Dissection Guide for
Human Anatomy

DISSECTION GUIDE FOR
Human Anatomy

David A. Morton, PhD

Department of Neurobiology and Anatomy
University of Utah School of Medicine
Salt Lake City, Utah

Kerry D. Peterson, LFP

Director, Body Donor Program
Department of Neurobiology and Anatomy
University of Utah School of Medicine
Salt Lake City, Utah

Kurt H. Albertine, PhD

Professor of Pediatrics, Medicine (Adjunct),
and Neurobiology and Anatomy (Adjunct)
Course Director, Human Gross Anatomy
Departments of Pediatrics, Medicine, and
Neurobiology and Anatomy
University of Utah School of Medicine
Salt Lake City, Utah

CHURCHILL LIVINGSTONE

An Imprint of Elsevier

CHURCHILL LIVINGSTONE
An Imprint of Elsevier
The Curtis Center
Independence Square West
Philadelphia, Pennsylvania 19106

DISSECTION GUIDE FOR HUMAN ANATOMY   ISBN 0-443-06627-2

Notice

Gross anatomy is an ever-changing field. Standard safety precautions must be followed but as new research and clinical experience broaden our knowledge, changes in treatment and drug therapy may become necessary or appropriate. Readers are advised to check the most current product information provided by the manufacturer of each drug to be administered to verify the recommended dose, the method and duration of administration, and contraindications. It is the responsibility of the treating physician, relying on experience and knowledge of the patient, to determine dosages and the best treatment for each individual patient. Neither the Publisher nor the author assumes any liability for any injury and/or damage to persons or property arising from this publication.

The Publisher

Printed in China

Last digit is the print number: 9 8 7 6 5 4 3 2 1

Acknowledgments

I express my appreciation to Tom Parks, Kerry Peterson, and Kurt Albertine for the opportunity to participate in the creation of this dissection guide. I am grateful to my parents (Gordon and Gabriella Morton), who taught me to appreciate the miracle of the human body. I also thank my brothers, my sister, my in-laws, the Templeman family, Brian, and Bo for their constant encouragement. I could not have done this without the support of my wonderful wife Celine and our children (Jared, Ireland and Gabriel). My wife and children are an inspiration in my life. This project has been one of the highlights of my schooling.

David A. Morton, Ph.D.

I express my sincere thanks to the body donors and their families who have taught me more about life than about anatomy. I am grateful to colleagues and students who have fostered my love for anatomy and its teaching. Foremost, I thank my wife (T.G.) for her unending support of my pursuits.

Kerry D. Peterson, L.F.P.

To health profession students who use this guide, gross anatomy is learned by dissecting the human body. This conviction, coupled with sage advice from my gross anatomy teacher, Professor Doctor Sigfried Zitzlsperger – who wrote in the *Gray's Anatomy* textbook he gave me many years ago, "Don't forget the many hours in the gross lab. That is where you learned your anatomy and that's where you should return to refresh your knowledge" – stimulated me to direct this composition. My hope is that our roadmap to human dissection facilitates learning in today's climate of compressed medical curricula.

On a personal note, I too extend sincere thanks to Dr. Parks (Chairman of Neurobiology and Anatomy) for his undaunted support, the body donors and their families (who are the real teachers of human anatomy and generosity), and Kerry Peterson, for expertly directing our body donor program, his superlative teaching, and his sense of humor. Thanks also are extended to the Dean's Office, particularly Dr. Betsy Allen, Dr. Larry Reimer, and former Dean Dr. Sam Shomaker, who provided institutional support and encouragement for our gross anatomy course. I am honored to have directed David Morton through this project, and I am proud to call him my colleague. I also appreciate the invaluable input provided by three classes of medical and dental students (to graduate in 2004–2006). Their constructive criticism and ideas improved the guide from their perspective, which is to dissect efficiently and learn what is expected. Lastly, I thank my wife (L.L.) for sticking by me despite the paucity of hours that I spend with her and our two children. To you, I dedicate my small part in this book.

Kurt H. Albertine, Ph.D.

A companion website presents cadaver dissection photos and outlines you can review before going in to the lab and to help plan your dissections. Visit:

http://evolve.elsevier.com/Morton/dissection/

The Key Idea Behind Writing the *Dissection Guide for Human Anatomy*

The key idea behind writing *Dissection Guide for Human Anatomy* was to produce a guide for efficient dissection of the entire human body in the reduced time that is allocated to basic science courses in medical schools. The key idea was not to produce a textbook or atlas of human gross anatomy. Therefore, our goal was to make the guide practical and easy to follow for medical, dental, and graduate students so that they would finish each dissection during the assigned laboratory period. Another goal was to provide students with clear expectations of the anatomic structures to be dissected and learned.

Accomplishing both goals required selecting a format that served each goal. We settled on a horizontal format that allowed us to position the text and corresponding illustration(s) on the same page. Thus each page is divided in half by a vertical bar, with text positioned on the left and the corresponding illustration(s) on the right. The book is spiral bound to display two pages simultaneously and to make the book as stable as possible on the book racks that are attached to dissection tables. In addition, we used boldface type for the names of anatomic structures the first time a structure is identified in the text. This design alerts the students that the named structure is to be dissected and identified.

The illustrations are simple black-and-white line drawings, with an occasional use of color. We chose this style of drawing because it focuses the reader's attention on a specific dissection task. Labels are kept to a minimum, and the labels that are used identify the anatomic structures

to be dissected. Red dashed lines highlight where cuts are to be made with a scalpel or saw.

Dissection Guide for Human Anatomy is organized into four regional units.

Unit 1: Back and Thorax
Unit 2: Abdomen, Pelvis, and Perineum
Unit 3: Neck and Head
Unit 4: Limbs

For each laboratory dissection, the guide begins with a table that identifies the structures to be dissected. The pages that follow contain the text and illustrations that direct the dissection. Each unit ends with a comprehensive table that identifies all the structures that were dissected and therefore are to be learned. We used *Terminologia Anatomica* (©1998; with permission from the Federative Committee on Anatomical Terminology) for anatomic nomenclature.

The tables in the *Dissection Guide for Human Anatomy* do not identify every anatomic structure that is present in a region. Because course hours for gross anatomy are fewer today than in the past, we chose not to make the tables all-inclusive and instead made editorial decisions on what to include. Our decisions for or against inclusion were based on practical experience. We recognize that our guide is a distillation of the total body of gross anatomic information, and we hope that students who desire more information and greater detail will invite their course instructors to embellish the students' knowledge and will independently seek more information by reading authoritative textbooks of human anatomy, such *Gray's Anatomy*, 38th edition, or the forthcoming 39th edition (Elsevier).

Some of the dissection approaches require cuts that destroy continuity of structures. Such cuts are made to improve access and/or visibility of anatomic structures. To minimize the impact of such cuts, we also provide instructions to spare one or more cadavers per dissection room, thereby preserving structural continuity. Students should confer with their instructors to determine whether the instructions to cut structures are to be modified or skipped.

Key:
m. – muscle
mm. – muscles
n. – nerve
nn. – nerves
a. – artery
aa. – arteries
v. – vein
vv. – veins
r. – right
l. – left
CN. – cranial nerve

Contents

Back and Thorax

Superficial Back

Prior to dissection, you should familiarize yourself with the following structures:

OSTEOLOGY
Skull
- Occipital b.
  - External occipital protuberance
  - Superior nuchal line
- Temporal b.
  - Mastoid process

Vertebral Column (33 vertebrae)
- Cervical vertebrae (7)
- Thoracic vertebrae (12)
- Lumbar vertebrae (5)
- Sacrum (5 fused vertebrae)
- Coccyx (4 fused vertebrae)

Vertebral Features
- Spinous process
- Transverse processes
- Transverse foramina (cervical only)
- Laminae
- Pedicles
- Body
- Articular processes (superior and inferior)
- Vertebral foramen
- Intervertebral foramina (notches)
- Intervertebral disc

Ilium
- Iliac crest
- Posterior superior iliac spine

Sacrum
Coccyx
Ribs
- Head
- Tubercle
- Angle

Scapula
- Spine
- Acromion
- Lateral margin
- Medial margin
- Superior angle
- Inferior angle

Miscellaneous
- Triangle of auscultation
- Lumbar triangle
- Midaxillary line

EXTRINSIC BACK MUSCLES
- Superficial layer
  - Trapezius m.
  - Latissimus dorsi m. (thoracolumbar fascia)
- Middle layer
  - Rhomboid major m.
  - Rhomboid minor m.
  - Levator scapulae m.
- Deep layer
  - Serratus posterior superior m.
  - Serratus posterior inferior m.

Other Muscles
- Splenius capitis and cervicis mm.

VESSELS AND NERVES
- Thyrocervical trunk
  - Transverse cervical a.
    - Superficial branch of transverse cervical a.
  - Dorsal scapular (deep branch) a.
- Posterior cutaneous neurovascular bundles
  - Dorsal rami of the spinal nn.
    - Posterior cutaneous nn. (medial and lateral branches)
  - Posterior intercostal aa.
    - Posterior cutaneous aa. (medial and lateral branches)
- Spinal accessory n. (CN XI)

TABLE 1–1   Extrinsic Back Muscles

Muscle	Proximal Attachment	Distal Attachment	Action	Innervation
Trapezius m.	Occipital bone, nuchal ligament, C7–T12 vertebrae	Lateral half of the clavicle, acromion and spine of the scapula	Elevates, retracts, depresses, and rotates the scapula	Spinal root of accessory n. (CN XI) and cervical nn. (C3–C4)
Latissimus dorsi m.	T7 vertebrae–sacrum, thoracolumbar fascia, iliac crest, inferior three ribs	Intertubercular groove of the humerus	Extends, adducts, and medially rotates the humerus; raises the body toward the arms during climbing	Thoracodorsal n. (C6–C8)
Levator scapulae m.	Transverse processes of C1–C4 vertebrae	Superior angle of the scapula	Elevates and rotates the scapula; inclines the neck to the same side of contraction	Cervical nn. (C3–C4) and dorsal scapular n. (C5)
Rhomboid major m.	Spinous processes of T2–T5 vertebrae	Medial margin of the scapula	Retract and rotate the scapula	Dorsal scapular n. (C4–C5)
Rhomboid minor m.	Nuchal ligament, spinous processes of C7–T1 vertebrae			
Splenius capitis m.	Nuchal ligament, spinous processes of C7–T4 vertebrae	Mastoid process of the temporal bone, occipital bone	*Acting alone:* laterally bends and rotates the head *Acting together:* extend the head and neck	Dorsal rami of the middle and lower cervical spinal nn.
Splenius cervicis m.	Spinous processes of T3–T6	Transverse processes of C2–C3 vertebrae		
Serratus posterior superior m.	C7–T3 vertebrae	Ribs 2–5	Elevates the ribs	Second to fifth intercostal nn.
Serratus posterior inferior m.	T11–L2 vertebrae	Ribs 8–12	Depresses the ribs	Ventral rami of 9th to 12th thoracic spinal nn.

Orientation and Surface Anatomy

■ Place the cadaver prone (face down)

■ Locate the following structures through the skin (Figure 1–1):

- **External occipital protuberance**
- **Spinous process of C7 (vertebra prominens)**
- **Acromion of the scapula**
- **Spine of the scapula**
- **Medial margin of the scapula**
- **Superior angle of the scapula**
- **Inferior angle of the scapula**
- **Midaxillary line** – imaginary line coursing vertically from the apex of the axilla along the lateral surface of the trunk
- **Iliac crests**
- **Sacrum**

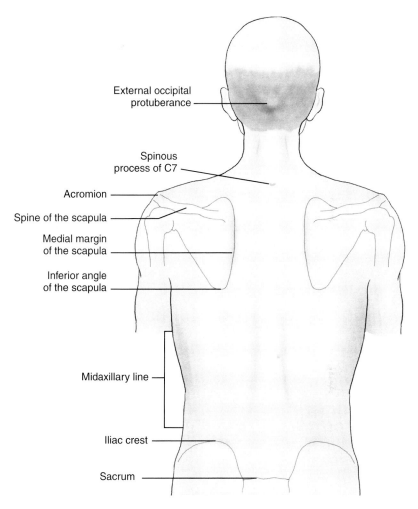

External occipital protuberance

Spinous process of C7

Acromion

Spine of the scapula

Medial margin of the scapula

Inferior angle of the scapula

Midaxillary line

Iliac crest

Sacrum

■ **Figure 1-1   Surface anatomy of the back**

Skin – Incisions

- The incisions should pass through the skin and superficial fascia, leaving the deep fascia and underlying muscle untouched; the depth of the incision will vary depending on the amount of fat in the superficial fascia

- Make the following incisions (Figure 1–2):

  - External occipital protuberance (*A*) to the sacrum (*B*)

  - Spinous process of C7 vertebra laterally to the acromion of each scapula (*D*)

  - Spinous process of (approximately) T6 vertebra (*C*) laterally to the midaxillary line (*E*)

  - Sacrum (*B*) laterally to (*F*)

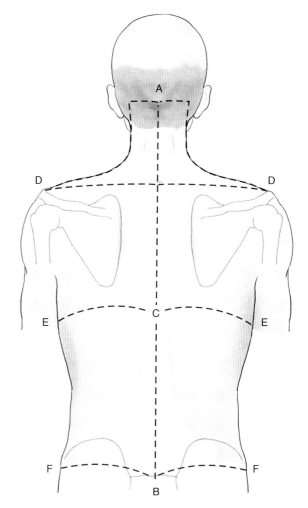

■ **Figure 1–2   Skin incisions of the back**

Skin – Reflection

■ To reflect the skin of the back (Figure 1–3):

- Grasp a corner of skin (using a hemostat or forceps) and pull superiorly (as shown in *A*)

- Keep pulling up while using a scalpel to cut (reflect) the skin and superficial fascia from the underlying deep fascia and muscles

- Once enough skin is reflected, use a scalpel to cut a "button-hole" through the skin; place a finger in the "button-hole" and pull up (as shown in *B*); this technique may be more effective than retracting with forceps

- Reflect the skin laterally from both sides of the back

■ *Note 1:* Look for neurovascular bundles that course between the deep fascia of the musculature and the superficial fascia being reflected

■ *Note 2:* Do not remove the skin; leave the flaps attached laterally to the body and replace them over the dissected areas when you are finished dissecting to prevent drying

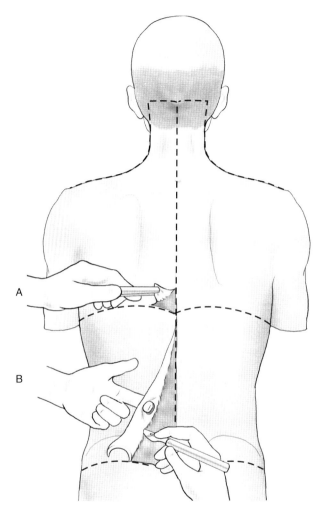

■ **Figure 1–3   Skin reflection of the back**

Nerves and Vessels of the Superficial Back

■ Identify the following (Figure 1–4):

- **Posterior cutaneous neurovascular bundles** – pass segmentally through fovea (small holes) in the deep fascia to enter the superficial fascia and skin

  - Located near **spinous processes** of the **cervical** and upper **thoracic** vertebrae

  - In the lower **thoracic** and **lumbar** regions, the bundles are located about three finger widths lateral to the midline

  - *Note:* While attempting to locate these bundles, only choose a small area of the back; you do not need to locate every segmental cutaneous neurovascular bundle

■ The posterior cutaneous nn. originate from **dorsal primary rami**

■ The posterior cutaneous aa. and vv. originate from **intercostal** and **lumbar vessels** (veins will often have a dark color)

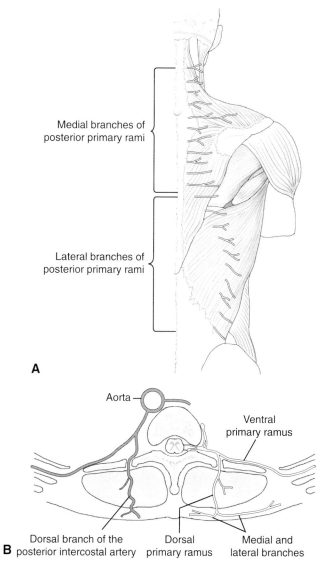

Medial branches of posterior primary rami

Lateral branches of posterior primary rami

A

Aorta

Ventral primary ramus

Dorsal branch of the posterior intercostal artery

Dorsal primary ramus

Medial and lateral branches

B

■ **Figure 1–4   Posterior cutaneous neurovascular bundles. *A*, Superficial view of the back illustrating posterior cutaneous neurovascular bundles. *B*, Cross-sectional view of the posterior cutaneous neurovascular bundles**

Extrinsic Back Muscles

■ Identify the following (Figure 1–5):

- **Trapezius m.** – identify its attachments to the base of the skull, nuchal ligament, spines of C7–T12 vertebrae, spines of both scapulae, and both acromial processes

  - Observe superior, medial, and inferior divisions of the trapezius m. (respective striations course in different directions)

- **Latissimus dorsi m.** – identify its attachments from T7 vertebra to the sacrum via the thoracolumbar fascia (the muscle inserts laterally in the intertubercular groove of the humerus; seen later)

- **Thoracolumbar fascia** – a thick layer of the deep fascia that is made thick by the fused aponeuroses of several muscles; attaches along the spinous processes from T7 vertebra to the sacrum

- **Triangle of auscultation** – bounded by the medial margin of the scapula and by the trapezius and latissimus dorsi mm.

- **Lumbar triangle** – bounded by the external oblique m., latissimus dorsi m., and iliac crest

■ Remove fascia to clearly demarcate the borders of each muscle (this is accomplished with a combination of scraping the muscle surfaces and blunt dissection)

■ *Note:* Extrinsic back muscles are located on the back but act on the upper extremity and will therefore be studied further in another laboratory session

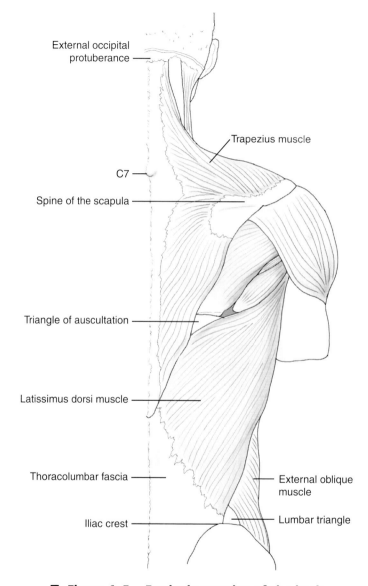

External occipital protuberance

Trapezius muscle

C7

Spine of the scapula

Triangle of auscultation

Latissimus dorsi muscle

Thoracolumbar fascia

External oblique muscle

Iliac crest

Lumbar triangle

■ **Figure 1–5   Extrinsic muscles of the back**

Dissection of the Trapezius Muscle

■ To dissect the trapezius m. (Figure 1–6):

- Grasp the most inferior attachment of the trapezius m. (spinous process of T12 vertebra) with a hemostat or forceps

- Free the deep surface of the trapezius m. from the underlying muscles using your fingers

- Cut (using a scalpel) the medial attachment of the trapezius m. in a superior direction from all vertebral spines (T12 toward C1)

- Do this for both sides of the body

- As you approach T1 vertebra, cut the trapezius m. from the spine of the scapula (*Do not* cut the rhomboid mm. deep to the trapezius m.)

- Reflect the trapezius m. laterally toward the shoulder

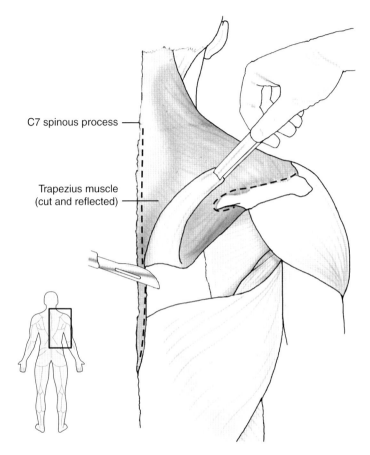

C7 spinous process

Trapezius muscle (cut and reflected)

■ **Figure 1–6   Reflection of the trapezius muscle**

Dissection of the Trapezius Muscle—cont'd

■ Identify the following (Figure 1–7):

- **Spinal accessory nerve (cranial nerve XI, CN XI)** – exits the skull via the jugular foramen and enters the deep surface of the trapezius m.

- **Superficial branch of the transverse cervical a.** – enters the deep superior surface of the trapezius m.

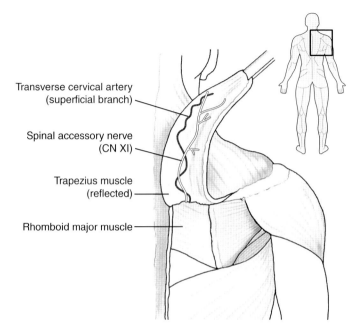

Transverse cervical artery (superficial branch)

Spinal accessory nerve (CN XI)

Trapezius muscle (reflected)

Rhomboid major muscle

■ **Figure 1–7   Spinal accessory nerve and transverse cervical artery**

Dissection of the Latissimus Dorsi Muscle

■ To dissect the latissimus dorsi m. (Figure 1–8):

- Cut both latissimus dorsi mm. (using a scalpel) in an inferior direction from the spinous processes from T7 vertebra to the sacrum

- Make horizontal incisions from the sacrum along the iliac crests

- Grasp the medial, inferior edge of the latissimus dorsi m. and begin to lift the muscle

  - Blunt dissection (using your hand or a blunt instrument) will allow lateral reflection of the latissimus dorsi m.

  - Do this for both sides of the body

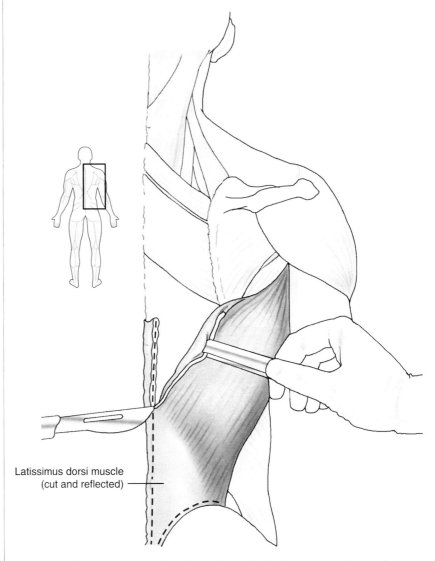

Latissimus dorsi muscle (cut and reflected)

■ **Figure 1–8   Reflection of the latissimus dorsi muscle**

Rhomboid and Levator Scapulae Muscles

■ Identify the following (Figure 1–9):

- **Rhomboid major m.** – observe its attachments along the spines of T2–T5 vertebrae and the medial margin of the scapula

- **Rhomboid minor m.** – observe its attachments along the spines of C7–T1 vertebrae and the medial margin of the scapula

  - The two may be fused and appear as one

■ Carry out the following instructions to reflect the rhomboid mm. (Figure 1–10):

- Cut the medial attachments of the rhomboid mm. (C7–T5 vertebrae) along all spinous processes and reflect the cut muscles laterally toward the scapula

■ Identify the following:

- **Dorsal scapular n. (C5)** – branches from the C5 root of the brachial plexus; courses along the deep surface of the levator scapulae and rhomboid mm.

- **Dorsal scapular (deep branch of the transverse cervical) a.** – enters the deep surface of the rhomboid musculature

- **Levator scapulae m.** – attaches to the superior angle of the scapula and transverse processes of C1–C4 vertebrae

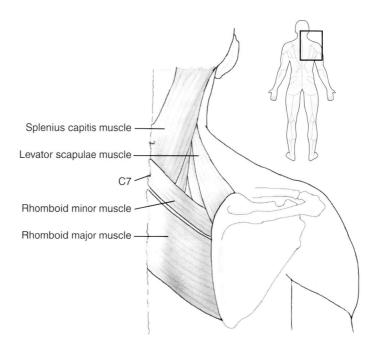

Splenius capitis muscle

Levator scapulae muscle

C7

Rhomboid minor muscle

Rhomboid major muscle

■ **Figure 1–9   Levator scapulae and rhomboid muscles**

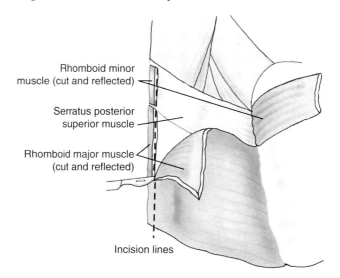

Rhomboid minor muscle (cut and reflected)

Serratus posterior superior muscle

Rhomboid major muscle (cut and reflected)

Incision lines

■ **Figure 1–10   Reflection of the rhomboid muscles**

Serratus Posterior Inferior and Superior Muscles

■ Identify the following (Figure 1–11):

- **Serratus posterior inferior m.** – located deep to the latissimus dorsi m. and attached along the spines of T11–L2 vertebrae and the ribs; may be attached to the deep surface of the latissimus dorsi m.

- **Serratus posterior superior m.** – deep to the levator scapulae and rhomboid mm. and superficial to the splenius capitis m.; attaches along the spines of C6–T2 vertebrae and the ribs

- **Erector spinae mm.** – located deep to the serratus posterior inferior and serratus posterior superior mm.

- **Splenius capitis and splenius cervicis mm.** – inferior attachments are visible deep to the serratus posterior superior m.

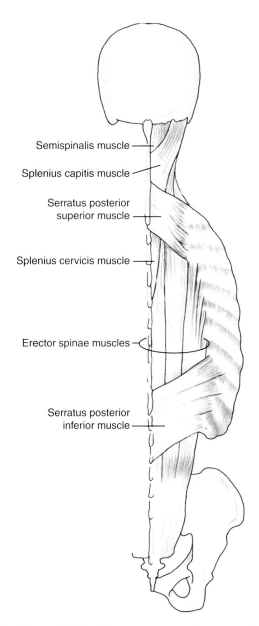

Semispinalis muscle

Splenius capitis muscle

Serratus posterior superior muscle

Splenius cervicis muscle

Erector spinae muscles

Serratus posterior inferior muscle

■ **Figure 1-11   Serratus posterior muscles**

Deep Back and Vertebral Column

Lab 2

Prior to dissection, you should familiarize yourself with the following structures:

INTRINSIC BACK MUSCLES

- Superficial layer
  - Erector spinae group
    - Iliocostalis m.
    - Longissimus m.
    - Spinalis m.

- Deep layer
  - Transversospinalis group
    - Semispinalis m.
    - Multifidus m.
    - Rotatores mm.
  - Intertransversarii mm.
  - Interspinales mm.
  - Levatores costarum mm.

VERTEBRAL LIGAMENTS

- Nuchal ligament
- Supraspinous ligament
- Interspinous ligament
- Ligamentum flavum
- Intertransverse ligament

TABLE 2–1   Intrinsic Muscles of the Back

Muscle	Proximal Attachment	Distal Attachment	Action	Innervation
Erector spinae group – series of muscles that extends from the sacrum to the skull				
Iliocostalis m.				
Iliocostalis lumborum	Iliac crest and sacrospinal aponeurosis	Thoracolumbar fascia and ribs 7–12		
Iliocostalis thoracis	Superior borders of ribs 6–12	Ribs 1–7 and transverse process of C7 vertebra		
Iliocostalis cervicis	Ribs 3–7	Transverse processes of C4–C6 vertebrae		
Longissimus m.				
Longissimus thoracis	Thoracodorsal fascia, transverse processes of T7–L2 vertebrae	Transverse processes of lumbar and thoracic vertebrae and inferior borders of ribs	*Bilateral:* extend the vertebral column *Unilateral:* laterally flex the vertebral column	Segmentally innervated by dorsal primary rami of spinal nn.
Longissimus cervicis	Transverse processes of T1–T6 vertebrae	Transverse processes of C2–C6 vertebrae		
Longissimus capitis	Transverse processes of C1–C4 vertebrae	Mastoid process of temporal bone		
Spinalis m.				
Spinalis thoracis	Spinous processes of T11–L2 vertebrae	Spinous processes of T2–T9 vertebrae		
Spinalis cervicis	Spinous processes of C6–T2 vertebrae	Spinous processes of C2–C4 vertebrae		

TABLE 2-1 Intrinsic Muscles of the Back—cont'd

Muscle	Proximal Attachment	Distal Attachment	Action	Innervation
Transversospinalis group				
Semispinalis m.				
Semispinalis thoracis	Transverse processes of T7–T12 vertebrae	Spinous processes of C6–T6 vertebrae	Extend the cervical and thoracic regions of vertebral column and rotate the vertebral column contralaterally	Segmentally innervated by dorsal primary rami of spinal nn.
Semispinalis cervicis	Transverse processes of T1–T6 vertebrae	Spinous processes of C2–C5 vertebrae		
Semispinalis capitis	Transverse processes of C7–T6 vertebrae	Between the superior and inferior nuchal lines on the occipital bone	Extends the head, and turns the face toward the opposite side	
Multifidus m.	Aponeurosis of the erector spinae mm., sacrum, and transverse processes of lumbar, thoracic, and cervical vertebrae	Spinous processes of lumbar, thoracic, and lower cervical vertebrae	Extend the vertebral column and rotate the vertebral column contralaterally	
Rotatores mm.	Transverse processes of C2 vertebra to the sacrum	Lamina immediately above vertebra of origin		
Interspinalis mm.	Spinous processes of cervical, upper thoracic, lower thoracic, and lumbar vertebrae	Inferior surfaces of spinous processes of vertebrae superior to the proximal attachment	Aid in extension and rotation of the vertebral column	
Intertransversarii mm.	Transverse processes of cervical, lower thoracic, and lumbar vertebrae	Transverse processes of adjacent vertebrae	Aid in lateral flexion of the vertebral column; acting bilaterally, stabilize the vertebral column	Dorsal rami of spinal nn.
Levatores costarum mm.	Tips of transverse processes of C7 and T1–T11 vertebrae	Pass inferolaterally and insert on a rib between its tubercle and angle	Elevate ribs, assisting inspiration; assist with lateral bending of the vertebral column	Dorsal rami of C8–T11 spinal nn.

Intrinsic Muscles of the Back – Superficial Group

■ **Erector spinae mm.** (dissect both sides)

  ● Identify each muscle in the **erector spinae group,** from lateral to medial (Figure 2–1):

    ● **Iliocostalis m.**

    ● **Longissimus m.**

    ● **Spinalis m.**

■ If not already done, cut the serratus posterior inferior m. and serratus posterior superior m. along their vertebral attachments; reflect both muscles laterally

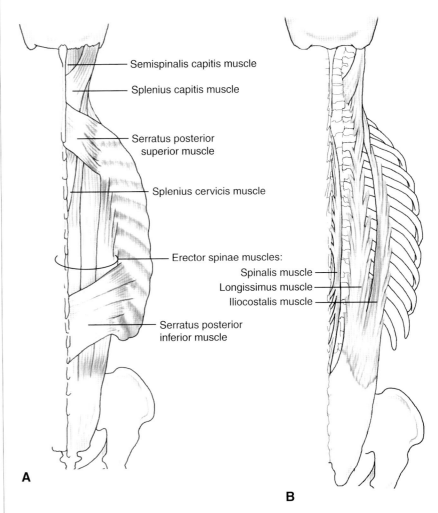

A

B

■ **Figure 2–1   Erector spinae muscles. *A*, Superficial view. *B*, Deep view**

Labels (Figure A):
Semispinalis capitis muscle
Splenius capitis muscle
Serratus posterior superior muscle
Splenius cervicis muscle
Erector spinae muscles:
Serratus posterior inferior muscle

Labels (Figure B):
Spinalis muscle
Longissimus muscle
Iliocostalis muscle

Intrinsic Muscles of the Back – Superficial Group—cont'd

- Dissect the erector spinae mm. (Figure 2–2):

  - Cut the inferior attachment of the erector spinae group, using a scalpel

- Reflect the erector spinae mm. (Figure 2–3):

  - Reflect the erector spinae group laterally from the midline of the back, exposing the deeper transversospinalis muscle group

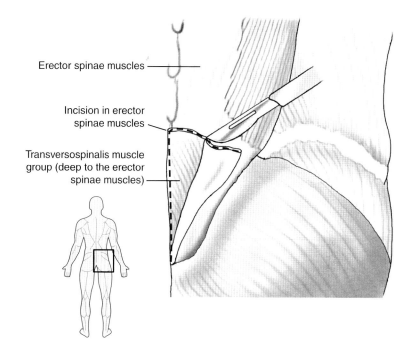

Erector spinae muscles

Incision in erector spinae muscles

Transversospinalis muscle group (deep to the erector spinae muscles)

■ **Figure 2–2   Dissection of the erector spinae muscles**

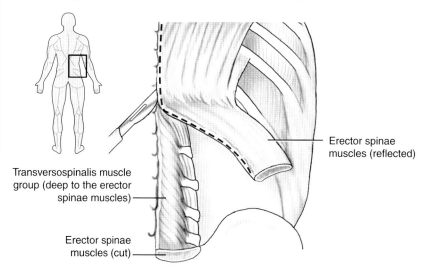

Erector spinae muscles (reflected)

Transversospinalis muscle group (deep to the erector spinae muscles)

Erector spinae muscles (cut)

■ **Figure 2–3   Reflection of the erector spinae muscles**

Intrinsic Muscles of the Back – Deep Group

■ Identify the following (Figure 2–4):

- **Transversospinalis group** – arise from a transverse process and ascend superiorly and medially to the spinous process of superior vertebrae

    - **Semispinalis m.** – muscle fibers span 5–6 vertebrae

    - **Multifidus m.** – muscle fibers span 3–4 vertebrae

    - **Rotatores m.** – muscle fibers span 1–2 vertebrae; deepest of the transversospinalis group

■ *Note:* Observe the muscle striations coursing from a transverse process below to spinous processes above (hence the name)

- **Levator costarum mm.** – between the transverse processes and ribs

- **Intertransversarii mm.** – span adjacent transverse processes

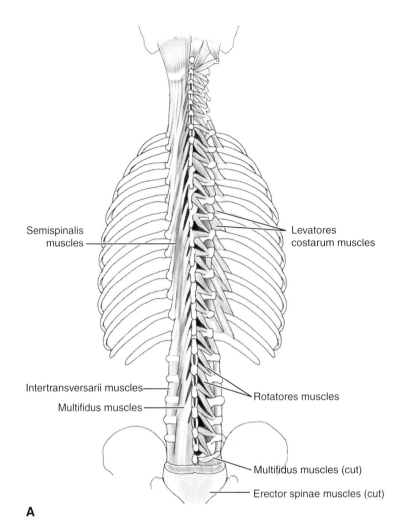

A

■ **Figure 2-4   Transversospinalis muscles. *A,* Deep view of the back illustrating the transversospinalis muscles. *B,* Schematic of the transversospinalis muscles**

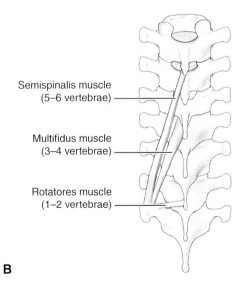

B

Ligaments of the Vertebrae

- To prepare for the laminectomy, you should have the vertebral lamina (area between the transverse and spinous processes) devoid of muscle tissue, from C7 to L5 vertebra

- *Cut* and *scrape* the vertebral laminae clean of muscle tissue because the bone saw works well on bones, not on muscles

- Identify the following ligaments (Figure 2–5):

  - **Supraspinous ligaments** – span the superficial tips of the vertebral spines

  - **Interspinous ligaments** – span the spines of adjacent vertebrae

  - **Intertransverse ligaments** – span the transverse processes of adjacent vertebrae

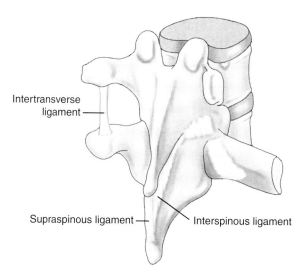

Intertransverse ligament

Supraspinous ligament

Interspinous ligament

■ **Figure 2–5   Vertebral ligaments**

Laminectomy

- Check that the saw switch is in the "off" position before the saw is plugged in; use care with the bone saw; keep your hands away from the blade

- Refer to Figures 2–6 and 2–7 for demonstration of the angle of the saw blade

- Turn on the saw and cut into the lamina of C7 vertebra

- As the saw blade cuts through the lamina, you will feel the blade drop (*Be careful;* avoid cutting the underlying **dura mater;** cut only as deep as the bone)

- Continue cutting to L5 vertebra

- Repeat for the other side of the vertebral column; it may be necessary to remove some vertebral spines to maintain a straight saw line

- Remove the cut portion of the vertebral column from the cadaver, freeing any remaining attachments with a mallet and chisel

  - Be careful around sharp edges of the remaining parts of the vertebrae

  - Use a mallet to hit (blunt) sharp bony edges

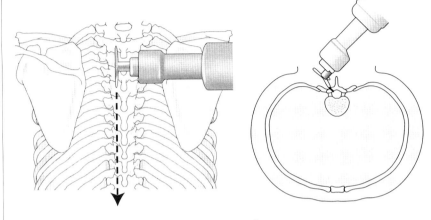

■ **Figure 2–6   Laminectomy**

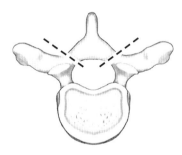

■ **Figure 2–7   Angle of the saw blade (the *arrows* indicate the plane of the saw blade)**

Inspection of the Removed Laminae

- Review the vertebral ligaments mentioned during this lab

- On the internal aspect of the removed laminae, observe the **ligamentum flavum** (Figure 2–8); it is yellow because of elastic fibers

- When review is complete, return the laminae to their anatomic position

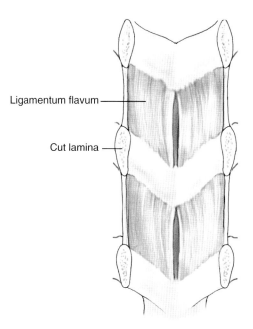

Ligamentum flavum

Cut lamina

■ **Figure 2–8   Internal view of the removed laminae**

Spinal Cord and Suboccipital Region

Lab 3

Prior to dissection, you should familiarize yourself with the following structures:

MENINGES
- Dura mater
  - Epidural space
  - Subdural space
  - Dural sac
- Arachnoid mater
  - Subarachnoid space
- Pia mater
  - Denticulate ligaments
  - Filum terminale

SPINAL CORD
- Posterior median sulcus
- Cervical spinal enlargement
- Lumbar spinal enlargement
- Conus medullaris

SPINAL NERVES
- Ventral roots (rootlets)
- Dorsal roots (rootlets)
  - Dorsal root ganglion
- Primary ventral rami
- Primary dorsal rami
- Cauda equina

OSTEOLOGY
- Superior nuchal line
- Inferior nuchal line

SUBOCCIPITAL REGION
Nuchal Ligament and Fascia
Muscles
- Splenius capitis m.
- Splenius cervicis m.
- Longissimus capitis m.
- Semispinalis capitis m.
- Suboccipital mm.
  - Rectus capitis posterior major m.
  - Rectus capitis posterior minor m.
  - Obliquus capitis superior m.
  - Obliquus capitis inferior m.

Vessels and Nerves
- Occipital a.
- Vertebral a.
- Suboccipital n. (dorsal ramus of C1)
- Greater occipital n. (dorsal ramus of C2)
- Dorsal ramus of C3

SUBOCCIPITAL TRIANGLE
Borders
- Rectus capitis posterior major m.
- Obliquus capitis superior m.
- Obliquus capitis inferior m.

Contents
- Vertebral a.
- Suboccipital n. (dorsal ramus of C1)

TABLE 3–1   Muscles of the Suboccipital Region

Muscle	Proximal Attachment	Distal Attachment	Action	Innervation
Rectus capitis posterior major	Spine of C2 vertebra (axis)	Lateral half of the inferior nuchal line	Extends and rotates the head	Dorsal primary ramus of C1 (suboccipital n.)
Rectus capitis posterior minor	Posterior tubercles of C1 vertebra (atlas)	Occipital bone below medial portion of inferior nuchal line	Extend the head	
Obliquus capitis superior	Transverse process of C1 vertebra (atlas)	Occipital bone between the superior and inferior nuchal line		
Obliquus capitis inferior	Spine of C2 vertebra (axis)	Transverse process of C1 vertebra (atlas)	Extends and rotates the head	

Spinal Canal Contents

- Remove the laminae to identify the following (Figures 3–1 and 3–2):

  - **Epidural fat** – you may notice the vertebral venous plexus in the epidural fat
  - **Dura mater** – positioned inside the **vertebral canal**

- Cut through the dura mater along its exposed length; carefully reflect and pin the dura mater laterally

  - **Subdural space** – region between the dura mater (externally) and the arachnoid mater (internally); use a probe to explore the subdural space
  - **Arachnoid mater** – the middle meninx; located immediately deep and loosely attached to the dura mater

- Cut through a region of the arachnoid mater

  - **Subarachnoid space** – space deep to the arachnoid mater; cerebrospinal fluid is contained within this space
  - **Pia mater** – intimately covers the **spinal cord** and cannot be grossly distinguished, with the exception of the **denticulate ligaments** and **filum terminale**
  - **Denticulate ligaments** – small bilateral projections of the pia mater that are attached laterally to the surrounding dura
  - **Ventral and dorsal roots** – observe that the denticulate ligaments separate them

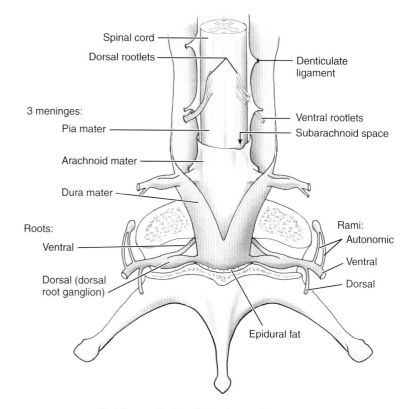

■ Figure 3–1 Spinal canal contents

Spinal Canal Contents—cont'd

■ Identify the following (see Figures 3–1 and 3–2):

- **Spinal nerves** – exit the lateral, external surface of the dura mater to traverse the intervertebral foraminae

- **Dorsal root ganglia** – a swelling along each dorsal root near the **pedicle** of the vertebra

- **Cervical spinal enlargement** – an enlargement of the spinal cord between vertebrae C3 and T2, housing nerve cell bodies for the upper limbs; only a portion of the cervical enlargement will be visible because the laminectomy stopped at the C7 vertebra

- **Lumbar spinal enlargement** – an enlargement of the spinal cord between vertebrae T9 and T12, housing nerve cell bodies for the lower limbs

- **Conus medullaris** – inferior region of the spinal cord where the spinal cord tapers into a cone (region of vertebra L2)

- **Cauda equina** (horse's tail) – the aggregate of nerve roots extending from the conus medullaris

- **Filum terminale** – the pia mater that continues inferiorly from the conus medullaris to the **coccyx** (usually silver in color)

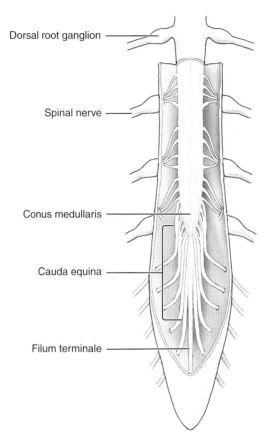

Dorsal root ganglion

Spinal nerve

Conus medullaris

Cauda equina

Filum terminale

■ **Figure 3–2   Spinal cord**

Suboccipital Triangle Dissection
(Right Side Only)

- Inclusion of this dissection will be indicated by your instructor

  - *Note:* Limit this dissection to the right side of the neck to avoid damage to the left posterior triangle of the neck (dissected later)

- Identify the following (Figure 3–3):

  - **Trapezius m.**

  - **Greater occipital n.** – dorsal ramus of C2 spinal n.; spend a few moments to find branches of the greater occipital n. and occipital a. in the nuchal fascia overlying the trapezius m.

- Dissect the trapezius m. from the following (see Figure 3–3):

  - **Spinous process of C7 to the occipital bone** – keep to the right side of the spinous processes and nuchal ligament

  - **Occipital bone** – along the superior nuchal line

- Reflect the trapezius m. inferiorly and laterally

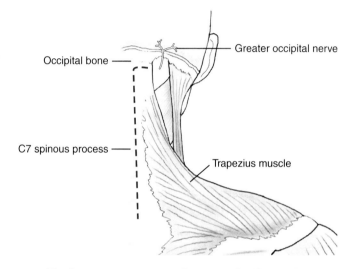

Occipital bone

Greater occipital nerve

C7 spinous process

Trapezius muscle

- **Figure 3-3   Trapezius muscle dissection**

Suboccipital Triangle Dissection (Right Side Only)—cont'd

■ Identify the following (Figure 3–4):

- **Splenius capitis m.** – observe that the striations course superiorly and laterally at an angle between the vertebrae and mastoid process of the temporal bone

- **Splenius cervicis m.** – deep to the splenius capitis m.

- **Semispinalis capitis m.** – a transversospinalis m.; deep to the splenius capitis m.; attached to the inferior nuchal line of the occipital bone

■ Dissect the following muscles from their superior attachments (stay close to the bone to spare the greater occipital n.):

- **Splenius capitis m.** – mastoid process and occipital bone

- **Semispinalis capitis m.** – inferior nuchal line (occipital bone)

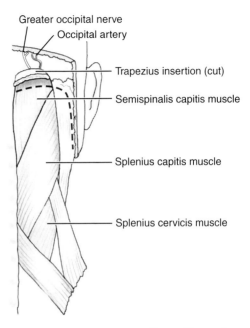

Greater occipital nerve
Occipital artery
Trapezius insertion (cut)
Semispinalis capitis muscle
Splenius capitis muscle
Splenius cervicis muscle

■ **Figure 3–4   Intermediate dissection of the posterior region of the neck**

Suboccipital Triangle Dissection
(Right Side Only)—cont'd

- Place a wood block under the superior part of the thorax to open the dissection field

- Identify the following muscles deep to the semispinalis and splenius capitis mm. (Figure 3–5):

  - **Rectus capitis posterior major m.** – courses from the spinous process of C2 vertebra to insert on the occipital bone just below the inferior nuchal ligament

  - **Obliquus capitis superior m.** – courses between the transverse process of C1 vertebra and the occipital bone

  - **Obliquus capitis inferior m.** – larger of the two oblique muscles; courses from the spinous process of C2 vertebra to the transverse process of C1 vertebra

  - **Longissimus capitis m.** – look for its attachment to the mastoid process, lateral to the splenius capitis m.

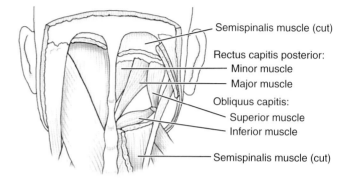

Semispinalis muscle (cut)

Rectus capitis posterior:
  Minor muscle
  Major muscle

Obliquus capitis:
  Superior muscle
  Inferior muscle

Semispinalis muscle (cut)

■ **Figure 3–5   Suboccipital triangle muscles**

Suboccipital Triangle Dissection (Right Side Only)—cont'd

■ Identify the following (Figure 3–6):

- **Greater occipital n.** – dorsal ramus of C2 spinal n.; exits the spinal cord between C1 and C2 vertebrae and emerges below the obliquus capitis inferior m.; courses in a superior direction

- **Rectus capitis posterior minor m.** – cut and reflect the superior attachment of the rectus capitis posterior *major* m.; the rectus capitis posterior *minor* m. is deep to the major m.

- **Suboccipital triangle** – borders:
  - **Rectus capitis posterior major m.**
  - **Obliquus capitis superior m.**
  - **Obliquus capitis inferior m.**

- **Suboccipital triangle** – contents
  - **Vertebral a.** – originates from the subclavian a. (seen later) and courses through the transverse foramina of the cervical vertebrae
    - ◆ Courses horizontally in the groove on the superior surface of the lamina of C1 vertebra (atlas)
  - **Suboccipital n.** – dorsal ramus of C1 spinal n.; courses superior to the lamina of C1 vertebra (atlas)

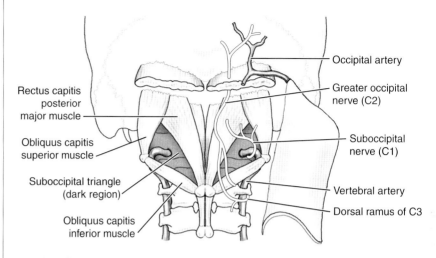

■ **Figure 3–6   Suboccipital triangle borders and contents**

Anterior Chest Wall and Breast

Lab 4

Prior to dissection, you should familiarize yourself with the following structures:

OSTEOLOGY

- Sternum
  - Jugular notch
  - Manubrium
  - Sternal angle
  - Body
  - Xiphoid process
- Clavicle
- Sternoclavicular joint
- Scapula
  - Acromion
  - Coracoid process

MUSCLES

- Pectoralis major m.
- Pectoralis minor m.
- Serratus anterior m.
- Subclavius m.
- Intercostal mm. (external, internal, innermost)
- Transversus thoracis m.

NERVES

- Ventral primary rami
  - Intercostal nn.
    - Lateral and anterior cutaneous nn.
  - Medial pectoral n.
  - Lateral pectoral n.

ARTERIES AND VEINS

- Thoracic aorta
  - Posterior intercostal vessels
    - Lateral cutaneous aa.
- Subclavian vessels
  - Internal thoracic vessels
    - Anterior intercostal vessels
      - Anterior cutaneous aa.
- Axillary a.
  - Thoracoacromial trunk
    - Deltoid, pectoral, acromial, and clavicular branches
  - Lateral thoracic vessels

Superficial Vein

- Cephalic v.

BREAST

- Nipple
- Areola
- Suspensory ligaments
- Lactiferous ducts
- Mammary gland
- Retromammary space

MISCELLANEOUS TERMS AND STRUCTURES

- Deltopectoral triangle
- Clavipectoral fascia
- Midaxillary line
- Parietal pleura
- Sternocostal joints
- Endothoracic fascia

Female cadavers – Turn to page 34 of the dissection guide.

Male cadavers – Turn to page 36 of the dissection guide.

TABLE 4-1 Muscles of the Anterior Thoracic Wall

Muscle	Proximal Attachment	Distal Attachment	Action	Innervation
Pectoralis major m.	Clavicular head: medial surface of the clavicle Sternocostal head: sternum, superior 6 costal cartilages	Lateral lip of the intertubercular groove (humerus)	Adducts, medially rotates the humerus Clavicular head: flexes the humerus Sternocostal head: extends the humerus	Medial and lateral pectoral nn.; clavicular head (C5–C6), sternocostal head (C7–T1)
Pectoralis minor m.	Ribs 3–5 (near their costal cartilages)	Coracoid process (scapula)	Stabilizes the scapula by drawing it inferiorly and anteriorly	Medial pectoral n. (C8–T1)
Subclavius m.	Rib 1 (near its costal cartilage)	Middle third of the clavicle (inferior surface)	Depresses and anchors the clavicle	Nerve to the subclavius (C5–C6)
External intercostal mm.	Inferior border of the ribs	Superior border of the ribs below	Stabilize the ribs External: elevate the ribs Internal: depress the ribs	Intercostal nn.
Internal intercostal mm.				
Innermost intercostal mm.				
Transversus thoracis m.	Posterior surface of the inferior portion of the sternum	Internal surface of costal cartilages 2–6	Depresses the ribs	

Female Cadaver – Orientation and Incisions

- Turn the cadaver supine (face up)

- Locate the following (Figure 4–1):

  - **Jugular notch**
  - **Sternal body**
  - **Xiphoid process**
  - **Sternoclavicular joint**
  - **Acromion of the scapula**
  - **Acromioclavicular joint**

- Make the following skin incisions (Figure 4–2):

  - Circular incision around each nipple
  - Jugular notch (*A*) to the xiphoid process (*B*)
  - Jugular notch (*A*) laterally along the clavicle to the lateral tip of the acromion (*C*)
  - Xiphoid process (*B*) laterally along the subcostal margin to the midaxillary line (*D*)

- Grasp a corner of skin with a hemostat and reflect the skin laterally

- Leave the skin flaps attached laterally

- *Leave* one breast attached to the cadaver

- *Note:* Look for neurovascular bundles that course between the deep fascia of the musculature and the superficial fascia

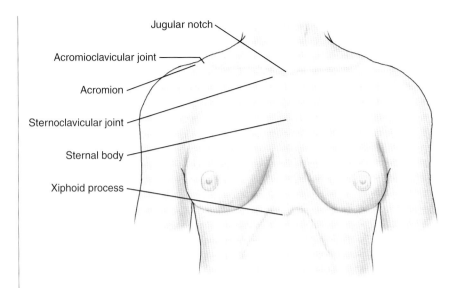

- **Figure 4–1   Anterior view of the anterior chest wall of the female**

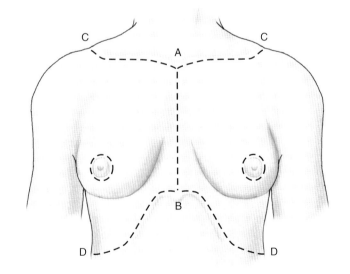

- **Figure 4–2   Skin incisions in the anterior chest wall of the female**

Female Cadaver – Breast Dissection

- The **breast** has no capsule and is located in the superficial fascia, superficial to the **pectoralis major m.**

- The glandular tissue is embedded in subcutaneous fat lobules

- Identify the following (Figure 4–3):

  - **Nipple**

  - **Areola**

- On one side, cut sagittally through the nipple, then continue the cut through the entire breast

- Identify the following (see Figure 4–3):

  - **Suspensory ligaments** of the breast – fibrous bands that connect the deep fascia to the skin

    - To better observe the suspensory ligaments, you may find it helpful to pull the nipple to put tension on (straighten) the ligaments

  - **Lactiferous ducts** – 15 to 20 ducts that drain milk from the glandular tissue

  - **Retromammary space** – loose connective tissue region between the deep fascia of the pectoralis major m, and the breast; insert your fingers into the retromammary space

- To continue the dissection, please turn to page 36

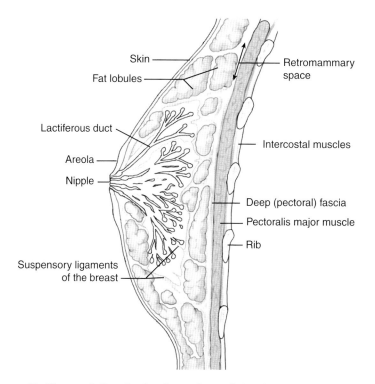

Skin
Fat lobules
Lactiferous duct
Areola
Nipple
Retromammary space
Intercostal muscles
Deep (pectoral) fascia
Pectoralis major muscle
Rib
Suspensory ligaments of the breast

■ **Figure 4–3   Sagittal section of the female breast**

Male Cadaver – Orientation and Incisions

- Turn the cadaver supine (face up)

- Locate the following structures (Figure 4–4):

  - **Jugular notch**
  - **Sternal body**
  - **Xiphoid process**
  - **Sternoclavicular joint**
  - **Acromion**
  - **Acromioclavicular joint**

- Make the following skin incisions (Figure 4–5):

  - Jugular notch (*A*) to the xiphoid process (*B*)
  - Jugular notch (*A*) laterally along the clavicle to the lateral tip of the acromion (*C*)
  - Xiphoid process (*B*) laterally along the subcostal margin to the midaxillary line (*D*)

- Grasp a corner of skin (with a hemostat) and reflect the skin laterally, removing the superficial fascia attached to the skin

- Repeat on both sides

- *Note:* Look for neurovascular bundles that course between the deep fascia of the musculature and the superficial fascia

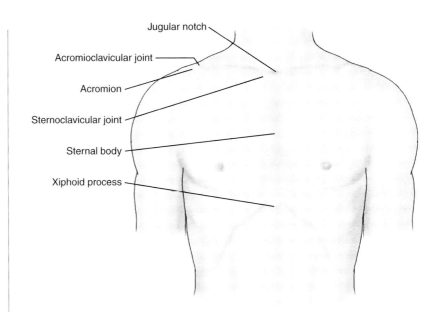

■ **Figure 4–4   Anterior view of the anterior chest wall of the male**

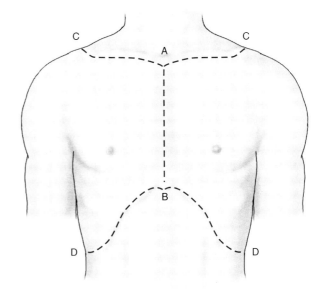

■ **Figure 4–5   Skin incisions in the anterior chest wall of the male**

Anterior Wall of the Thorax – Superficial Fascia (Female and Male)

■ Identify the following (Figures 4–6 and 4–7):

- **Anterior cutaneous neurovascular bundles –** branches of the **intercostal neurovascular bundles**

  • Upper five anterior cutaneous neurovascular bundles emerge just lateral to the sternum through slips in the pectoralis major m.

  • Lower six anterior cutaneous neurovascular bundles run inferiorly to the anterior abdominal wall (studied further in another laboratory session)

- **Lateral cutaneous neurovascular bundles –** branches of the **intercostal neurovascular bundles**

  • Emerge between slips of the serratus anterior m.

■ *Note:* Look for cutaneous vv. (which are dark); when you find a cutaneous v., the corresponding a. and n. will accompany it

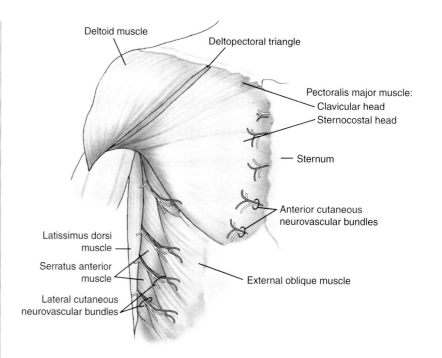

Deltoid muscle
Deltopectoral triangle
Pectoralis major muscle:
Clavicular head
Sternocostal head
Sternum
Anterior cutaneous neurovascular bundles
Latissimus dorsi muscle
Serratus anterior muscle
External oblique muscle
Lateral cutaneous neurovascular bundles

■ **Figure 4–6   Muscles of the pectoral region**

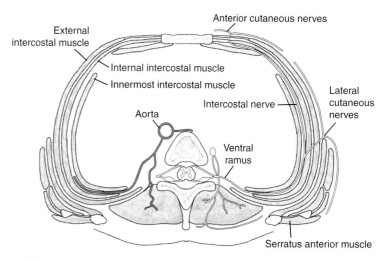

External intercostal muscle
Anterior cutaneous nerves
Internal intercostal muscle
Innermost intercostal muscle
Lateral cutaneous nerves
Intercostal nerve
Aorta
Ventral ramus
Serratus anterior muscle

■ **Figure 4–7   Cross-section through the thoracic wall**

Anterior Wall of the Thorax

■ Identify the following (Figure 4–8):

- **Pectoralis major m.** – observe the clavicular and sternal heads, which are named for their bony attachments

- **Deltoid m.** – you will only be able to see the anterior half of the deltoid m. attached to the clavicle

- **Deltopectoral triangle** – bordered by the clavicle, clavicular head of pectoralis major m., and anterior border of the deltoid m.

   - **Cephalic v.** – contained within the deltopectoral triangle

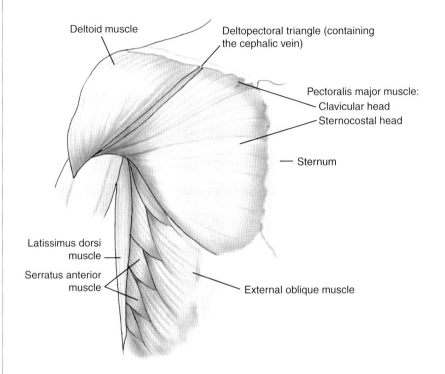

Deltoid muscle

Deltopectoral triangle (containing the cephalic vein)

Pectoralis major muscle:
Clavicular head
Sternocostal head

Sternum

Latissimus dorsi muscle

Serratus anterior muscle

External oblique muscle

■ **Figure 4-8   Anterior thoracic wall**

Anterior Wall of the Thorax—cont'd

■ Follow these instructions to dissect the pectoralis major m. from the thoracic wall (Figure 4–9):

- Gently insert your fingers deep to the inferior border of the pectoralis major m.
- Using your scalpel, cut along the sternal and clavicular attachments of the pectoralis major m.
- Reflect the pectoralis major m. laterally
- Watch for the medical-pectoral n.

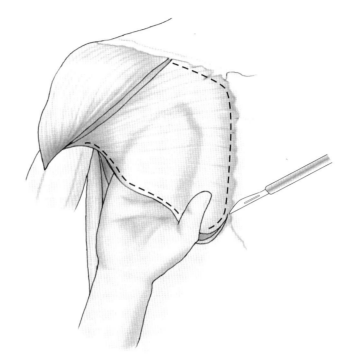

■ **Figure 4–9   Incision to dissect the pectoralis major muscle**

Anterior Wall of the Thorax – Deltopectoral Triangle

■ Identify the following (Figures 4–10 and 4–11):

- **Clavipectoral fascia** – originates on the clavicle and envelopes the subclavius and pectoralis minor mm.; extends between the two muscles as a single sheet

  - **Suspensory ligament of the axilla** – continuation of the clavipectoral fascia into the axillary region

- **Subclavius m.** – attached to the inferior surface of the clavicle and rib 1

- **Pectoralis minor m.** – attached to the coracoid process and ribs 3–5

■ Dissect the clavipectoral fascia between the pectoralis minor and subclavius mm.

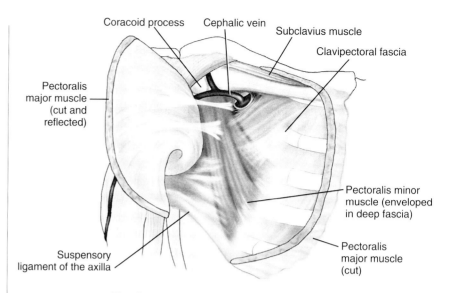

Coracoid process · Cephalic vein · Subclavius muscle · Clavipectoral fascia

Pectoralis major muscle (cut and reflected)

Pectoralis minor muscle (enveloped in deep fascia)

Suspensory ligament of the axilla

Pectoralis major muscle (cut)

■ **Figure 4–10   Clavipectoral fascia**

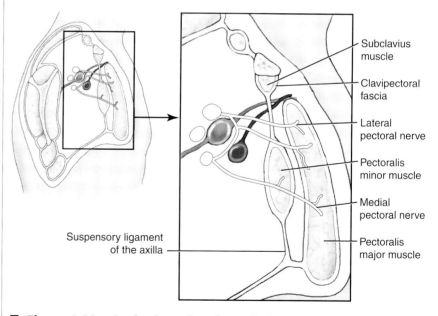

Subclavius muscle

Clavipectoral fascia

Lateral pectoral nerve

Pectoralis minor muscle

Medial pectoral nerve

Pectoralis major muscle

Suspensory ligament of the axilla

■ **Figure 4–11   Sagittal section through the axilla to demonstrate the clavipectoral fascia**

Anterior Wall of the Thorax – Deltopectoral Triangle Contents

■ Identify the following structures (Figures 4–12 and 4–13):

- **Lateral pectoral n.** – originates from the lateral cord of brachial plexus; enters deep surface of pectoralis major m.

- **Medial pectoral n.** – originates from the medial cord of the brachial plexus and passes through the pectoralis minor m. to enter the deep surface of the pectoralis major m.

  - *Note:* The names "medial" and "lateral" pectoral n. are derived from the cords of the brachial plexus from which each nerve arises, not their anatomic position

- **Cephalic v.** – drains into the **axillary v.**

- **Thoracoacromial trunk/artery** – originates from the **axillary a.**; prior to the axillary a. coursing deep to the pectoralis minor m., it divides into four branches according to their respective fields of distribution:

  - **Pectoral branch** – follows the pectoral nerves

  - **Deltoid branch** – may arise from the acromial branch; accompanies the cephalic v. in the deltopectoral triangle

  - **Acromial branch** – crosses the coracoid process deep to the deltoid m.

  - **Clavicular branch** – ascends medially between the pectoralis major and clavipectoral fascia

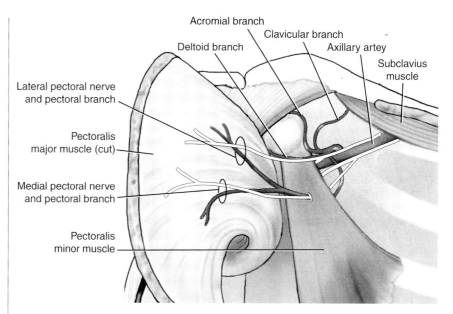

■ **Figure 4-12   Branches of the thoracoacromial trunk**

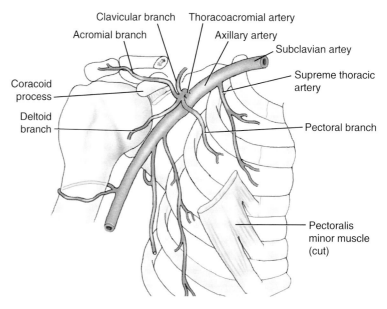

■ **Figure 4-13   Branches of the axillary artery**

Intercostal Structures

■ Identify the following (the upper intercostal spaces are usually best to demonstrate the following details) (Figure 4–14):

● **External intercostal mm.**

- The muscle fibers course in an oblique fashion (superior and lateral to inferior and medial)

- The muscle belly fills the intercostal spaces posteriorly and laterally; anteriorly it is a membrane

● **Internal intercostal mm.**

- Internal intercostal m. fibers run at a right angle and deep to the external intercostal mm.

- The muscle belly fills the intercostal spaces anteriorly and laterally; posteriorly it is a membrane

● **Innermost intercostal mm.**

- Innermost intercostal m. follows the same course as the internal intercostal mm.

- The muscle belly fills the intercostal spaces laterally; anteriorly and posteriorly it is a membrane

● **Intercostal v., a.,** and **n.** – cut along the inferior border of the external and internal intercostal mm.; separate the internal intercostal from the innermost intercostal mm. to identify the intercostals v., a., and n.

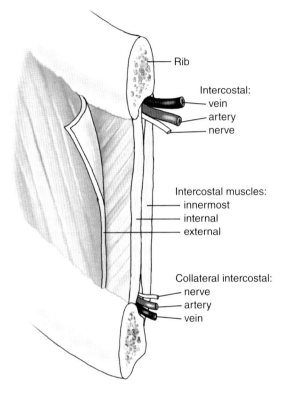

■ **Figure 4-14   Intercostal structures**

Removal of the Anterior Thoracic Wall

- Cut a window through the intercostal mm. just lateral to the sternum, in the first intercostal space (Figure 4–15)

  - Identify and transect the **internal thoracic vessels** transversely in the first intercostal space on both sides of the sternum

- Reflect the serratus anterior, pectoralis major, and pectoralis minor mm. laterally

- Using a scalpel, mark the following on the anterior thoracic wall (Figure 4–16):

  - Sternal angle
  - The first intercostal space (inferior to rib 1)
  - Midaxillary line
  - Along eighth or ninth intercostal space between midaxillary line and sternum
  - The sternoxiphoid joint

- *Bone saw or rib cutter:* Cut through the sternum and ribs (remember to only cut through the bone, *not* deeper to avoid cutting the parietal pleura and the lungs)

- *Scalpel:* Cut through the intercostal muscles and other soft tissues

- Lift the rib cage away from the body

- Use your fingers to open the space in the **endothoracic fascia** (a thin fascial layer between the internal intercostal mm./membrane and **parietal pleura**) while separating the rib cage from the parietal pleura

- *Note:* Many pathologic conditions cause adhesions of the parietal pleura to the visceral pleura; you may not be able to open the pleural space without shredding the pleurae

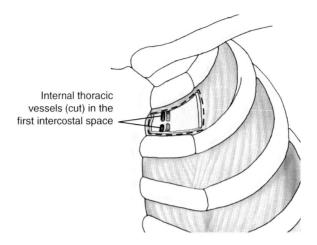

Internal thoracic vessels (cut) in the first intercostal space

■ **Figure 4–15   Intercostal incision**

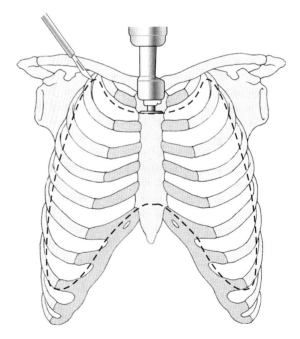

■ **Figure 4–16   Rib cage incisions and saw cuts**

Examination of the Removed Anterior Thoracic Wall

■ Identify the following structures on the internal surface of the removed anterior thoracic wall (Figure 4–17):

- **Internal thoracic vessels** – originate from the **sub-clavian a.,** just posterior to the **sternoclavicular joint**

  - The internal thoracic vessels are the anterior contribution to the intercostal circulation

  - Both internal thoracic aa. terminate by dividing into **superior epigastric** and **musculophrenic aa.** at about rib 6

- **Transversus thoracis mm.** – observe that the muscle is attached to the sternal body and costal cartilage of ribs 1–5

- On the cut ends of the ribs, identify the following:

  - **External intercostal m.**

  - **Internal intercostal m.**

  - **Innermost intercostal m.**

  - **Intercostal v., a., and n.**

■ Return the dissected rib cage to its anatomic position on the cadaver

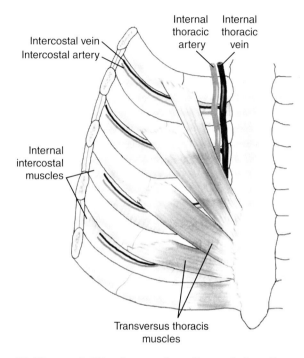

Intercostal vein
Intercostal artery
Internal thoracic artery
Internal thoracic vein
Internal intercostal muscles
Transversus thoracis muscles

■ **Figure 4-17   Internal surface of the rib cage and sternum**

Thoracic Situs and Lungs

Lab 5

Prior to dissection, you should familiarize yourself with the following structures:

PLEURAL SPACE
- Endothoracic fascia
- Parietal pleura
  - Costal pleura
  - Mediastinal pleura
  - Diaphragmatic pleura
  - Cervical pleura
- Pleural space
- Visceral pleura
  - Same regions as the parietal pleura

PLEURAL RECESSES
- Costomediastinal recess
- Costodiaphragmatic recess

ROOT OF THE LUNGS
- Pulmonary aa.
- Pulmonary vv.
- Bronchi
- Pulmonary ligaments
- Bronchial aa.
- Lymph nodes

NERVES AND VESSELS
- Phrenic nn.
- Vagus nn.
  - Recurrent laryngeal nn.
- Pericardiacophrenic vessels

LEFT LUNG
- Apex
- Costal, mediastinal, and diaphragmatic surfaces
- Hilum
  - Pulmonary ligament
  - Bronchi, pulmonary aa. and vv.
- Superior (upper) and inferior (lower) lobes
- Oblique fissure
- Cardiac notch
- Lingula
- Contact impressions after fixation
  - Cardiac impression
  - Aortic groove impression
  - Descending aorta impression

- Bronchi
  - Primary
  - Secondary
  - Tertiary
- Bronchopulmonary (hilar) lymph nodes

RIGHT LUNG
- Apex
- Costal, mediastinal, and diaphragmatic surfaces
- Hilum
  - Pulmonary ligament
  - Bronchi, pulmonary aa. and vv.
- Superior (upper), middle, and inferior (lower) lobes
- Oblique and horizontal fissures
- Contact impressions after fixation
  - Cardiac impression
  - Esophageal groove
- Bronchi
  - Primary
  - Secondary
  - Tertiary
- Bronchopulmonary (hilar) lymph nodes

Pleural Spaces

■ Identify the following (Figure 5–1):

- **Pleural spaces** – left and right spaces between the parietal and visceral pleurae

- **Middle mediastinum** – contains the **pericardial sac** and **heart**

- **Parietal pleura** – four regions:

  - **Costal parietal pleura** – lines the internal surface of the rib cage

  - **Mediastinal parietal pleura** – lines the mediastinum

  - **Diaphragmatic parietal pleura** – lines the superior surface of the diaphragm

  - **Cervical parietal pleura** – extends into the thoracic inlet, superior to rib 1

■ Make a longitudinal incision through the parietal pleura (if it is still intact) from the apex of the chest to the diaphragm, then reflect the parietal pleura laterally

- **Visceral pleura** – attached directly to the lung's surface and as such cannot be distinguished from the lung itself, except at the pulmonary ligaments, where the visceral and parietal pleurae are contiguous

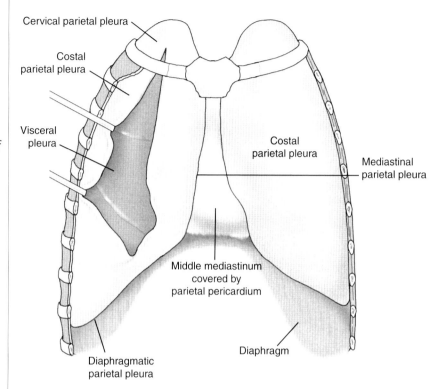

■ **Figure 5-1   Pleural sacs**

Pleural Recesses

- **Pleural recesses** – regions where the parietal pleura reflects so sharply that two layers of parietal pleura are in contact with each other during expiration; lung tissue fills this potential space during inspiration

- Identify the following (Figures 5–2 and 5–3):

  - **Costomediastinal recess** – the parietal pleura reflects from the ventral thoracic wall onto the mediastinum

  - **Costodiaphragmatic recess** – the parietal pleura reflects from the lateral thoracic wall to the diaphragm

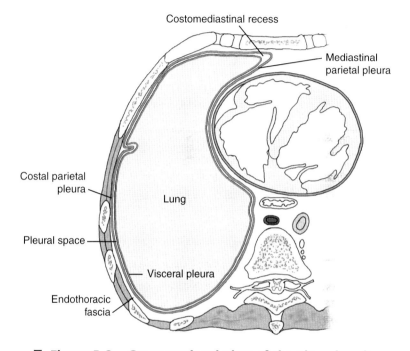

■ **Figure 5–3   Cross-sectional view of the pleural cavity**

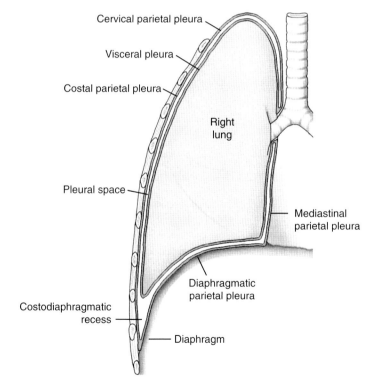

■ **Figure 5–2   Coronal view of the pleural cavity and recesses**

Phrenic Nerves and Pericardiacophrenic Vessels

■ Identify the following (Figure 5–4):

- **Left phrenic n.** – located between the subclavian a. and v. deep to the clavicle; courses anterior to the root of the left lung to enter the diaphragm

- **Right phrenic n.** – lateral to the superior vena cava; courses anterior to the root of the right lung

- **Pericardiacophrenic a.** and **v.** – located with the phrenic n. between the mediastinal parietal pleura and the fibrous pericardium; you will most likely have to cut through the mediastinal parietal pleura to get to the pericardiacophrenic vessels and phrenic n.

- *Do not* cut the phrenic n. or the pericardiacophrenic a. or v. when making the subsequent incisions through the root of lungs

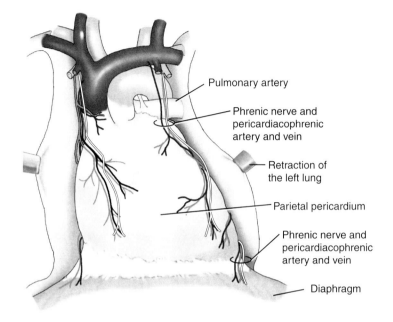

Pulmonary artery

Phrenic nerve and pericardiacophrenic artery and vein

Retraction of the left lung

Parietal pericardium

Phrenic nerve and pericardiacophrenic artery and vein

Diaphragm

■ **Figure 5–4   Phrenic nerves and pericardiacophrenic vessels**

Removal of the Lungs from the Thoracic Cavity

■ To remove the lungs, follow these instructions (Figure 5–5):

- Expose the root of a lung; with your hand placed between the lung and pericardium, pull the lung laterally

- Using a scalpel, *carefully* transect the root of the lung between the lung and pericardium, (do not cut your fingers or the heart)

- Repeat this procedure for the other lung

■ *Note:* Some pathologic conditions cause lung adhesions to surrounding structures; break the adhesions with fingers or cut them with scissors

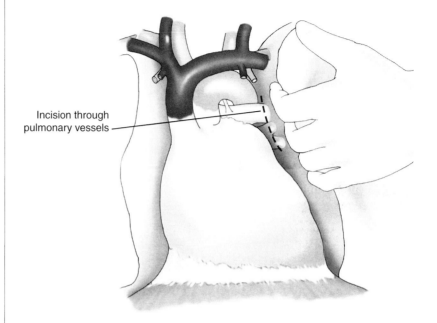

Incision through pulmonary vessels

■ **Figure 5–5   Incisions at the lungs' roots**

Left Lung

■ Examine the excised lungs; if available, use a hand air pump or compressed air hose to inflate bronchopulmonary segments

■ Identify the following (Figures 5–6 and 5–7):

- **Apex**
- **Costal, mediastinal,** and **diaphragmatic surfaces**
- **Hilum**
  - **Pulmonary ligament**
    - ◆ Transition of parietal pleura to visceral pleura
    - ◆ Hangs inferiorly from the hilum
    - ◆ The parietal pleura forms a double membrane (pulmonary ligament) that anchors the lung to the mediastinum
  - **Bronchi** – generally located posteriorly
  - **Pulmonary aa.** – generally located superiorly
  - **Pulmonary vv.** – generally located inferiorly
    - ◆ *Note:* These positional terms depend on where the hilum is cut
  - **Bronchial aa.** and numerous **bronchopulmonary lymph nodes**
- **Superior and inferior lobes** – separated by the **oblique fissure**
- **Cardiac notch** – gap in the superior lobe where the heart lies
- **Lingula** – most ventral and inferior portion of superior lobe
- **Contact impressions** – cardiac impression, aortic arch groove, and descending aorta groove

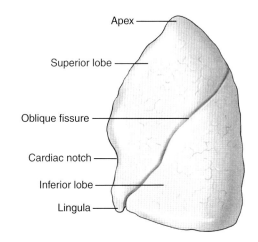

■ Figure 5–6   Lateral view of the left lung

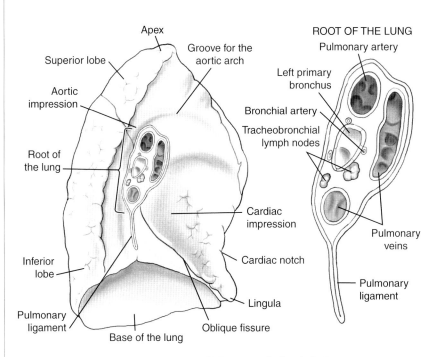

■ Figure 5–7   Medial view of the left lung

Right Lung

■ Identify the following (Figures 5–8 and 5–9):

- **Apex** and **costal, mediastinal,** and **diaphragmatic surfaces**
- **Hilum**
  - **Pulmonary ligament**
  - **Bronchi** (posterior); **pulmonary aa.** (superior); **pulmonary vv.** (inferior)
    - ◆ *Note:* These positional terms depend on where the hilum is cut
  - **Bronchial aa.** and numerous **bronchopulmonary lymph nodes**
- **Superior lobe** – separated from the **middle lobe** via the **horizontal fissure**
- **Inferior lobe** – separated from the superior lobe and middle lobe via the **oblique fissure**
- **Contact impressions – cardiac impression** and **esophageal groove**
- **Bronchi** – follow the bronchi into one lung, using blunt dissection; find two or three lobar segments.

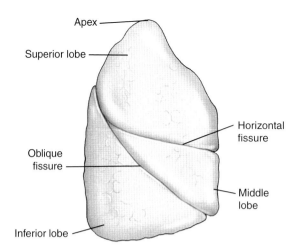

■ **Figure 5–8   Lateral view of the right lung**

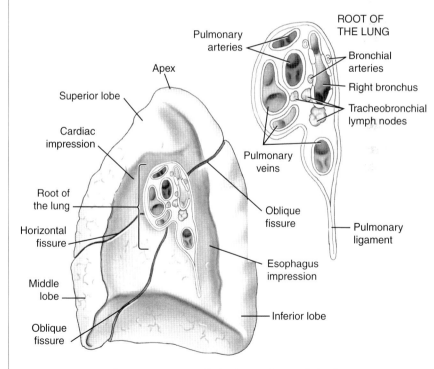

■ **Figure 5–9   Medial view of the right lung**

Heart

Prior to dissection, you should familiarize yourself with the following structures:

ANTERIOR MEDIASTINUM
- Thymus

MIDDLE MEDIASTINUM

Pericardial Sac
- Parietal pericardium
  - Fibrous and serous layers
- Pericardial cavity
- Visceral pericardium (epicardium)
- Transverse pericardial sinus
- Oblique pericardial sinus
- Left vagus n.
  - Recurrent laryngeal n.
- Ligamentum arteriosum
- Phrenic n.

Heart
- Right and left atria
- Right and left ventricles
- Atrioventricular (AV) sulcus

- Interventricular groove
- Apex
- Base

CORONARY ARTERIES
- Left coronary a.
  - Anterior interventricular branch
  - Circumflex branch
    - Left (obtuse) marginal branch
- Right coronary a.
  - Atrial a.
    - Sinuatrial branch
  - Right (acute) marginal branch
  - Posterior interventricular branch

CORONARY VEINS
- Coronary sinus
  - Great cardiac v.
  - Middle cardiac v.
  - Small cardiac v.
  - Anterior cardiac vv.

RIGHT HEART

Right Atrium (Auricle)
- Pectinate mm.
- Sulcus terminalis
- Crista terminalis
- Sinus venarum
- Superior and inferior venae cavae
- Coronary sinus
- Fossa ovalis
- Tricuspid valve
- *SA and AV nodes

Right Ventricle
- Tricuspid valve (right AV valve)
- Anterior, posterior, and septal cusps
- Chordae tendineae
- Papillary mm.
- Trabeculae carneae
- Supraventricular crest

*SA – sinuatrial; AV – atrioventricular

- Septomarginal trabecula (moderator band)
- Conus arteriosus
- Orifice for the pulmonary trunk

Pulmonary Trunk

- Pulmonary valve (right, left, and anterior cusps)

LEFT HEART

Left Atrium (Auricle)

- Pulmonary vv. (4)

- Bicuspid orifice
- Foramen ovale and its valve
- Bicuspid valve (mitral or left AV valve)

Left Ventricle

- Bicuspid valve
- Anterior and posterior cusps
- Chordae tendineae
- Papillary mm.
- Trabeculae carneae

- Orifice for the aorta
- Interventricular septum

Aorta

- Aortic valve (right, left, and posterior cusps)

Orientation

■ Identify the following (Figure 6–1):

- **Thymus gland** – directly behind the manubrium (in the anterior mediastinum), overlies the aortic arch, left brachiocephalic v., trachea, and the pericardium (may be absent)

- **Aorta** – arches superiorly over the pulmonary trunk

- **Pulmonary trunk (a.)** – bifurcates into the right and left pulmonary aa. inferior to the aortic arch

- **Superior vena cava** – courses along the right side of the proximal portion of the aortic arch

- **Ligamentum arteriosum** – fibrosed arterial connection between the pulmonary trunk and aortic arch; probe the area between the bifurcation of the pulmonary trunk and the aorta to find it

- **Left vagus n.** – courses inferiorly immediately lateral to the ligamentum arteriosum

  - The left vagus n. courses anterior to the aortic arch and then posterior to the left pulmonary a. and root of the left lung

  - **Left recurrent laryngeal n.** – branches from the left vagus n. at the level of the ligamentum arteriosum; loops around the aortic arch to ascend to the larynx

- **Right vagus n.** – passes between the **right brachiocephalic a.** and **v.** and posterior to the hilum of the right lung

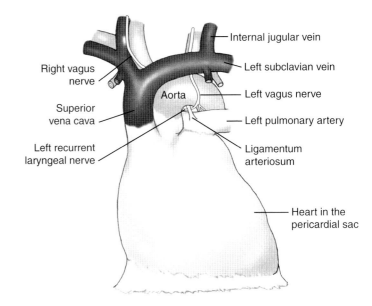

■ **Figure 6–1   Heart in situ**

Labels in figure:
- Internal jugular vein
- Right vagus nerve
- Left subclavian vein
- Aorta
- Left vagus nerve
- Superior vena cava
- Left pulmonary artery
- Left recurrent laryngeal nerve
- Ligamentum arteriosum
- Heart in the pericardial sac

Open the Pericardial Sac

- Identify the following (Figure 6–2):

  - **Pericardial sac** – fibroserous membrane surrounding the heart; the sac must be opened before the heart can be studied and removed from the middle mediastinum

- Pinch a fold of the pericardial sac with forceps

- Cut the fold with a scalpel or scissors; make a longitudinal incision

- Fold back the **parietal pericardium** and identify the following:

  - **Fibrous layer** – the external surface of the parietal pericardium

  - **Serous layer** – the internal surface of the parietal pericardium

  - **Pericardial cavity (space)** – the space between the parietal pericardium and the visceral pericardium

  - **Visceral pericardium** – surrounds the adipose (fat) tissue enveloping the heart

- Dissect the entire ventral portion of the pericardial sac to better investigate the heart; do not dissect the dorsal portion of pericardial sac at this time

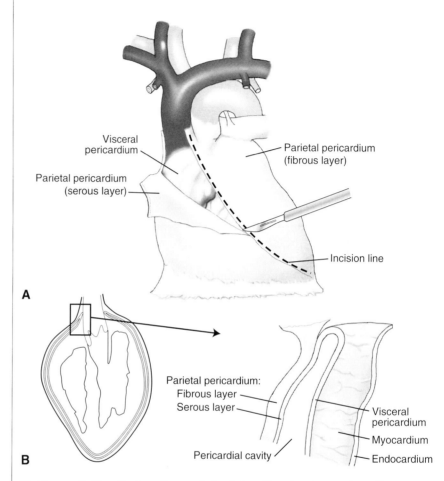

- **Figure 6–2   Pericardium. *A*, Incision line in the pericardium. *B*, Coronal section through the heart to demonstrate the pericardial sac**

Pericardial Sinuses

■ Identify the following (Figure 6–3):

- **Transverse pericardial sinus**

  - Push your right index finger posterior to the **pulmonary trunk,** from the cadaver's left to right sides of the heart

  - Continue to push your finger between the ascending aorta and superior vena cava, immediately superior to the pulmonary vv.

  - This space is the **transverse pericardial sinus**

- **Oblique pericardial sinus** (see Figure 6–5)

  - On the left border of the cadaver's heart, push your fingers behind the aorta into the space between the **inferior vena cava** and the four pulmonary vv. (the heart will be cupped by your palm)

  - This area is the **oblique pericardial sinus**

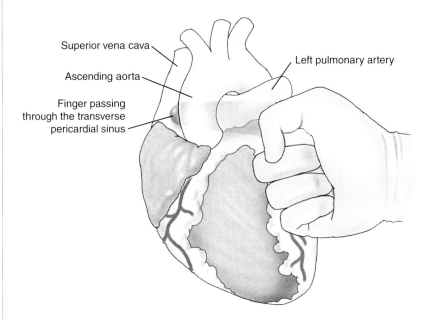

Superior vena cava

Ascending aorta

Finger passing through the transverse pericardial sinus

Left pulmonary artery

■ **Figure 6–3   Transverse pericardial sinus**

Removal of the Heart from the Middle Mediastinum

- Using a scalpel, transect the following (Figure 6–4):

  - **Pulmonary trunk** – across the pulmonary trunk and ascending aorta where the transverse pericardial sinus is located

  - **Inferior vena cava (IVC)** – transect the IVC as inferiorly as possible

  - **Superior vena cava (SVC)** – transect the SVC about 2 cm superior to its union with the right atrium (inferior to the **azygos v.**)

  - **Pulmonary vv.** – lift the apex of the heart ventrally and superiorly to stretch the four pulmonary vv. from the remaining pericardial sac and transect each vein

- The only thing holding the heart in place at this time is the connective tissue between the transverse and oblique sinuses

- Use blunt dissection to remove the heart from the pericardial sac

- Identify the following (Figure 6–5):

  - **Pericardial sac** – examine the dorsal aspect of the pericardial sac after the heart has been removed

  - **Transverse pericardial sinus**

  - **Oblique pericardial sinus**

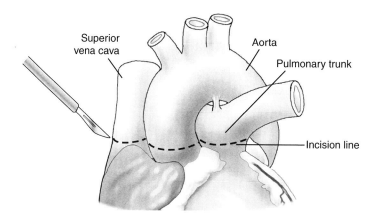

■ **Figure 6–4   Heart incisions for removal of the heart**

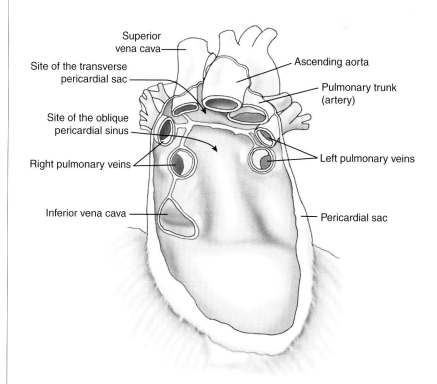

■ **Figure 6–5   Pericardial sac (empty)**

Inspection of the Removed Heart

■ Identify the following surfaces (Figure 6–6):

- **Anterior (sternocostal) surface** – formed mostly by the right ventricle

- **Inferior (diaphragmatic) surface** – formed mostly by the left ventricle

- **Left (pulmonary) surface** – formed mostly by the left ventricle, which creates the cardiac notch in the left lung

- **Superior** and **inferior venae cavae**

- **Pulmonary trunk** – bifurcates inferior to the aortic arch

- **Ascending aorta** – the right atrium may overlap the ascending aorta

- **Pulmonary vv.** – four located directly posterior and inferior to the pulmonary trunk; two right pulmonary vv. positioned between the superior and inferior venae cavae

- **Coronary sulcus** – oblique groove separating the atria from the ventricles; filled by cardiac vessels and adipose tissue

- **Anterior/posterior interventricular grooves** – filled by the left anterior and right posterior interventricular (descending) coronary aa. and adipose tissue

- **Apex** – the inferolateral portion of left ventricle; lies posterior to the fourth or fifth intercostal space, about one hand width from the midsagittal line

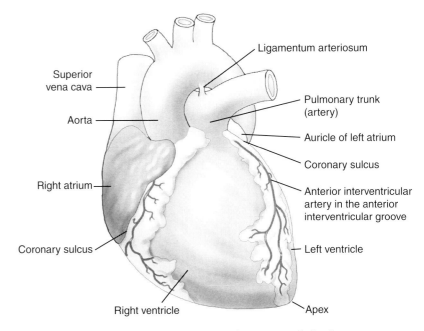

■ **Figure 6–6   Superficial features of the heart**

Coronary Arteries

■ Identify the following (Figure 6–7):

- **Right coronary a.** – arises from the aortic arch, directly superior to the right cusp of the aortic valve

  - **Atrial branch** – first branch from the right coronary a.; supplies the right atrium

    ◆ **Sinuatrial branch** – division of the atrial branch; supplies the sinuatrial (SA) node

  - **Acute marginal a.** – supplies the right surface of the right ventricle; courses toward the apex along the inferior border of the heart

  - **Posterior interventricular a.** (posterior descending a.) – descends in the posterior interventricular groove toward the apex

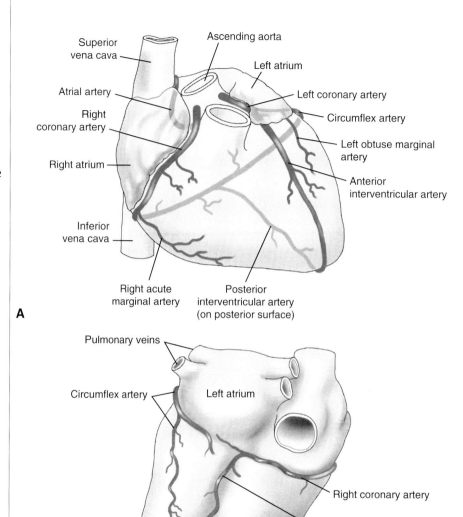

A

B

■ **Figure 6–7   Coronary arteries. *A*, Anterior view of the heart. *B*, Posterior view of the heart**

Coronary Arteries—cont'd

■ Identify the following (Figure 6–8):

- **Left coronary artery** – arises from the aortic arch, directly superior to the left cusp of the aortic valve; divides at the left atrium into the following two branches:

  - **Anterior interventricular branch** (left anterior descending; LAD branch) – courses in the anterior interventricular groove to the apex

    ◆ May anastomose with the posterior interventricular branch of the right coronary a. at the apex

  - **Circumflex branch** – follows the coronary sulcus around the left border of the heart and gives off the marginal aa.

    ◆ May anastomose with the posterior interventricular branch of the right coronary a. in the coronary sulcus

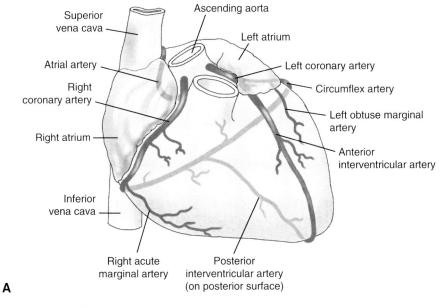

A

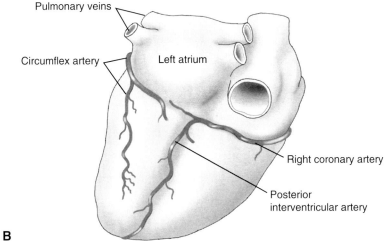

B

■ **Figure 6–8   Coronary arteries. *A*, Anterior view of the heart. *B*, Posterior view of the heart**

Cardiac Veins

■ Identify the following (Figures 6–9 and 6–10):

- **Coronary sinus** – large tributary v. that courses along the posterior coronary sulcus; receives venous blood from all of the cardiac vv. (except the anterior cardiac vv., which drain directly into the right atrium); the coronary sinus returns blood to the right atrium

  - **Great cardiac v.** – originates on the anterior surface of the heart, near its apex, and ascends with the left anterior interventricular (descending) a.

  - **Middle cardiac v.** – posterior surface of the heart, in the posterior interventricular groove; ascends with the posterior interventricular a.

  - **Small cardiac v.** – anterior surface of the heart, along the inferior margin of the right ventricle; ascends with the acute marginal a.

  - **Anterior cardiac vv.** – drain the wall of the right atrium and empty directly into the right atrium

■ *Note:* Often, two of the above identified veins flank the corresponding coronary artery

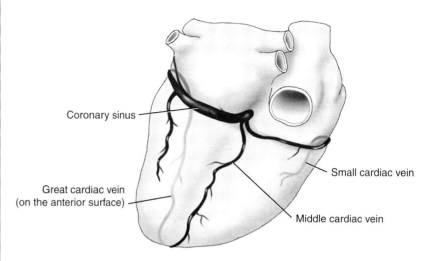

■ **Figure 6–9   Posterior view of the heart to demonstrate the cardiac veins**

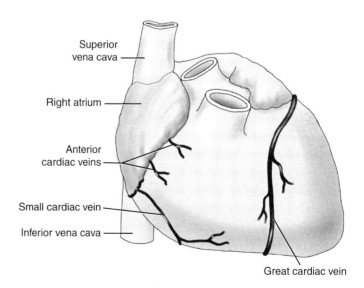

■ **Figure 6–10   Anterior view of the heart to demonstrate the cardiac veins**

Right Atrial Incisions

- Make the following incisions through the wall of the right atrium (Figure 6–11):

  - Horizontally from *A* to the anterior border (*B*)
  - Longitudinally from *B* to the inferior border (*C*)
  - Horizontal from *C* to the posterior-inferior border (*D*; not shown)

- Open the cut flap, exposing the internal contents; use forceps to remove blood clots; wash the atrium with water

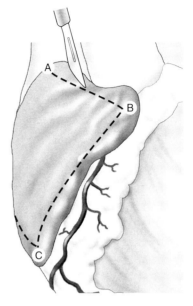

■ **Figure 6-11   Right atrial incisions**

Right Atrial Structures

■ This chamber receives venous blood from the superior and inferior venae cavae, and coronary sinus

■ Identify the following within the right atrium (Figure 6–12):

- **Pectinate mm.** – rough, muscular anterior wall region

- **Sinus venarum** – smooth, thin-walled posterior region

- **Crista terminalis** – separates the smooth and rough parts of the atrial wall *internally*

- **Sulcus terminalis** – a groove on the external surface of the right atrium directly over the crista terminalis

- **Sinuatrial (SA) node** – located where the crista terminalis meets the superior vena cava; located in the atrial wall, between the sulcus terminalis and crista terminalis; *difficult to identify*

- **Coronary sinus** – opens into right atrium, between the tricuspid valve and the opening of the inferior vena cava

- **Interatrial septum** – separates the two atria

- **Fossa ovalis** – a depression in the interatrial septum, superior to the inferior vena cava (formerly the foramen ovale in a fetus)

- **Tricuspid valve** – observe the superior border of the three cusps

- **Atrioventricular (AV) node** – located in the atrioventricular septum, above the opening of the coronary sinus; *difficult to identify*

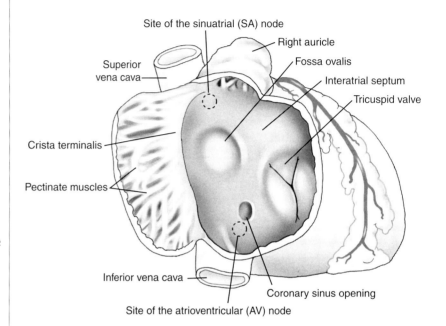

Site of the sinuatrial (SA) node
Right auricle
Fossa ovalis
Interatrial septum
Tricuspid valve
Superior vena cava
Crista terminalis
Pectinate muscles
Inferior vena cava
Coronary sinus opening
Site of the atrioventricular (AV) node

■ **Figure 6–12   Right atrial structures**

Right Ventricular Incisions

■ Push a blunt instrument through the pulmonary trunk into the right ventricle

■ Make the following incisions through the wall of the right ventricle (Figure 6–13):

● Horizontally from *A* to *B*, inferior to the pulmonary valve (use a blunt instrument as a guide)

● Make another incision about 1 cm medial and parallel to the coronary groove toward the inferior border (*C*)

● Make an additional incision about 2 cm medial and parallel to the anterior interventricular groove toward the inferior border (*D*) of the right ventricle

■ Open the cut ventricular flap inferiorly; use forceps to remove blood clots; wash the ventricle with water

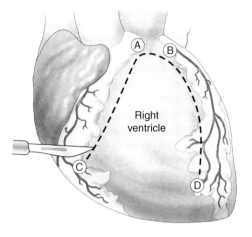

■ **Figure 6–13   Right ventricular incisions**

Right Ventricular Structures

■ Identify the following (Figure 6–14):

- **Trabeculae carneae** – irregular muscular elevations
- **Papillary mm.** – myocardium attached to the **tricuspid valve** via **chordae tendineae** (attach the tricuspid valve to the papillary mm.)
  - **Anterior papillary m.** (largest) – arises from the anterior, inner surface; attaches to the anterior and posterior cusps of the tricuspid valve
  - **Posterior papillary m.** – arises from the inferior surface; attaches to the posterior and septal cusps
  - **Septal papillary m.** – arises from the interventricular septum; attaches to the anterior and septal cusps
- **Septomarginal trabecula** (moderator band) – muscular bundle that runs between the interventricular septum and the anterior papillary m. (present in about 60% of hearts)
- **Conus arteriosus** – smooth surface of the right ventricle immediately inferior to the pulmonary trunk
- **Supraventricular crest** – separates the smooth conus arteriosus from the muscular ventricle
- **Pulmonary valve** – observe the anterior, left, and right cusps

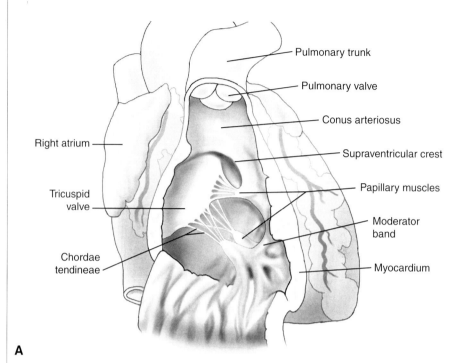

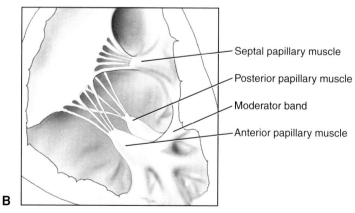

■ **Figure 6–14   Right ventricular structures. *A*, Right ventricle open. *B*, Papillary muscles of the right ventricle**

Left Atrial Incisions and Structures

■ Make an inverted U shaped incision through the wall of the left atrium, between the pulmonary vv. (Figure 6–15)

  ● Fold the flap of the atrium inferiorly; remove blood clots; wash with water

■ Identify the following:

  ● **Four pulmonary vv.** (two on each side) – enter the lateral aspects of the left atrium

  ● **Bicuspid (mitral) valve**

  ● **Pectinate mm.**

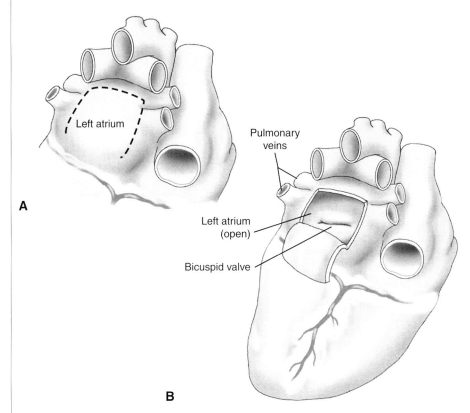

A

B

■ **Figure 6–15** **Left atrium.** *A,* **Left atrial incisions.** *B,* **Left atrium open**

Left Ventricular Incisions and Structures

■ Make the following incision through the wall of the left ventricle (Figure 6–16):

- *Note:* Some coronary vessels will be cut

- Start at the ascending aorta (inferior to the right and left coronary cusps) and continue the incision laterally and parallel to the interventricular groove

- Continue the incision through the length of the left ventricle to its apex

■ Open the left ventricle; remove blood clots; wash with water

■ Observe that the myocardium is thicker for the left ventricle than the right ventricle

■ Identify the following in the left ventricle (see Figure 6–16):

- **Bicuspid (mitral) (left AV) valve** – two papillary mm. and chordae tendineae are attached to the cusps of the bicuspid valve

- **Trabeculae carneae** – irregular muscular elevations

- **Interventricular septum** – observe the muscular and membranous regions

- **Aortic valve** – has right, left, and posterior cusps; fibrous thickenings in the middle edge of each cusp are called **nodules**

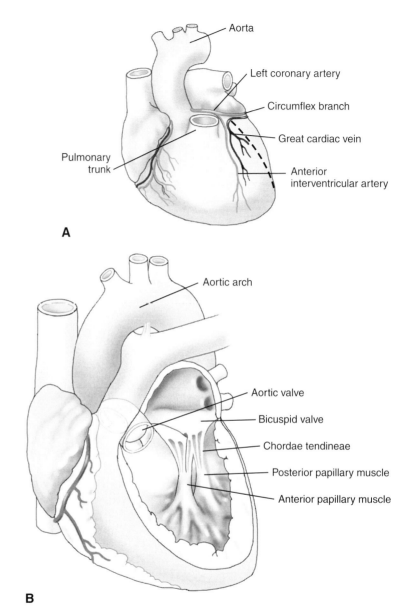

A

B

■ **Figure 6–16   Left ventricular structures.** *A,* Left ventricular incisions. *B,* Left ventricle open

Valves of the Heart

■ Identify the following cusps of the valves from a superior view (Figure 6–17):

- **Tricuspid (right AV) valve**
  - Anterior cusp
  - Septal (medial) cusp
  - Posterior cusp
- **Pulmonary semilunar valve**
  - Anterior cusp
  - Right cusp
  - Left cusp
- **Bicuspid (left AV; mitral) valve**
  - Anterior cusp
  - Posterior cusp
- **Aortic semilunar valve**
  - Right cusp – associated with the right coronary a.
  - Left cusp – associated with the left coronary a.
  - Posterior cusp

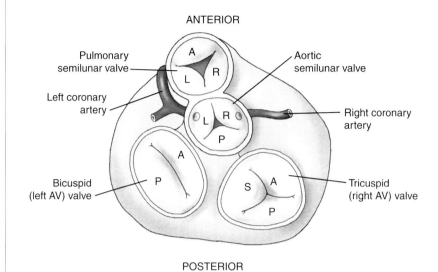

■ **Figure 6-17   Valves of the heart**

Mediastina

Prior to dissection, you should familiarize yourself with the following structures:

VEINS
- Azygos v.
- Hemiazygos v.
- Accessory hemiazygos v.
- Brachiocephalic vv.
- Intercostal vv.

ARTERIES
- Aortic arch
  - Brachiocephalic a.
  - Left common carotid a.
  - Left subclavian a.
- Descending aorta
  - Intercostal aa.
  - Bronchial aa.
  - Esophageal aa.

NERVES
- Thoracic sympathetic trunks

- Greater splanchnic nn.
- Lesser splanchnic nn.
- Cardiac plexus
- Pulmonary plexus
- Esophageal plexus
- Vagal trunks (anterior and posterior)
- Intercostal nn.

MUSCLES
- Innermost intercostal mm.
- Subcostal mm.

MISCELLANEOUS
- Trachea
  - Carina
- Esophagus
- Thoracic duct

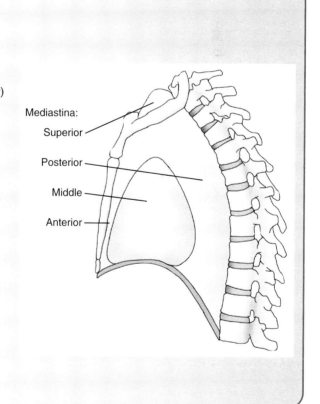

Mediastina:
Superior
Posterior
Middle
Anterior

Posterior Mediastinum

■ Identify the borders of the posterior mediastinum (Figure 7–1):

- **Vertebral bodies T5–T12**
- **Posterior surface of the pericardial sac**

■ Remove the posterior aspect of the pericardial sac (leave the pericardium attached to the diaphragm)

■ Identify the following:

- **Esophagus** – anterior and slightly to the right of the aorta
- **Descending aorta** – posterior and slightly to the left of the esophagus
- **Vagus nn.**
    - **Right vagus n.** – posterior to the root of the right lung; medial to the azygos v.
    - **Left vagus n.** – posterior to the root of the left lung; anterior to the aortic arch
        - ◆ **Left recurrent laryngeal n.** – branch from the left vagus n.; immediately left and posterior to the ligamentum arteriosum
- **Esophageal plexus** – the vagus n. branches and becomes a plexus on the esophagus (you may need to clean the esophagus to identify the plexus)
    - **Anterior vagal trunk** – the esophageal plexus on the *left* side converges to form the anterior vagal trunk at the distal end of the esophagus
    - **Posterior vagal trunk** – the esophageal plexus on the *right* side converges to form the posterior vagal trunk at the distal end of the esophagus

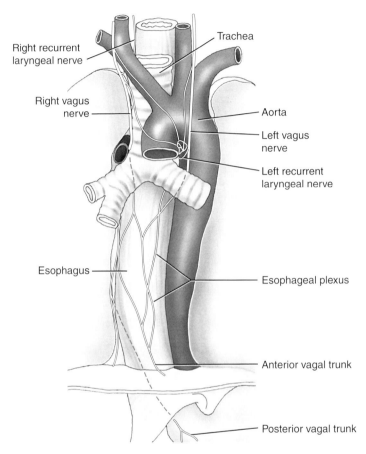

■ **Figure 7–1   Esophageal plexus and vagus nerves**

The Trachea and Related Structures

■ Identify the following (Figure 7–2):

- **Trachea** – the windpipe; located anterior to the esophagus

  - **Tracheal rings** – palpate the trachea to identify the tracheal C-shaped rings (they are not complete rings)

- **Carina** – the bifurcation of the trachea into the right and left primary bronchi; used as a landmark during bronchoscopy

- **Primary bronchi** – the trachea bifurcates into two primary bronchi

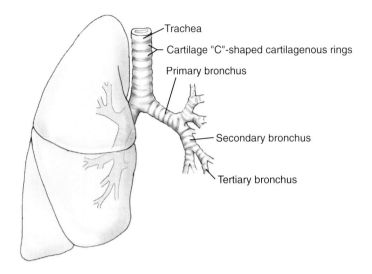

■ **Figure 7–2   Trachea**

Azygos System of Veins

■ Identify the following (Figures 7–3 and 7–4):

- **Azygos v.** – drains the right side of the thoracic wall (expect variation)

  - Arches over the superior border of the right primary bronchus to drain into the SVC

  - Follow to the posterior wall and inferiorly to the diaphragm along the right side of vertebral bodies T4–T12

    - **Right intercostal vv.** – drain into the azygos v.

- **Hemiazygos v.** – drains the left posterior, inferior portion of the thoracic wall

  - Ascends on the left side of the vertebral column, posterior to the descending aorta

  - At about T9 vertebra, crosses to the right side, posterior to the aorta, thoracic duct, and esophagus, to join the azygos v.

- **Accessory hemiazygos v.** – drains the left posterior, superior portion of the thoracic wall

  - Originates at the fourth or fifth intercostal space and descends on the left side of the vertebrae to about T8

  - At about T7 or T8 vertebra, crosses to the right side, posterior to the aorta, thoracic duct, and esophagus, to join the azygos v.

■ *Note:* Expect variation with the hemiazygos venous system; the superior and supreme intercostal vv. usually drain into the brachiocephalic vv.

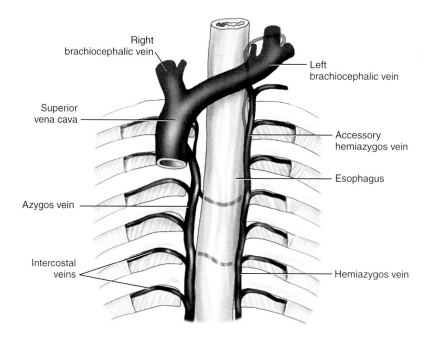

■ **Figure 7–3   Azygos venous system**

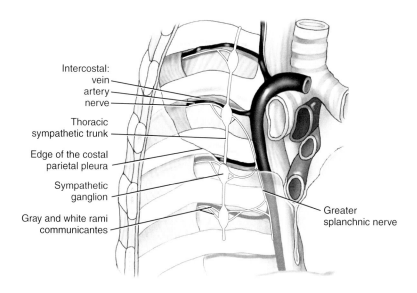

■ **Figure 7–4   Lateral view of the azygos vein**

Lymphatic System in the Thorax

■ Identify the following (Figure 7–5):

- **Thoracic duct** – the largest lymphatic vessel in the body; to identify it, pull the esophagus to the right side

  - Located between the descending aorta on its left and esophagus on its right; 3–4 mm in diameter and may look dull white

  - Terminates superiorly by emptying into the junction of the left internal jugular v. and left subclavian v. (studied further in another laboratory session)

- **Tracheobronchial lymph nodes** – located on either side of the tracheal bifurcation

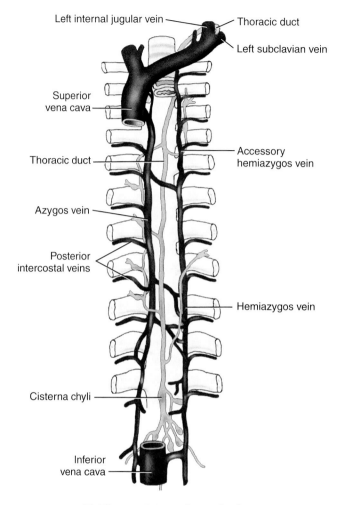

Left internal jugular vein

Thoracic duct

Left subclavian vein

Superior vena cava

Accessory hemiazygos vein

Thoracic duct

Azygos vein

Posterior intercostal veins

Hemiazygos vein

Cisterna chyli

Inferior vena cava

■ **Figure 7–5   Thoracic duct**

Nerves in the Posterior Mediastinum

■ Peel the parietal pleura and endothoracic fascia from the posterior rib cage

■ Identify the following (Figure 7–6):

- **Thoracic sympathetic trunks** – located bilaterally:
  - Heads of the ribs in the superior part of the thorax
  - Costovertebral joints in the midthorax
  - Vertebral bodies in the inferior part of the thorax
- **Rami communicantes** – located between the sympathetic trunks and intercostal nn.
- **Greater splanchnic nn.** (spinal levels T5–T9) – located on the lateral surface of the vertebral bodies T6–T12
- **Lesser splanchnic nn.** (spinal levels T10–T11) – located on the lateral surface of the vertebral bodies T10–T11
- **Least splanchnic nn.** (spinal level T12) – located on the lateral surface of T12 vertebral body
- **Intercostal neurovascular bundles (v.a.n.)** – review them

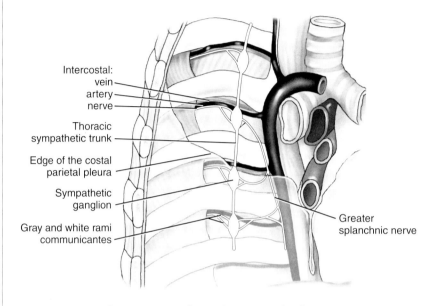

Intercostal:
vein
artery
nerve

Thoracic
sympathetic trunk

Edge of the costal
parietal pleura

Sympathetic
ganglion

Gray and white rami
communicantes

Greater
splanchnic nerve

■ **Figure 7–6   Thoracic sympathetic nerves**

Superior Mediastinum

- Identify the inferior border of the superior mediastinum:

  - A horizontal plane from the sternal angle to the T4–T5 intervertebral disc

- To gain wider access to this region, remove the inferior portion of the manubrium with a bone saw or rib cutter (leave the sternoclavicular and sternocostal joints for rib 1 intact)

- Cut the superior vena cava transversely, immediately superior to the azygos v., and reflect the cut end superiorly

- Identify the following (Figure 7–7):

  - **Brachiocephalic vv.** – unite to form the superior vena cava

  - **Aortic arch** – located deep to the brachiocephalic vv.; arches over the left bronchus, becomes the descending thoracic aorta (the thread-sized nerves in this area are part of the **cardiac plexus**)

  - Branches of the aortic arch:

    - **Brachiocephalic trunk** – first branch; 1–2 inches long; bifurcates to form the right common carotid and right subclavian aa.

    - **Left common carotid a.** – second branch

    - **Left subclavian a.** – third branch

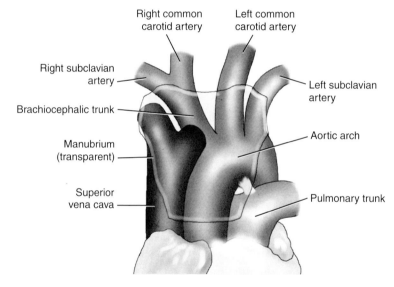

■ **Figure 7-7   Aortic arch branches**

Superior Mediastinum—cont'd

- Branches of the descending thoracic aorta (Figure 7–8):

  - **Esophageal aa.** – very small and hard to identify

  - **Bronchial aa.** – small; look at the hilum of either lung to identify these vessels

  - **Intercostal aa.** – originate from the descending thoracic aorta; easiest to find

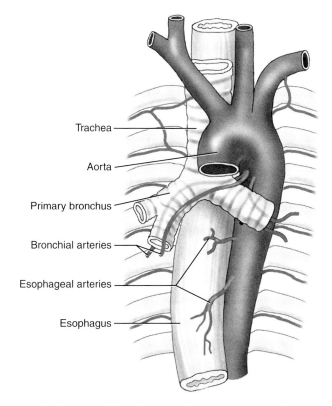

Trachea

Aorta

Primary bronchus

Bronchial arteries

Esophageal arteries

Esophagus

■ **Figure 7–8   Descending aorta branches**

Unit 1   Back and Thorax Overview

At the end of Unit 1 you should be able to identify the following structures on cadavers, skeletons, and/or radiographs:

Osteology

- Occipital bone
  - External occipital protuberance
- Vertebral column (33 vertebrae)
  - Cervical vertebrae (7)
  - Thoracic vertebrae (12)
  - Lumbar vertebrae (5)
  - Sacrum (5 fused vertebrae)
  - Coccyx (4 fused vertebrae)
- Vertebral structures
  - Spinous and transverse processes
  - Laminae and pedicles
  - Bodies and intervertebral discs
  - Articular processes (superior and inferior)
  - Vertebral foramina
  - Intervertebral foramina (notches)
- Ilium
  - Iliac crest
  - Posterior superior iliac spine
- Ribs
  - Head, tubercle, angle
- Scapula
  - Spine and acromion
  - Coracoid process
  - Lateral and medial margins
  - Superior and inferior angles

- Sternum
  - Manubrium and jugular notch
  - Sternal angle
  - Body and xiphoid process
- Clavicle
- Sternoclavicular joint

Muscles

- Extrinsic muscles of the back
  - Superficial layer
    - Trapezius m.
    - Latissimus dorsi m.
  - Middle layer
    - Rhomboid major m.
    - Rhomboid minor m.
    - Levator scapulae m.
  - Deep layer
    - Serratus posterior superior m.
    - Serratus posterior inferior m.
- Splenius capitis and cervicis mm.
- Intrinsic muscles of the back
  - Superficial layer
    - Erector spinae group
      - Iliocostalis m.
      - Longissimus m.
      - Spinalis m.
  - Deep layer
    - Transversospinalis group

- Semispinalis m.
- Multifidus m.
- Rotatores m.
      - Intertransversarii and interspinales mm.
      - Levatores costarum mm.
- Suboccipital mm.
  - Rectus capitis posterior major m.
  - Rectus capitis posterior minor m.
  - Obliquus capitis superior m.
  - Obliquus capitis inferior m.
- Pectoralis major and minor mm.
- Subclavius m.
- Intercostal mm. (external, internal, innermost)
- Transversus thoracis m.
- Subcostal mm.

Vertebral Ligaments

- Nuchal ligament and nuchal fascia
- Supraspinous
- Interspinous
- Ligamentum flavum
- Intertransverse

Miscellaneous

- Triangle of auscultation

- Lumbar triangle
- Thoracolumbar fascia
- Deltopectoral triangle
- Clavipectoral fascia
- Midaxillary line
- Parietal pleura
- Sternocostal joints
- Trachea
- Carina
- Esophagus
- Thoracic duct
- Thymus

Breast
- Nipple, lactiferous ducts, and areola
- Suspensory ligaments
- Retromammary space
- Mammary gland

Vessels (arteries and veins)
- Superficial/deep branches of transverse cervical a.
- Intercostal vessels
  - Posterior, lateral, and anterior cutaneous aa. and vv.
- Cephalic v.
- Subclavian a. and v.
  - Vertebral a.
  - Internal thoracic a. and v.
- Axillary a.
  - Thoracoacromial trunk
    - Deltoid, pectoral, acromial, and clavicular branches
- Superior vena cava
  - Brachiocephalic vv.
  - Azygos v.

- Hemiazygos v.
- Accessory hemiazygos v.
- Aortic arch
  - Brachiocephalic a.
  - Left common carotid a.
  - Left subclavian a.
- Descending thoracic aorta
  - Intercostal aa.
  - Bronchial aa.
  - Esophageal aa.

Nerves
- Dorsal ramus of C1
- Greater occipital n. (dorsal ramus of C2)
- Dorsal ramus of C3
- Spinal accessory n. (CN XI)
- Intercostal nerves
- Posterior, lateral, and anterior cutaneous nn.
- Lateral and medial pectoral nn.
- Phrenic nn.
- Vagus nn.
  - Recurrent laryngeal nn.

Spinal Cord
- Meninges
  - Dura mater
    - Epidural and subdural spaces
    - Dural sac
  - Arachnoid mater
    - Subarachnoid space
  - Pia mater
    - Denticulate ligaments
    - Filum terminale
- Spinal cord
  - Posterior median sulcus
  - Cervical and lumbar spinal enlargements
  - Conus medullaris

- Ventral and dorsal roots
  - Spinal nerves
    - Primary ventral rami
    - Primary dorsal rami
      - Dorsal root ganglion
    - Cauda equina

Lung
- Endothoracic fascia
- Pleural cavity
  - Parietal pleura
    - Costal, mediastinal, diaphragmatic, and cervical pleurae
  - Pleural space
  - Visceral pleura (same regions as the parietal pleura)
- Pleural recesses
  - Costomediastinal recess
  - Costodiaphragmatic recess
- Left lung
  - Apex
  - Costal, mediastinal and diaphragmatic surfaces
  - Hilum
    - Pulmonary ligament
    - Bronchi, bronchial aa., lymph nodes, pulmonary aa. and vv.
  - Superior/inferior lobes and oblique fissure
  - Cardiac notch and lingula
  - Contact impressions
    - Cardiac impression
    - Aortic groove impression
    - Descending aorta impression
  - Bronchi
    - Primary, secondary, and tertiary
  - Hilar lymph nodes
- Right lung
  - Apex
  - Costal, mediastinal, and diaphragmatic surfaces

- Hilum
  - Pulmonary ligament
  - Bronchi, bronchial aa., pulmonary aa. and vv.
- Superior, middle, and inferior lobes
- Oblique and horizontal fissures
- Contact impressions
  - Cardiac impression
  - Esophageal groove
- Bronchi
  - Primary, secondary, and tertiary
- Hilar lymph nodes

Heart

- Pericardial sac
  - Parietal pericardium
    - Fibrous and serous layers
  - Pericardial space
  - Visceral pericardium (epicardium)
  - Transverse and oblique pericardial sinuses
  - Ligamentum arteriosum
- Heart
  - Right and left atria and ventricles
  - Atrioventricular (AV) sulcus
  - Interventricular groove
  - Apex and base

- Coronary arteries
  - Left coronary a.
    - Anterior interventricular branch
    - Circumflex branch
      - Obtuse marginal branch
  - Right coronary a.
    - Atrial a.
      - Sinuatrial (SA) branch
    - Acute marginal branch
    - Posterior interventricular branch
- Coronary veins
  - Coronary sinus
    - Great, middle, and small cardiac vv.
    - Anterior cardiac vv.
- Right atrium (auricle)
  - Pectinate mm.
  - Sulcus and crista terminalis
  - Sinus venarum
  - Superior and inferior venae cavae
  - Coronary sinus
  - Fossa ovalis and tricuspid (right AV) valve
  - SA and AV nodes
- Right ventricle
  - Tricuspid valve (right AV valve)
    - Anterior, posterior, and septal cusps

- Chordae tendineae and papillary muscles
- Trabeculae carneae
- Septomarginal trabecula (moderator band)
- Conus arteriosus
- Supraventricular crest
- Pulmonary trunk
  - Pulmonary valve (right, left, and anterior cusps)
- Left atrium (auricle)
  - Pulmonary vv. (4)
  - Bicuspid orifice
  - Foramen ovale / fossa ovalis
- Left ventricle
  - Bicuspid valve (mitral or left AV valve)
  - Anterior and posterior cusps
  - Chordae tendineae
  - Papillary mm.
  - Trabeculae carneae
  - Orifice for the aorta
  - Interventricular septum
- Aorta
  - Aortic valve (right, left, and posterior cusps)

Abdomen, Pelvis, and Perineum

2

Anterior Abdominal Wall and Inguinal Canal

Lab 8

Prior to dissection, you should familiarize yourself with the following structures:

OSTEOLOGY

- Ribs (costal margin)
- Xiphoid process
- Pubis
  - Pubic symphysis
  - Pubic crest and tubercle
  - Pecten pubis
- Ilium
  - Iliac crest
  - Anterior superior iliac spine

MUSCLES

- Rectus abdominis m.
  - Tendinous intersections
  - Rectus sheath
    - Anterior and posterior layers
    - Arcuate line
- Pyramidalis m.
- External oblique m.
  - Inguinal ligament
    - Lacunar ligament
    - Superficial inguinal ring
- Internal oblique m.

- Transversus abdominis m.
  - Conjoint tendon (inguinal falx)
  - Deep inguinal ring

NERVES AND VESSELS

- Internal thoracic a. and v.
  - Superior epigastric aa. and vv.
  - Musculophrenic aa. and vv.
- Subcostal n. (T12)
- Iliohypogastric n. (L1)
- Ilioinguinal n. (L1)
- Genital branch of the genitofemoral n. (L1–L2)
- Lower intercostal and lumbar v.a.n.
  - Lateral cutaneous v.a.n.
  - Anterior cutaneous v.a.n.
- External iliac a.
  - Inferior epigastric a. and v.
  - Deep circumflex iliac a. and v.
- Femoral a.
  - Superficial circumflex iliac a. and v.
  - Superficial epigastric a. and v.

- Lateral thoracic v.
  - Thoracoepigastric v.
- Paraumbilical vv.

SUPERFICIAL INGUINAL RING

- Medial and lateral crura
- Intercrural fibers
- Spermatic cord/round ligament of the uterus

ANTERIOR ABDOMINAL WALL (INTERNAL)

- Median umbilical fold (obliterated urachus)
- Medial umbilical folds (obliterated umbilical aa.)
- Lateral umbilical folds (inferior epigastric aa. and vv.)
- Superior epigastric aa. and vv.
- Ligamentum teres (obliterated umbilical v.)
- Falciform ligament (liver)
  - Ligamentum teres hepatis

Continued

Layers of the Anterolateral Abdominal Wall

- Skin
- Superficial fascia
  - Superficial (fatty) layer (Camper's fascia)
  - Deep (membranous) layer (Scarpa's fascia)

- External oblique m./aponeurosis
- Internal oblique m./aponeurosis
- Transversus abdominis m./aponeurosis
- Transversalis fascia
- Extraperitoneal fat
- Parietal peritoneum

OTHER STRUCTURES

- Umbilicus
- Linea alba
- Linea semilunaris
- Inguinal triangle
- Inguinal canal

TABLE 8-1　Muscles of the Anterior Abdominal Wall

Muscle	Proximal Attachments	Distal Attachments	Action	Innervation
External oblique m.	External surface of ribs 5–12, latissimus dorsi m., and thoracolumbar fascia	Linea alba, pubic tubercle, iliac crest	Compress and support abdominal viscera (micturition, defecation, forced expiration) Rectus abdominis, external oblique, and internal oblique: flex the trunk External and internal oblique: lateral flexion and rotation of the trunk	Intercostal nn. 8–12 and subcostal n. (internal oblique and transversus abdominis mm. also include first lumbar n.)
Internal oblique m.	Thoracolumbar fascia, anterior two thirds of the iliac crest, and lateral half of the inguinal ligament	Cartilages of ribs 10–12, linea alba, and the pecten pubis by way of the conjoint tendon		
Transversus abdominis m.	Costal cartilages 7–12, thoracolumbar fascia, iliac crest, and lateral third of the inguinal ligament	Linea alba with the aponeurosis of internal oblique m., and the pecten pubis (conjoint tendon)		
Rectus abdominis m.	Pubic crest and pubic symphysis	Costal cartilages 5–7, xiphoid process		
Pyramidalis m.	Pubis	Linea alba midway between the umbilicus and pubis	Tenses the linea alba	Subcostal n.

TABLE 8–2 Layers of the Abdominal Wall and Scrotum (Male)

Layers of the Abdominal Wall	Corresponding Layers in the Scrotum

1. Skin
Superficial fascia
   2. Fatty (Camper's)
 	
   3. Membranous (Scarpa's)
4. External oblique m.
5. Internal oblique m.
6. Transversus abdominis m.
7. Transversalis fascia
8. Extraperitoneal fatty tissue
9. Peritoneum
	1. Skin
	
	2. Disappears because the scrotum contains no fat
	3. Dartos muscle and fascia
	4. External spermatic fascia
	5. Cremaster muscle and fascia
	6. Cremaster muscle and fascia
	7. Internal spermatic fascia
	8. Areolar tissue with a little fat
	9. Tunica vaginalis

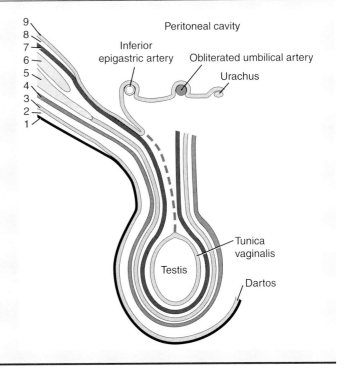

Orientation

■ Locate the following surface structures (Figure 8–1):

- **Xiphoid process**
- **Costal margin**
- **Iliac crest**
- **Anterior superior iliac spine**
- **Inguinal ligament**
- **Pubic crest and tubercle**
- **Umbilicus**

■ During this dissection, you will explore nine layers of tissue that make up the anterolateral abdominal wall

1. Skin

2. Camper's fascia (fatty layer)

3. Scarpa's fascia (membranous layer)

4. External oblique m. – enveloped by deep fascia

5. Internal oblique m. – enveloped by deep fascia

6. Transversus abdominis m. – enveloped by deep fascia

7. Transversalis fascia

8. Extraperitoneal fat

9. Parietal peritoneum

■ Refer to Figure 8–2 throughout this dissection as reference

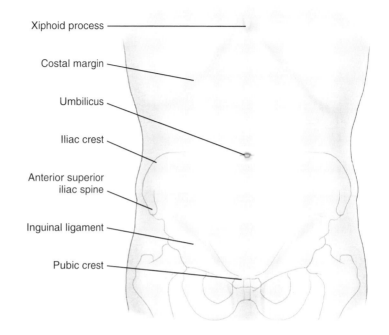

Xiphoid process

Costal margin

Umbilicus

Iliac crest

Anterior superior iliac spine

Inguinal ligament

Pubic crest

■ **Figure 8–1   Abdominal surface anatomy**

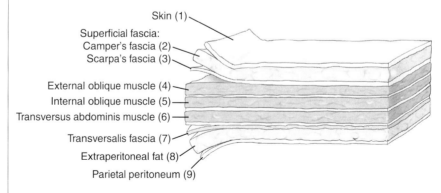

Skin (1)

Superficial fascia:
Camper's fascia (2)
Scarpa's fascia (3)

External oblique muscle (4)

Internal oblique muscle (5)

Transversus abdominis muscle (6)

Transversalis fascia (7)

Extraperitoneal fat (8)

Parietal peritoneum (9)

■ **Figure 8–2   Layers of the anterolateral abdominal wall**

Abdominal Skin Incisions

■ Make the following incisions to remove the skin, but leave the superficial fascia (fat) intact (Figure 8–3):

- Xiphoid process (*A*) to the pubic symphysis (*B*) (cut around the umbilicus)
- Pubic symphysis (*B*) laterally to the iliac crest (*C*) (1 cm inferior to the inguinal ligament)
- Xiphoid process (*A*) laterally to the midaxillary line

■ After the incisions are made, reflect the skin laterally to expose the superficial fascia

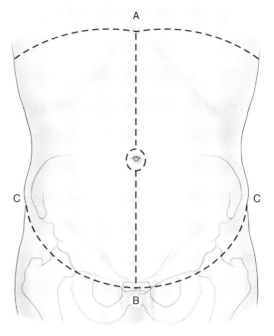

■ **Figure 8–3   Abdominal skin incisions**

Superficial Fascia

■ Identify the two planes of the subcutaneous layer (superficial fascia) of the anterolateral abdominal wall (Figure 8–4):

- **Fatty (superficial) layer (Camper's fascia)**

    - Fat may be several centimeters thick

- **Membranous (deep) layer (Scarpa's fascia)**

    - In the lower abdominal area, make an incision in the superficial fascia, as shown in Figure 8–4

    - Reflect the margins of the cut fascia to identify Camper's fascia externally and Scarpa's fascia internally

■ Explore the space between Scarpa's fascia and the deep fascia

- Insert a finger into the space between Scarpa's fascia and the rectus sheath

- Push your finger in all directions (especially inferiorly into the scrotal sac/labia majora and toward the perineum)

- Note the **fundiform ligament** superior to the penis/clitoris, in the midline

- *Note:* You cannot push your finger into the thigh because Scarpa's fascia is fused to the fascia lata (deep fascia of the thigh) 2.5 cm inferior to the inguinal ligament

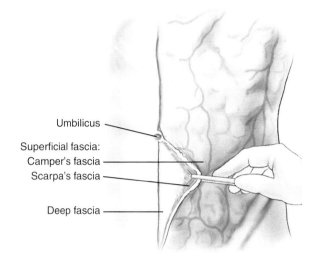

Umbilicus
Superficial fascia:
Camper's fascia
Scarpa's fascia
Deep fascia

■ **Figure 8–4   Superficial fascia**

Cutaneous Nerves

■ Identify the following (Figure 8–5):

- **Thoracoepigastric v.** – tributary of the lateral thoracic v., which is a tributary of the axillary v.

- **Superficial epigastric v.** – probe through the fat between the umbilicus and inguinal region; tributary to the femoral v.

- **Superficial circumflex iliac v.** – inferior and parallel to the inguinal ligament; tributary to the femoral v.

- **Paraumbilical vv.** – tributary to the hepatic portal v.

- **Anterior cutaneous nn.** – immediately lateral to the linea alba; pierce the anterior wall of the rectus sheath

  - **T7, T8,** and **T9 nn.** supply the region superior to the umbilicus

  - **T10 n.** supplies skin around the umbilicus

  - **T11, T12 (subcostal n.),** and **L1 (iliohypogastric and ilioinguinal nn.)** supply the region inferior to the umbilicus; the iliohypogastric n. is about 4 cm superior to the pubic symphysis, just superior to the superficial inguinal ring

- **Lateral cutaneous nn.** – emerge through the serratus anterior and external oblique mm.

- **T12 (subcostal) nn.** – superior to the iliac crest and descend to supply that region of the skin

- **L1 (iliohypogastric** and **ilioinguinal) nn.** – superior to the iliac crest (just inferior to the subcostal n.) and descend to supply that region of the skin

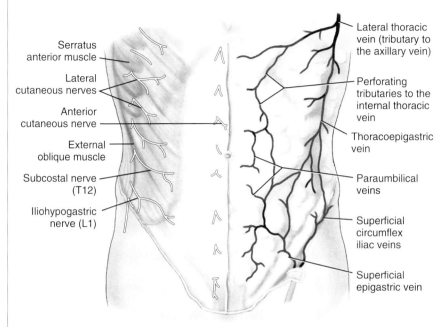

■ **Figure 8–5   Cutaneous nerves**

Labels in figure:
- Serratus anterior muscle
- Lateral cutaneous nerves
- Anterior cutaneous nerve
- External oblique muscle
- Subcostal nerve (T12)
- Iliohypogastric nerve (L1)
- Lateral thoracic vein (tributary to the axillary vein)
- Perforating tributaries to the internal thoracic vein
- Thoracoepigastric vein
- Paraumbilical veins
- Superficial circumflex iliac veins
- Superficial epigastric vein

Anterior Abdominal Wall Musculature

- Remove all remains of superficial fascia

- Identify the following (Figure 8–6):

  - **External oblique m.** – observe the direction of its striations, which terminate near the plane of the anterior superior iliac spine

  - **External oblique aponeurosis** – the anterior tendinous membrane of the external oblique m.; forms part of the anterior lamina of the **rectus sheath**

  - **Rectus sheath** – a tendinous sheath formed by the aponeuroses of the external and internal oblique and transversus abdominis mm.; surrounds the rectus abdominis m.

  - **Linea semilunaris** – superficial crease formed by the lateral border of the rectus sheath

  - **Tendinous intersections** – superior, middle, and inferior tendinous slips that anchor the rectus abdominis m. to the anterior layer of the rectus sheath

  - **Linea alba** – strong tendinous seam between the two rectus sheaths

  - **Pyramidalis m.** (often absent) – located deep to the anterior layer of the rectus sheath and anterior to the inferior-most portion of the rectus abdominis m.

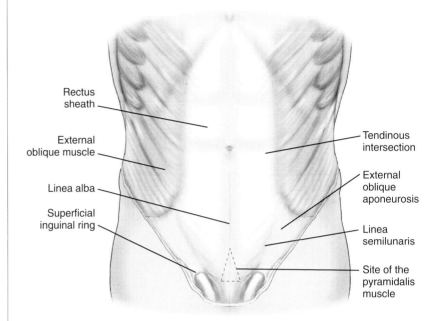

Rectus sheath

External oblique muscle

Linea alba

Superficial inguinal ring

Tendinous intersection

External oblique aponeurosis

Linea semilunaris

Site of the pyramidalis muscle

■ **Figure 8–6   External oblique muscle**

Superficial Inguinal Ring

- Identify the following (Figure 8–7):

  - **Superficial inguinal ring** – triangular passage through the external oblique aponeurosis, 2.5 cm superolateral to the pubic tubercle; traversed by the spermatic cord (male) or round ligament of the uterus (female); borders of the ring are the following:

    - **Medial crus**
    - **Intercrural fibers**
    - **Lateral crus**

  - **Spermatic cord/round ligament of the uterus** – passes through the superficial inguinal ring

  - **Inguinal ligament** – the inferior border of the external oblique aponeurosis between the anterior superior iliac spine and the pubic tubercle

- *Note:* A thin layer of connective tissue overlays the superficial inguinal ring; you will need to remove it from the crura to get a clear view of the ring

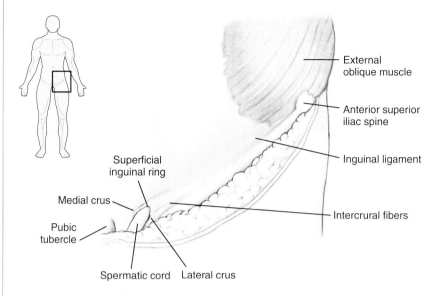

- **Figure 8–7   Superficial inguinal ring**

Abdominal Wall Incisions

- Make the following abdominal incisions (*do not* cut into the underlying viscera) through the anterolateral abdominal wall (Figure 8–8):

  - Xiphoid process (*A*) to just above the umbilicus (*B*), along the linea alba

  - Extend the incision from *B* to about 2 cm superior to the anterior superior iliac spine (*C*) on both sides

- *Note:* For the following pages, panels 1, 2, and 3 from Figure 8–8 will be used as reference for regional study of the anterior abdominal wall

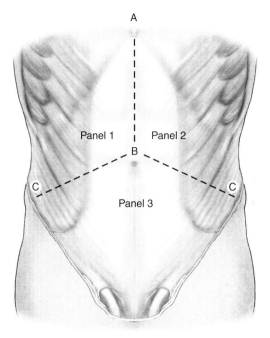

■ **Figure 8–8   Abdominal wall incisions**

Muscles of the Anterior Abdominal Wall

- Use a probe to separate the layers in the cut edges of panels 1 and 2 of the abdomen

- Identify the following (Figure 8–9):

  - **External oblique m.** – superficial muscle layer

  - **Internal oblique m.** – intermediate muscle layer; the striations course at right angles to the external oblique m.

  - **Transversus abdominis m.** – deep muscle layer; the striations parallel those of the internal oblique m.

  - **Rectus abdominis m.** – courses longitudinally from the sternum and costal margin to the pubis; the paired muscles are separated by the linea alba

    - Dissect the anterior layer of the rectus sheath from the rectus abdominis m.; the rectus sheath is thicker inferiorly

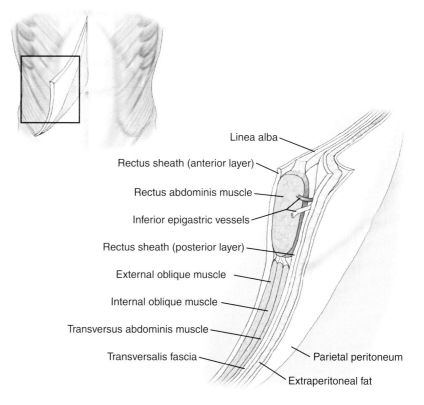

Linea alba

Rectus sheath (anterior layer)

Rectus abdominis muscle

Inferior epigastric vessels

Rectus sheath (posterior layer)

External oblique muscle

Internal oblique muscle

Transversus abdominis muscle

Transversalis fascia

Parietal peritoneum

Extraperitoneal fat

■ **Figure 8–9   Anterior abdominal wall muscles**

Anterior Abdominal Wall – Posterior Layer of the Rectus Sheath

■ Identify the following (Figures 8–10 and 8–11):

- **Rectus sheath** – the posterior layer is formed by the aponeuroses of the internal oblique and transversus abdominis mm.

  - Below the **arcuate line,** located between the umbilicus and the pubic symphysis, the posterior layer passes anterior to the rectus abdominis m.

    - ◆ At the level of arcuate line, the inferior epigastric vessels enter the rectus sheath

- **Transversalis fascia** – a thin aponeurotic membrane between the transversus abdominis m., rectus abdominis m., and the extraperitoneal fat

- **Extraperitoneal fat** – located between the transversalis fascia and parietal peritoneum

- **Parietal peritoneum** – serous lining on the internal surface of the abdominal wall

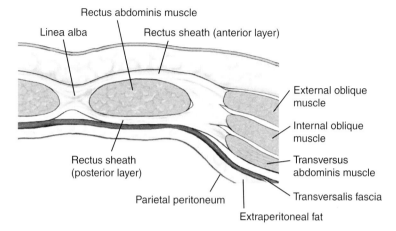

■ **Figure 8–10   Cross-section of the rectus sheath superior to the arcuate line**

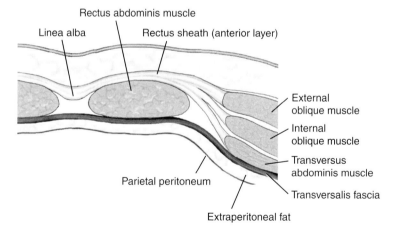

■ **Figure 8–11   Cross-section of the rectus sheath inferior to the arcuate line**

Epigastric Vessels Deep to the Rectus Abdominis Muscle

■ Separate the rectus abdominis m. from the posterior layer of the rectus sheath

■ Identify the following (Figure 8–12):

- **Lower intercostal nn.** and **vessels** – not shown

- **Subcostal and lumbar vessels** – course between the internal oblique and transversus abdominis mm. (not shown)

- **Superior epigastric vessels** – terminal branches of the internal thoracic vessels

- **Inferior epigastric vessels** – branches of the external iliac vessels, just superior to the inguinal ligament; course superiorly in the transversalis fascia to enter the rectus sheath below the arcuate line

  - These vessels anastomose on the deep surface of the rectus abdominis m.

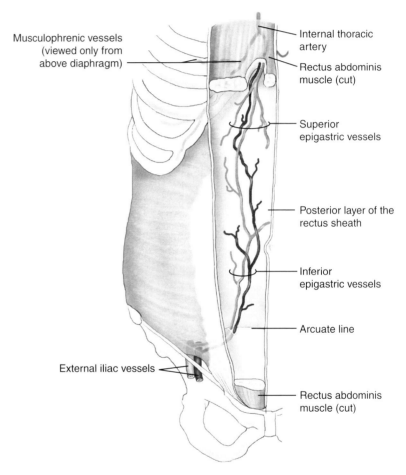

Musculophrenic vessels (viewed only from above diaphragm)

Internal thoracic artery

Rectus abdominis muscle (cut)

Superior epigastric vessels

Posterior layer of the rectus sheath

Inferior epigastric vessels

Arcuate line

External iliac vessels

Rectus abdominis muscle (cut)

■ **Figure 8-12   Epigastric vessels**

Anterior Abdominal Wall – Internal Surface

■ Identify the following structures on the internal surface of the abdominal wall (Figure 8–13):

- **Med*ian* umbilical fold (ligament)** – unpaired fold in the midsagittal plane; formed by the urachus (obliterated allantois)

- **Med*ial* umbilical fold*s* (ligaments)** – paired folds lateral to the median umbilical fold; formed by the obliterated umbilical arteries

- **Lateral umbilical fold*s* (ligament)** – paired folds lateral to the medial umbilical folds; formed by the **inferior epigastric vessels**

- **Falciform ligament** – mesenteric fat that spans between the umbilicus and the liver (studied further in another laboratory session)

  - **Ligamentum teres hepatis** – ligamentous cord in the free margin of the falciform ligament; formed by the obliterated umbilical v.

- **Deep inguinal ring** – appears as an oval defect on the internal surface of the abdominal wall; located midway between the anterior superior iliac spine and the pubic symphysis, superior to the inguinal ligament and lateral to the inferior epigastric vessels

- **Ductus deferens (male)**

- **Round ligament of the uterus (female)**

- **Inguinal triangle** – bounded by the rectus abdominis m. medially, inguinal ligament inferiorly, and inferior epigastric vessels laterally

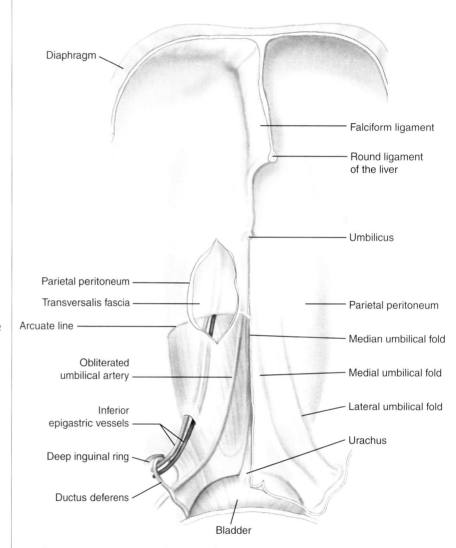

■ **Figure 8–13   Internal view of the anterior abdominal wall with the peritoneum removed from the left side**

Inguinal Canal – Borders

- **Inguinal canal** – the passageway through the inferior part of the abdominal wall that lies parallel and superior to the inguinal ligament; located between the superficial and deep inguinal rings

- Make a cut through the external abdominal aponeurosis, about 5 cm above the inguinal ligament, from the superficial to deep inguinal ring (Figure 8–14*A*)

- Identify the following borders of the canal (Figure 8–14*B*, *C*):

  - **External opening** – superficial inguinal ring

  - **Roof** – fascia of the internal oblique and transversus abdominis mm.

  - **Floor** – formed by the lacunar and inguinal ligaments

    - **Lacunar ligament** – formed by the external oblique aponeurosis; wraps underneath and medial to the contents of the inguinal canal; attaches along the pectineal line

    - **Pectineal ligament** – fibers of the lacunar ligament that continue laterally along the pecten pubis

  - **Anterior wall** – aponeurosis of the external oblique m.

  - **Posterior wall** – transversalis fascia and conjoint tendon

    - **Conjoint tendon** – formed by the fusion of the inferior, medial fibers of both the internal oblique and transversus abdominis mm.; attaches to the pubic crest and pecten pubis

  - **Internal opening** – deep inguinal ring

A

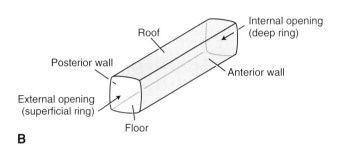

B

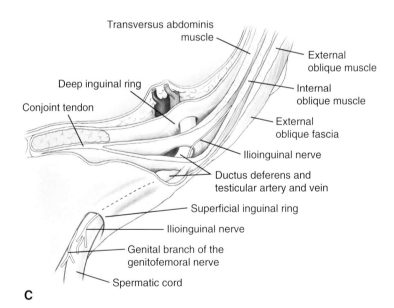

C

- **Figure 8-14   Inguinal canal borders. *A*, Incision lines. *B*, Schematic of the inguinal canal. *C*, Inguinal region of the anterior abdominal wall on the left side**

BOTH SEXES

Inguinal Canal – Contents

■ Identify the following (Figure 8–15):

- **Ilioinguinal n. (L1)** – descends into the inguinal canal between the internal and external oblique mm.

  - **Male** – traverses the inguinal canal, coursing along the inferior aspect of the spermatic cord to the superficial inguinal ring, where it contributes to the upper thigh scrotum and penis.

  - **Female** – same course but supplies the labia majora

- **Genital branch** of the **genitofemoral n.** (L1–L2) – enters the inguinal canal through the deep inguinal ring

  - **Male** – supplies the cremaster m. and the scrotal skin

  - **Female** – accompanies the round ligament of the uterus and ends within the skin of the labia majora and mons pubis

- **Round ligament of the uterus** – main occupant in the female; courses from the uterus (pelvic cavity), through the deep inguinal ring, through the inguinal canal; exits via the superficial inguinal ring and spreads into fibrous strands within the labia majora

- **Spermatic cord** – main occupant in the male; begins at the deep inguinal ring lateral to the inferior epigastric vessels and ends in the scrotum at the posterior border of the testis

■ The inguinal canal also contains blood and lymphatic vessels

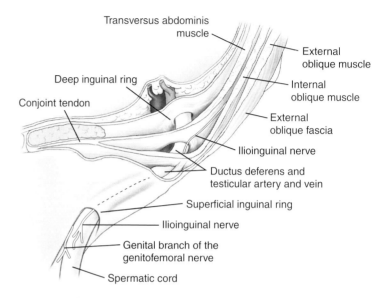

■ **Figure 8–15   Contents of the inguinal canal**

Peritoneum and Foregut

Prior to dissection, you should familiarize yourself with the following structures:

PERITONEUM

- Mesentery
  - Root of the mesentery
- Mesocolon
  - Transverse mesocolon
  - Sigmoid mesocolon
  - Mesoappendix
- Lesser omentum
  - Hepatophrenic ligament
  - Hepatogastric ligament
  - Hepatoduodenal ligament
- Greater omentum
  - Gastrosplenic ligament
  - Splenorenal ligament
- Peritoneal attachments of the liver
  - Coronary ligament
    - Falciform ligament
    - Triangular ligaments
    - Hepatorenal ligament
- Recesses, fossae, and folds
- Peritoneal cavity
  - Greater sac

- Lesser sac (omental bursa)
  - Omental (epiploic) foramen
- Paravesical fossa
- Rectovesical pouch
- Vesicouterine pouch
- Rectouterine pouch

ARTERIES

- Celiac trunk
  - Left gastric a.
    - Esophageal a.
  - Splenic a.
    - Left gastroepiploic a.
    - Short gastric aa.
  - Common hepatic a.
    - Hepatic a. proper
      - Right gastric a.
      - Left and right hepatic aa.
      - Cystic a.
    - Gastroduodenal a.
      - Supraduodenal a.
      - Anterior/posterior superior pancreaticoduodenal a.
      - Right gastroepiploic a.

STOMACH

- Greater and lesser curvatures
- Pylorus and pyloric sphincter
- Body, cardia, and fundus
- Gastric rugae

SMALL INTESTINE

- Duodenum
  - Duodenojejunal junction
  - Suspensory ligament of the duodenum
- Jejunum and ileum
- Ileocecal junction

LARGE INTESTINE

- Cecum
  - Vermiform appendix
- Ascending colon
- Transverse colon
- Descending colon
- Sigmoid colon
- Rectum
- Anus

LIVER
- Diaphragmatic and visceral surfaces
- Left and right lobes
- Quadrate and caudate lobes
- Falciform and round ligaments (ventral mesentery)

- Porta hepatis
- Subphrenic recess
- Bare area
- Common hepatic duct
  - Cystic duct
  - Common bile duct

OTHER ABDOMINAL ORGANS
- Gallbladder
- Spleen
- Pancreas

Peritoneal Cavity – Greater Omentum

■ As a result of incisions already made through the abdominal wall (previous lab), you have cut through the anterior layer of the parietal peritoneum

■ Identify the following (Figures 9–1 and 9–2):

• **Peritoneal cavity** (greater sac) – the potential space between the visceral and parietal layers of peritoneum; the peritoneal cavity is divided into two regions—the greater and lesser sacs:

  ● **Greater sac** – the space between the parietal and visceral peritoneum, not including the sac deep to the stomach (lesser sac)

• **Greater omentum** – apron of adipose and connective tissue covering the abdominal viscera; consists of a double sheet that folds upon itself to create four layers of peritoneum:

  ● One sheet descends from the anterior surface of the stomach, the other from the posterior surface of the stomach

  ● The sheets fuse at the **greater curvature** of the stomach and descend, in front of the small intestine, toward the pelvis

  ● This membrane turns upon itself and ascends; at the **transverse colon** this membrane splits to surround the transverse colon

    ◆ *Note:* In most adults, the four layers are fused

  ● The left border of the greater omentum is continuous with the **gastrosplenic ligament;** the right border extends to the **duodenum;** expect variations

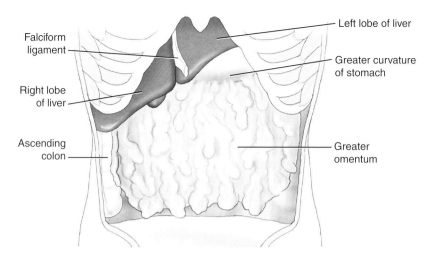

■ Figure 9–1   Greater omentum in situ

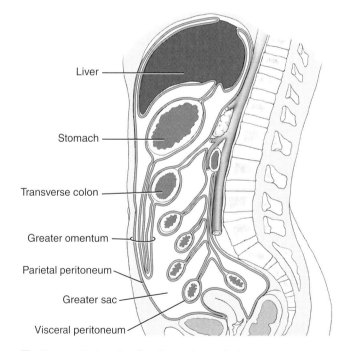

■ Figure 9–2   Sagittal section of the abdominopelvic cavity

Peritoneal Cavity – Lesser Omentum

■ Identify the following (Figures 9–3 and 9–4):

● **Lesser omentum** – two layers of peritoneum that course from the anterior and posterior surfaces of the stomach and proximal duodenum, fuse at the **lesser curvature** of the stomach, and ascend as a double fold to the liver; two ligaments comprise the lesser omentum:

  ● **Hepatogastric ligament** – the part of the lesser omentum between the liver and the stomach

  ● **Hepatoduodenal ligament** – the part of the lesser omentum between the liver and the duodenum

    ◆ **Porta hepatis** – enclosed in the free border of the hepatoduodenal ligament; comprised of the common bile duct, hepatic a., and portal v.

  ● **Epiploic foramen** – located on the right side of the midline, at the free margin of the hepatoduodenal ligament; push your finger through the foramen into the **lesser sac**

  ● **Lesser sac** (omental bursa) – space behind the stomach and lesser omentum

    ◆ The greater and lesser sacs communicate through the epiploic foramen

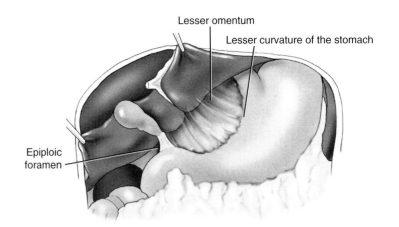

■ Figure 9–3   Lesser omentum in situ

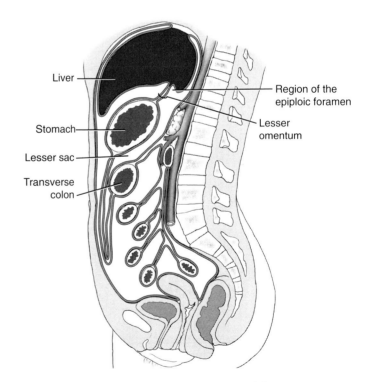

■ Figure 9–4   Lesser omentum and lesser sac

Overview of the Abdominal Viscera

■ Identify the following within the abdominal cavity (Figure 9–5):

- **Liver** – located in the right upper quadrant

- **Gallbladder** – attached to the inferior surface of the liver

- **Spleen** – left upper quadrant, posterior and lateral to the stomach; touches the diaphragm and the tail of the pancreas

- **Stomach** – left upper quadrant, inferior to the left lobe of the liver

■ Reflect the greater omentum superiorly to expose the intestine, and identify the following:

- Most of the **duodenum** and all of the **pancreas** are retroperitoneal and cannot be seen at this time

- **Jejunum** – second segment of the small intestine

- **Ileum** – third segment of the small intestine

- **Ileocecal junction** – union of the ileum to the cecum

- **Cecum** and the **vermiform appendix** – the appendix is located on the inferior border of the cecum

- **Ascending, transverse, descending,** and **sigmoid colon**

- **Kidneys** and **adrenal glands** – retroperitoneal; the right kidney is inferior to the liver; the left kidney is slightly higher than the right kidney because of the liver; palpate but do not dissect the kidneys at this time; the kidneys will be studied further in another laboratory session

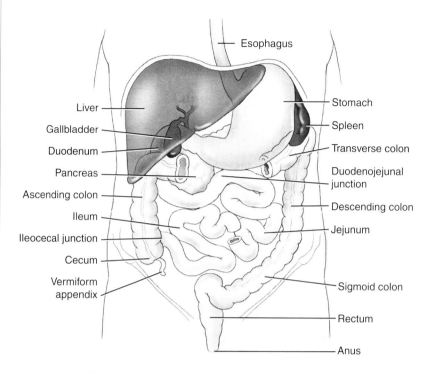

■ **Figure 9–5   Abdominal viscera in situ**

Pelvic Peritoneum

■ Gently lift loops of the small intestine out of the pelvic basin to identify the following (Figures 9–6 and 9–7):

- **Male:**

  - **Paravesical fossae** – peritoneal recesses lateral to the bladder

  - **Pararectal fossae** – peritoneal recesses lateral to the rectum

  - **Rectovesical pouch** – a peritoneal recess between the rectum and the bladder

  - **Sigmoid mesocolon** – at the level of S3 vertebra, the peritoneum becomes the mesentery of the sigmoid colon, which is intraperitoneal

- **Female:**

  - **Vesicouterine pouch** – a peritoneal recess between the bladder and uterus

  - **Rectouterine pouch** – a peritoneal recess between the rectum and uterus

  - **Paravesical fossae** – peritoneal recesses lateral to the bladder

  - **Pararectal fossae** – peritoneal recesses lateral to the rectum

  - **Sigmoid mesocolon** – at the level of S3 vertebra, the peritoneum becomes the mesentery of the sigmoid colon, which is intraperitoneal

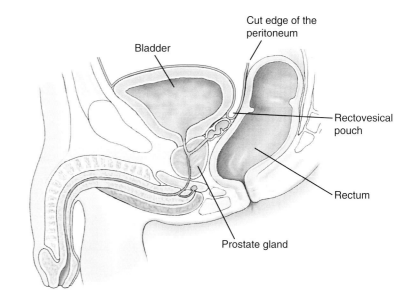

■ **Figure 9–6   Male pelvic peritoneum**

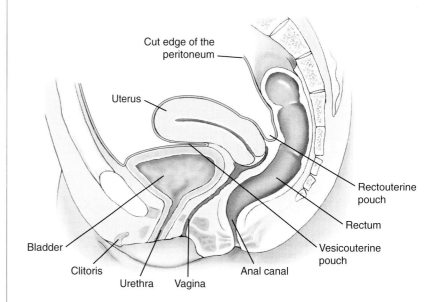

■ **Figure 9–7   Female pelvic peritoneum**

Foregut – Stomach

■ Identify the following (Figures 9–8 and 9–9):

- **Esophagus** – the terminal end empties into the cardia of the stomach; this is the location of the **cardiac (esophageal) sphincter,** formed by the right crus of the diaphragm

- **Cardia** – territory of the cardiac sphincter; region where the esophagus enters the stomach

- **Greater curvature** – inferior border of the stomach

- **Lesser curvature** – superior border of the stomach

- **Fundus** – dilated, superior part of the stomach related to the left dome of the diaphragm

- **Body** – region between the fundus and pylorus

- **Pylorus** – funnel-shaped, terminal region of the stomach; region surrounding the **pyloric sphincter**

- **Gastric rugae** – make an incision through the anterior wall of the stomach to reveal the longitudinal gastric rugae (folds)

- **Pyloric sphincter** – make a sagittal incision through the pylorus to observe the pyloric sphincter (band of smooth muscle)

- **Gastric canal** – the direct communication between the cardia and pylorus of the stomach

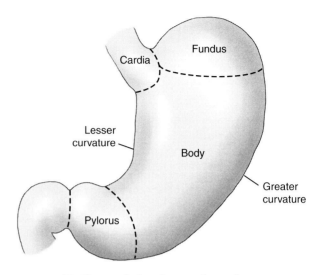

■ **Figure 9–8   Stomach regions**

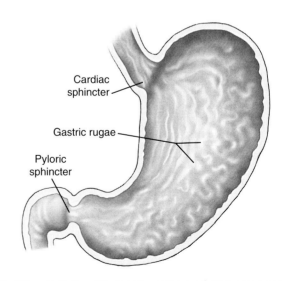

■ **Figure 9–9   Coronal section of the stomach**

Vasculature – Celiac Trunk

- **Celiac trunk** – the principal arterial supply to the foregut; it is the first unpaired branch from the abdominal aorta, at the level of T12 vertebra (Figure 9–10)

  - To identify the branches of the celiac trunk, you will dissect the mesentery to reveal the arteries and veins to the foregut organs

  - To avoid cutting the vessels, do *not* use a scalpel or scissors

- You will identify the three primary branches of the celiac trunk, along with their successive branches:

  - **Common hepatic a.** – branch to the liver

  - **Left gastric a.** – branch to the stomach

  - **Splenic a.** – branch to the spleen

- The origin of the celiac trunk will be dissected later, once the organs that overlay the abdominal aorta are dissected

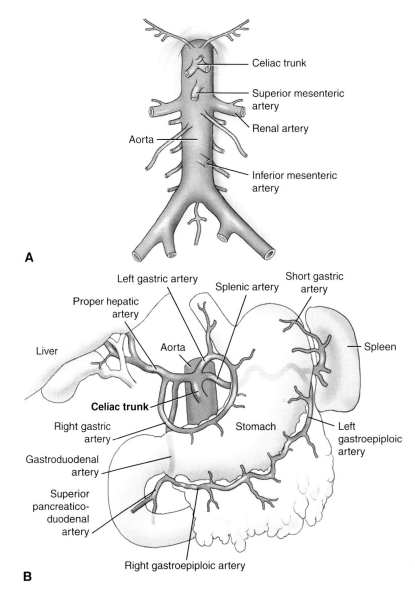

A

B

- Figure 9–10   The celiac trunk. *A*, Abdominal aorta. *B*, Branches of the celiac trunk

Celiac Trunk – Common Hepatic Artery and Its Branches

■ Identify the following branches from the common hepatic a. (Figure 9–11):

● **Common hepatic a.** – branch to the liver and other organs

  ● **Hepatic a. proper** – free the hepatic a. proper, located medial to the bile duct; it supplies the liver and gallbladder

   ◆ **Right gastric a.** – courses to the lesser curvature of the stomach on the righthand side; remove the remaining lesser omentum to reveal the arterial anastomosis with the left gastric a.

   ◆ **Left and right hepatic aa.** – course to the left and right sides of the liver, respectively

     ◆ **Cystic a.** – a small branch of the right hepatic a. to the gallbladder

  ● **Gastroduodenal a.** – posterior to the duodenum; find its terminal branches:

   ◆ **Right (omental) gastroepiploic a.**

   ◆ **Anterior superior pancreaticoduodenal aa.**

   ◆ **Posterior superior pancreaticoduodenal aa.**

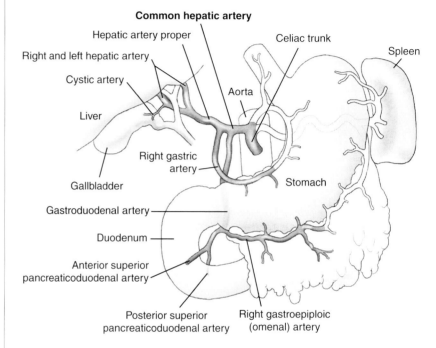

Common hepatic artery

Hepatic artery proper

Right and left hepatic artery

Cystic artery

Liver

Celiac trunk

Spleen

Aorta

Right gastric artery

Stomach

Gallbladder

Gastroduodenal artery

Duodenum

Anterior superior pancreaticoduodenal artery

Posterior superior pancreaticoduodenal artery

Right gastroepiploic (omenal) artery

■ **Figure 9–11   Common hepatic artery branches of the celiac trunk**

Celiac Trunk – Left Gastric and Splenic Arteries

■ Identify the remaining two branches of the celiac trunk (Figure 9–12):

- **Left gastric a.** – superior branch of the celiac trunk; follow to the left lesser curvature of the stomach, where the artery anastomoses with the right gastric a.

  - **Esophageal a.** – attempt to identify an arterial branch from the left gastric a. to the esophagus

- **Splenic a.** – left branch of the celiac trunk; shaped like a corkscrew as it courses over/through/under the pancreas to reach the spleen; courses within the splenorenal ligament

  - **Short gastric aa.** – to the fundus of the stomach

  - **Left (omental) gastroepiploic a.** – to the left side of the greater curvature of stomach; courses in the gastrosplenic ligament; anastomoses with the right gastroepiploic a.

■ Trim, if necessary, the greater omentum from the greater curvature of the stomach; leave the gastroepiploic (omental) vessels attached to the stomach and leave the greater omentum attached to the transverse colon

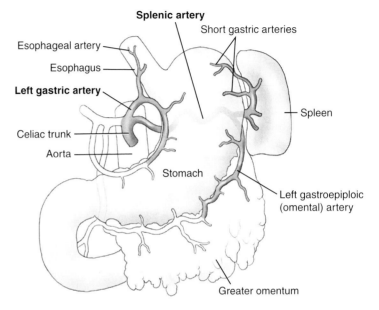

■ **Figure 9–12   Left gastric and splenic branches of the celiac trunk**

Foregut – Duodenum

■ The duodenum is the first segment of the small intestine; it is C-shaped and about 12 inches long; curves around the head of the pancreas; most of the duodenum is retroperitoneal (see Figure 9–2)

■ To locate the **duodenum:**

- Lift the transverse colon

- The duodenum is revealed by cutting through the transverse mesocolon

■ Identify the following (Figures 9–13 and 9–14):

- **Superior (first) part** – 5 cm long; lies anterolateral to L1 vertebra; the proximal region is mobile but the distal region and remainder of the duodenum are retroperitoneal

- **Descending (second) part** – 7–10 cm long; descends to the right of L1–L3 vertebrae; curves around the head of the pancreas; the bile and pancreatic ducts enter its posteromedial wall

- **Horizontal (third) part** – 6–8 cm long; crosses L3 vertebra; crossed by the superior mesenteric a. and v.

- **Ascending (fourth) part** – 5 cm long; ascends on the left of L3 to L2 vertebrae

- **Lesser omentum** – comprised of two regions/ligaments:

  - **Hepatogastric ligament** – connects the liver and stomach

  - **Hepatoduodenal ligament** – connects the liver and duodenum

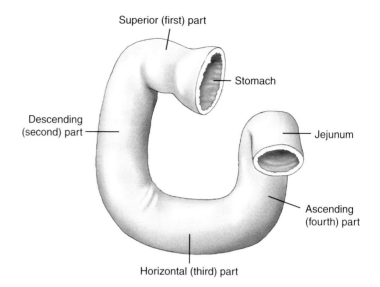

■ Figure 9–13   Regions of the duodenum

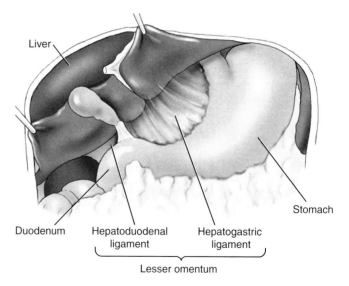

■ Figure 9–14   Lesser omentum

Duodenojejunal Junction

■ Reflect the greater omentum and transverse colon superiorly

■ Grasp and move the small intestine to the left side of the abdomen; follow the intestinal tube proximally (see Figure 9–15)

■ Identify the following (Figures 9–15 and 9–16):

• **Duodenojejunal junction** – region where the immobile (retroperitoneal) duodenum ends and the mobile (intraperitoneal) jejunum begins; demarcated by the suspensory ligament of the duodenum

• **Suspensory ligament of the duodenum** – attaches the duodenum to the diaphragm

• **Duodenal fossa** – fossa formed by folds of serous membrane created by the retroperitoneal-to-intraperitoneal transition of the first two parts of the small intestine; insert your finger into the fossa to feel the ascending (fourth) portion of the duodenum

• **Transverse mesocolon** – mesentery of the transverse colon; bluntly dissect the transverse mesocolon at the duodenal fossa to reveal the retroperitoneal part of the duodenum and the pancreas

• *Note:* Be aware of, but do *not* cut, the large vessels in the transverse mesocolon

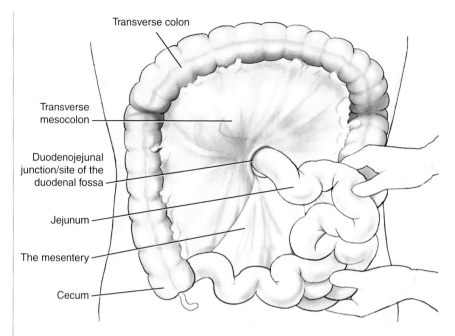

■ **Figure 9–15   Duodenojejunal junction**

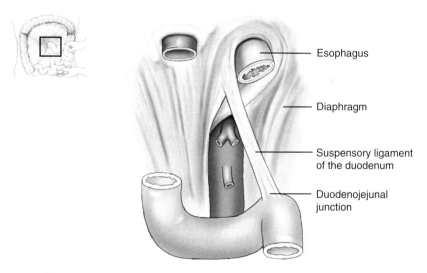

■ **Figure 9–16   Suspensory ligament of the duodenum**

Abdominal Viscera – Spleen and Associated Structures

■ Identify the following (Figures 9–17 and 9–18):

- **Spleen** – located intraperitoneally in the left upper quadrant; rests just above the left colic flexure; the tail of the pancreas touches the spleen at its hilum

  - **Hilum** – peritoneum covers the entire surface of the spleen, except at its hilum, where the splenic vessels exit and enter

- **Gastrosplenic ligament** – on the cadaver's left side, push your right hand beneath the diaphragm and grasp the spleen

  - Push a finger of your left hand into the lesser sac until your fingers make contact

  - The barrier that prevents your fingers from touching is the **gastrosplenic ligament;** contains the left gastroepiploic vessels

- **Splenorenal ligament** – stretches between the spleen and left kidney; contains the splenic vessels

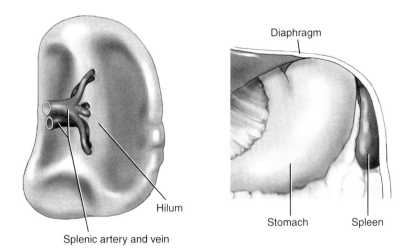

■ **Figure 9–17   Spleen**

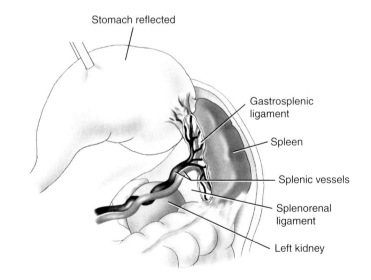

■ **Figure 9–18   Splenic ligaments**

Abdominal Viscera – Liver

- The liver is the largest gland in the body, weighing about 1.5 kg; located in the upper right quadrant of the abdomen

- To access the liver, follow these instructions (Figure 9–19):

  - Cut the sixth and seventh right costal cartilages near the xiphisternal junction; you may partially incise the diaphragm

- Identify the following (Figure 9–20):

  - **Diaphragmatic surface** – region of the liver in contact with the diaphragm

  - **Visceral surface** – region of the liver in contact with viscera – (stomach, duodenum, colon, right kidney)

  - **Right lobe of the liver** – largest of the liver's four lobes; located to the right of the falciform and round ligaments; from its inferior surface hangs the gallbladder

  - **Left lobe of the liver** – located to the left of the falciform and round ligaments

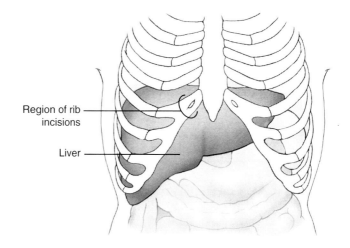

■ Figure 9–19   Liver topography

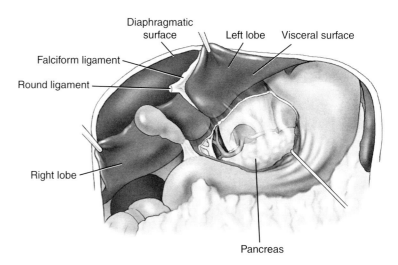

■ Figure 9–20   Anterior surface of the liver

Liver – Mesentery and Ligaments

■ Identify the following (Figures 9–21 and 9–22):

- **Subphrenic recess** – slip your hands between the right and left lobes of the liver and the diaphragm; your hands fill the subphrenic recess; you cannot touch your fingers together because of ligaments (reflections of the peritoneum from the liver to the diaphragm)

- **Coronary ligaments** – reflections of peritoneum from the diaphragm to superior and posterior surfaces of the liver

  - **Triangular ligaments** – the left and right triangular ligaments are sharp lateral margins of the coronary ligaments

- **Bare area** – region on the superior surface of the liver between the coronary and triangular ligaments; lacks peritoneum

- **Falciform ligament** – reflections of the parietal peritoneum around the obliterated umbilical v.; attaches the liver to both the diaphragm and anterior abdominal wall

  - The free (inferior) border of the falciform ligament contains the **round ligament of the liver** (obliterated umbilical v.)

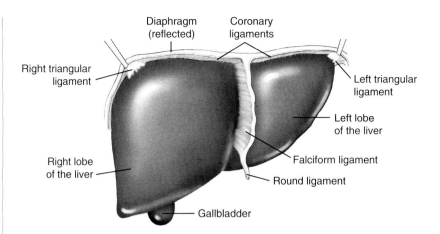

■ **Figure 9–21   Anterior view of the liver (diaphragm reflected)**

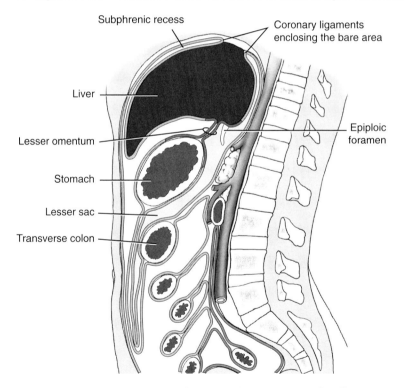

■ **Figure 9–22   Sagittal section through the liver**

Liver – Inferior (Visceral) Surface

■ Identify the following structures on the inferior surface of the liver (Figure 9–23):

- **Right lobe**

- **Left lobe**

- **Quadrate lobe** – located anteriorly; bordered by the gallbladder, porta hepatis, and round ligament of the liver

- **Caudate lobe** – located posteriorly; bordered by the ligamentum venosum, groove for inferior vena cava, and porta hepatis

■ Cut a wedge off the left lobe of the liver to identify the following structures:

- **Intrahepatic portal triads** – continuation of the structures in the porta hepatis; branches of the bile duct, portal v., and hepatic a.

- **Intrahepatic vv.** – deliver contents to the inferior vena cava

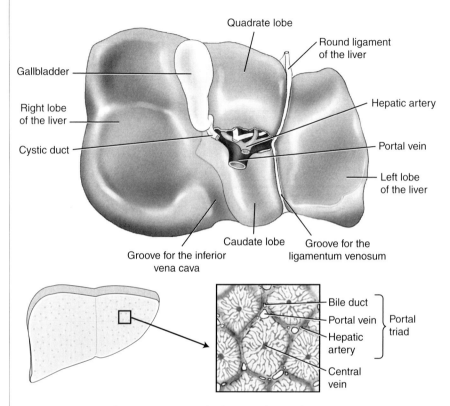

Figure 9–23 Inferior surface of the liver

Porta Hepatis

- **Porta hepatis** – located between the left and right lobes of the liver, within the hepatoduodenal ligament (the free edge of the lesser omentum); contains the common bile duct, hepatic a. proper, and the portal v.

- Remove the peritoneum from the **hepatoduodenal ligament** to identify the following contents (Figures 9–24 and 9–25):

  - **Common bile duct** – green and usually on the superficial right side of the porta hepatis

    - **Cystic duct** – duct from the gallbladder, which joins the **common hepatic duct** to form the common bile duct

      - ◆ Follow the common hepatic duct to the liver to locate the **right** and **left hepatic ducts**

  - **Hepatic a. proper** – branch from the hepatic a., after the latter gives rise to the gastroduodenal a., on the superficial left side of the porta hepatis

  - **Portal v.** – deep and posterior to the common bile duct and hepatic a. proper; the largest of the three structures

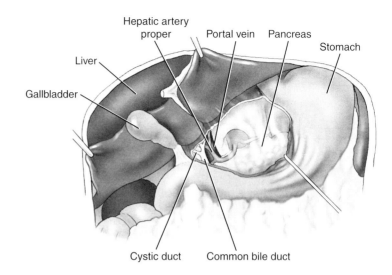

■ **Figure 9–24   Porta hepatis in situ**

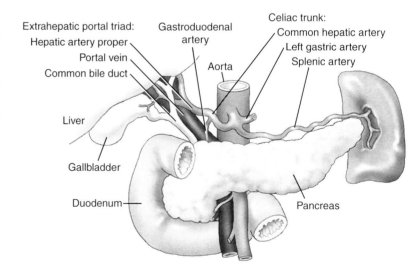

■ **Figure 9–25   Structures associated with the porta hepatis**

Vasculature – Celiac Trunk

■ Identify the following structures (Figure 9–26):

- **Celiac ganglia** – a bilateral, tough network of autonomic nerve fibers that surrounds the celiac trunk; each ganglian is approximately the size of a quarter

  - Site of synapse of the preganglionic sympathetic fibers from the greater splanchnic nerves (T5–T9 spinal cord segments)

  - You will have to cut the plexus of nerves to see the origin of the celiac trunk

- **Celiac trunk** – principal artery to the foregut; the first unpaired branch from the abdominal aorta, at the level of T12 vertebra; identify the three primary branches, along with their successive branches:

  - **Common hepatic a.** – branch to the liver

  - **Left gastric a.** – branch to the stomach

  - **Splenic a.** – branch to the spleen

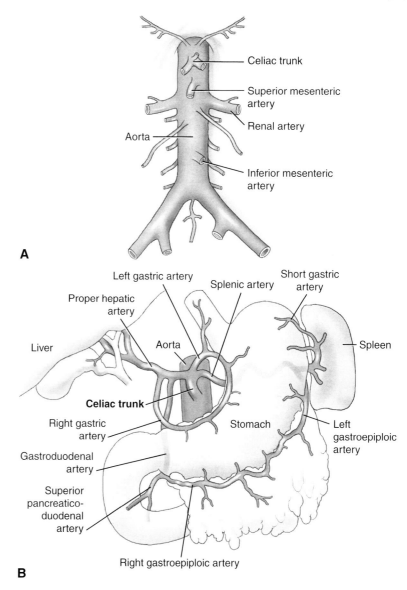

■ **Figure 9-26   The celiac trunk. *A*, Abdominal aorta. *B*, Branches of the celiac trunk**

Midgut and Hindgut

Prior to dissection, you should familiarize yourself with the following structures:

PANCREAS
- Head
- Uncinate process
- Neck
- Body
- Tail
- Main pancreatic duct
- Accessory pancreatic duct

MIDGUT

(Supplied by the Superior Mesenteric Vessels)

- Duodenum (4 parts)
  - Major and minor duodenal papillae (inside the duodenum)
    - Proximal to the major papilla – foregut
    - Distal to the major papilla – midgut
  - Pancreaticoduodenal ampulla (outside the duodenum)
  - Duodenal sphincter (surrounds the ampulla)
- Jejunum
  - Circular folds
- Ileum
  - Ileocecal junction/valve
  - Peyer's patches of lymphoid tissue

- Cecum
  - Vermiform appendix
- Ascending colon
- Right colic (hepatic) flexure
- Transverse colon (proximal 2/3)

HINDGUT

(Supplied by the Inferior Mesenteric Vessels)

- Transverse colon (distal 1/3)
- Left colic (splenic) flexure
- Descending colon
- Sigmoid colon
  - Semilunar folds

Features of the Colon
- Omental (epiploic) appendices
- Haustra
- Taenia coli mm.
- Rectum

PORTAL SYSTEM
- Portal v
  - Splenic v.
  - Superior mesenteric v.

- Gastric vv.
- Inferior mesenteric v.
- Gastroepiploic vv.
- Superior rectal v.

ARTERIES AND LYMPHATICS
- Aorta
  - Superior mesenteric a.
    - Anterior and posterior inferior pancreaticoduodenal aa.
    - Jejunal aa.
      - Vasa recta
    - Ileal aa.
      - Arcades
    - Ileocolic a.
      - Appendicular a.
    - Right colic a.
    - Middle colic a.
    - Marginal a.
    - Associated lymph nodes and lymphatics
  - Inferior mesenteric a.
    - Left colic a.
    - Sigmoid aa.
    - Marginal a.
    - Superior rectal a.
    - Associated lymph nodes and lymphatics

Duodenum

- Identify the following (Figure 10–1):

  - **Duodenum** – if not already done, locate by lifting the transverse colon; the duodenum is revealed by cutting through the transverse mesocolon

    - Follow the common bile duct to the posterior surface of the head of the pancreas

    - Carefully dissect the common bile duct from the surrounding pancreatic tissue; estimate the extent of the descending (second) part of the duodenum

  - Open the duodenum opposite to the junction with the common bile duct; wipe the lumen and wall clean to identify

    - **Circular folds** – distinct invaginations folds of the mucosal lining

    - **Major duodenal papilla** – region where the pancreatic and common bile ducts enter the duodenal lumen

    - **Minor duodenal papilla** (may be absent) – 2 cm superior to the major duodenal papilla; receives the accessory pancreatic duct (persistence of the proximal duct of the dorsal pancreatic bud)

  - **Pancreaticoduodenal ampulla** – outside of the duodenum; the ampulla is the bulbous junction of the common bile and pancreatic ducts; it opens into the lumen of the duodenum via the major duodenal papilla

  - **Duodenal sphincter** – circularly arranged smooth muscle fibers in the wall of the pancreaticoduodenal ampulla

  - **Main pancreatic duct** – use the ampulla as your guide to separate the pancreatic tissue to reveal the main pancreatic duct

    - **Accessory pancreatic duct** – superior and parallel to the main pancreatic duct; usually smaller in diameter than the main pancreatic duct

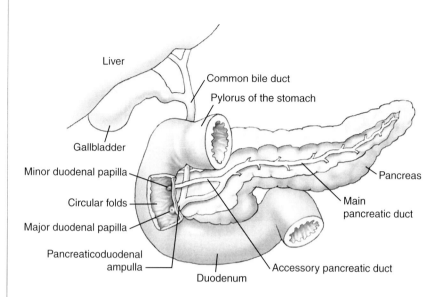

■ **Figure 10–1   Duodenum**

Midgut – Jejunum and Ileum

■ Identify the following (Figures 10–2 and 10–3):

- **Jejunum** – begins at the duodenojejunal junction and ends at the ileum

  - Generally lies coiled in the upper left quadrant of the peritoneal cavity, below the transverse mesocolon

  - Proximal three fifths of the small intestine

- **Ileum** – generally lies in the lower right part of the peritoneal cavity and pelvis

  - Distal two fifths of the small intestine

- **The mesentery** – coils of jejunum and ileum are attached to the posterior abdominal wall by a fan-shaped, double-fold of peritoneum known as the mesentery

  - Provides support for the small intestine and contains the mesenteric vessels (arteries, veins, and lymphatics) and nerves from the posterior abdominal wall to the small intestine

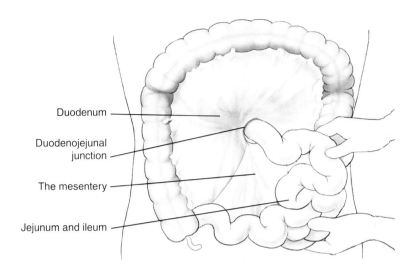

Duodenum
Duodenojejunal junction
The mesentery
Jejunum and ileum

■ **Figure 10–2   Jejunum and ileum**

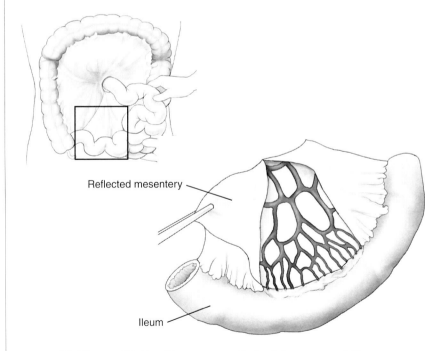

Reflected mesentery

Ileum

■ **Figure 10–3   Mesentery of the jejunum and ileum**

Midgut – Jejunum and Ileum—cont'd

■ Cut open and clean segments (several centimeters) of the jejunum, proximal ileum, and distal ileum

■ Identify the following internal features (Figures 10–4 and 10–5):

● **Circular folds** – folds in the mucosal layer of the gut that increase its surface area to accommodate the passage of digesting food

  ● The jejunum may be distinguished from the ileum by the appearance of the **circular folds:**

    ◆ **Jejunum** – circular folds are numerous

    ◆ **Ileum** – circular folds in the proximal ileum are less numerous; they gradually disappear toward the distal ileum

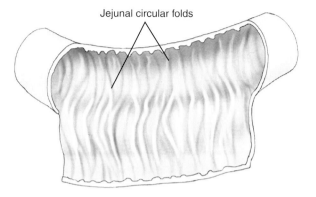

Jejunal circular folds

■ **Figure 10–4   Internal features of the jejunum**

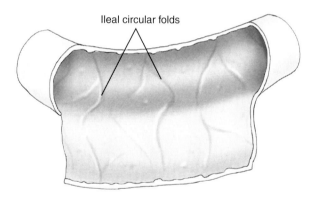

Ileal circular folds

■ **Figure 10–5   Internal features of the ileum**

Midgut – Large Intestine

■ Identify the following (Figures 10–6 and 10–7):

- **Ileocecal junction** – the union of the ileum and cecum; internally a sphincter muscle separates the small and large intestines

- **Cecum** – beginning of the large intestine; extends inferiorly at the ileocecal junction into the right iliac fossa; intraperitoneal

- **Vermiform appendix** – opens into the cecum inferior to the ileocecal junction; the appendix occupies many different positions and varies in size; intraperitoneal

- **Ascending colon** – extends from the cecum to the transverse colon; ends at the **right colic (hepatic) flexure;** retroperitoneal

- **Transverse colon** – extends from the right colic flexure to the left colic (splenic) flexure

  - **Phrenicocolic ligament** – mesentery attaching the left colic flexure to the diaphragm

  - Suspended by the transverse mesocolon so it is intraperitoneal

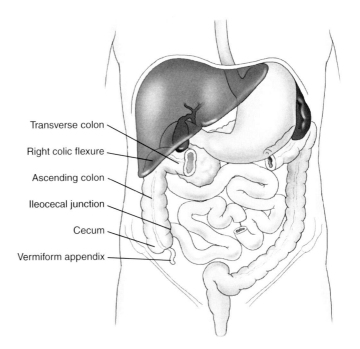

■ **Figure 10–6   Large intestine**

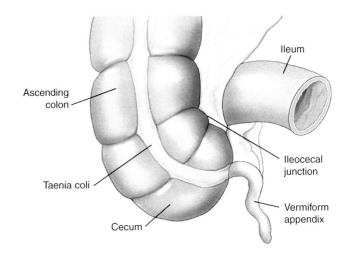

■ **Figure 10–7   Cecum and appendix**

Hindgut – Large Intestine

■ Identify the following (Figures 10–8 and 10–9):

- **Transverse colon** – extends from the right colic flexure to the **left colic (splenic) flexure**

- **Descending colon** – extends from the left colic flexure to the pelvic brim; retroperitoneal

- **Sigmoid colon** – extends from the pelvic brim to the rectum; attached to the lateral pelvic wall via the sigmoid mesocolon, hence it has considerable mobility; intraperitoneal

- **Rectum** – extends from the sigmoid colon to the anal canal; partially covered in peritoneum (studied further in another laboratory session)

- Features of the large intestine (midgut and hindgut)

  - **Taenia coli mm.** – three longitudinal bands of smooth muscle; the anterior and posterior bands are easy to identify on the colon (the middle muscle is deep to the mesentery and more difficult to observe, except at the cecum)

  - **Haustra** – bulges formed by contraction of the taenia coli mm.; located throughout the length of the colon

  - **Omental (epiploic) appendices** – sacs (appendages) of fat that are located along the length of the colon, from the ascending to sigmoid colon

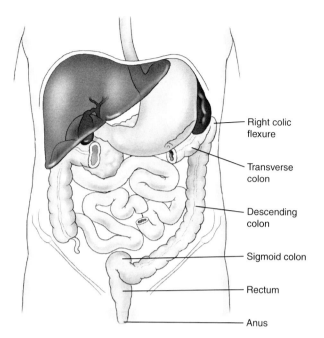

■ **Figure 10–8   Large intestine**

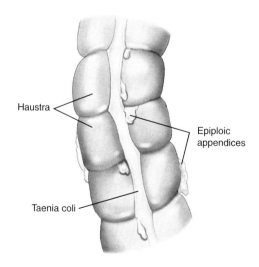

■ **Figure 10–9   Structures of the large intestine**

Midgut – Superior Mesenteric Artery

■ Identify the following (Figure 10–10):

- **Superior mesenteric a.** – the second unpaired artery arising from the abdominal aorta, just inferior to the celiac trunk at the level of L1 vertebra

  - Lift the transverse colon over the costal margin; pull the small intestine to the left to stretch the mesentery; palpate to the right of the duodenojejunal junction for the superior mesenteric a. and v.

  - Bluntly dissect the mesentery and observe that the superior mesenteric a. crosses anterior to the left renal v. and the third part of the duodenum

  - **Anterior** and **posterior inferior pancreaticoduodenal aa.** – branches of the superior mesenteric a. to the head of the pancreas and midgut-derived portion of the duodenum; anastomose with the anterior and posterior superior pancreaticoduodenal aa. from branches of the celiac trunk

- **Superior mesenteric v.** – courses along the right side of the superior mesenteric a.

- **Superior mesenteric plexus** – plexus of autonomic nerves that surround the superior mesenteric a.

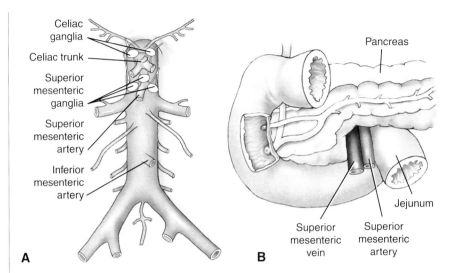

A

B

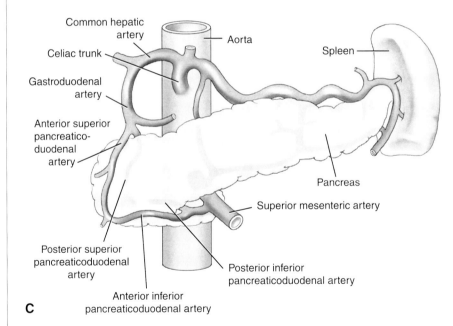

C

■ **Figure 10–10   Superior mesenteric artery. A, Abdominal aorta. B, Superior mesenteric artery and vein in situ. C, Arteries associated with the pancreas**

Midgut – Superior Mesenteric Artery—cont'd

- To expose the branches of the mesenteric vessels, reflect the anterior layer of the mesentery (see Figure 10–11); keep the vessels intact

- The names of the branches of the superior mesenteric a. correspond to the structures that are supplied by it

- Identify the following branches to the small intestine (Figure 10–11):

  - **Jejunal** and **ileal aa.** – 15 to 18 arterial branches of the superior mesenteric a.; supply the jejunum and ileum; pass between the two layers of the mesentery

  - **Arterial arcades** – arteries form arcades (arches) within the mesentery that eventually form straight terminal branches **(vasa recta)** to the jejunum and ileum

- *Note:* Compared to the jejunum, the ileum has more arterial arcades that are stacked upon one another; consequently, the vasa recta of the ileum are shorter than those of the jejunum

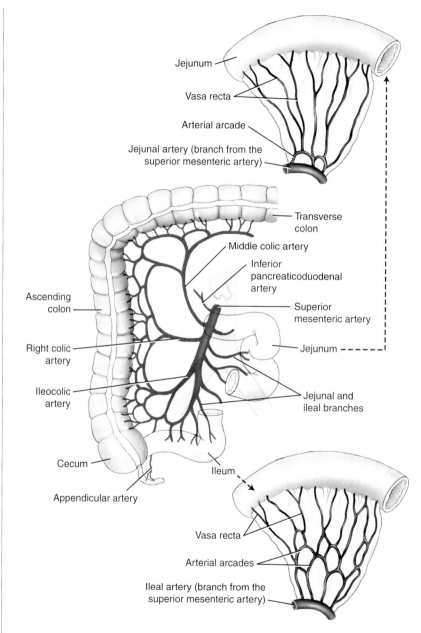

■ **Figure 10–11 Jejunal and ileal arteries**

Midgut – Superior Mesenteric Artery—cont'd

- Identify the following branches to the proximal portion of the colon (Figure 10–12):

  - **Ileocolic a.** – terminal branch of the superior mesenteric a.; supplies the cecum, appendix, and distal portion of the ileum; anastomoses with ileal branches from the right colic a.

    - **Appendicular a.** – terminal branch of the ileocolic a.; supplies the appendix

  - **Right colic a.** – origin is variable, in that it may arise from the superior mesenteric a., the ileocolic a., or the middle colic a.; supplies the ascending colon; anastomoses with the ileocolic and middle colic aa.

  - **Middle colic a.** – supplies the right half of the transverse colon; anastomoses with the right colic a. and the left colic a. (branch of the inferior mesenteric a.)

  - **Marginal a.** – anastomoses of the colic aa.

  - *Note:* Tributaries of the superior mesenteric v. drain the same regions supplied by the superior mesenteric a.

  - **Superior mesenteric lymph nodes** – look for lymph nodes in the mesentery; named according to the organ that they drain

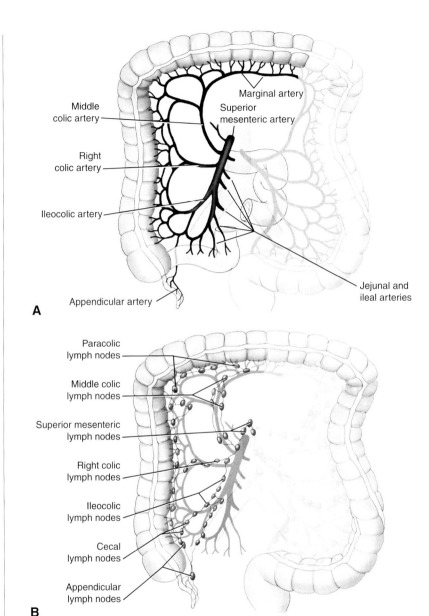

A

B

- **Figure 10–12 Superior mesenteric artery and its branches.** *A*, Branches of the superior mesenteric artery. *B*, Superior mesenteric lymph nodes

Hindgut – Inferior Mesenteric Artery

- The inferior mesenteric a. is the third unpaired vessel arising from the abdominal aorta, at the level of L3 vertebra

- Identify the following branches to the distal portion of the colon (Figure 10–13):

- **Inferior mesenteric a.** – originates about 3 cm superior to the aortic bifurcation, usually to the left of the midline of the aorta

  - **Left colic a.** – supplies the descending colon and the left portion of the transverse colon; anastomoses with the middle colic a. (from the superior mesenteric a.)

  - **Sigmoid aa.** – usually 4–5 branches that form arches; supply blood to the sigmoid colon

  - **Superior rectal a.** – supplies the proximal portion of the rectum; usually a large vessel (the terminal branch of the inferior mesenteric a.)

  - **Marginal a. (of Drummond)** – anastomoses of the colic aa. around the mesenteric margin of the large intestine

    - ◆ Begins at the ileocecal junction, where it anastomoses with the cecal branches of the superior mesenteric a., and ends at the superior rectal a.

- **Inferior mesenteric lymph nodes** – look for lymph nodes in the mesentery; named according to the organ that they drain

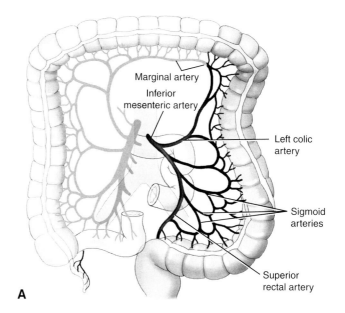

Marginal artery

Inferior mesenteric artery

Left colic artery

Sigmoid arteries

Superior rectal artery

A

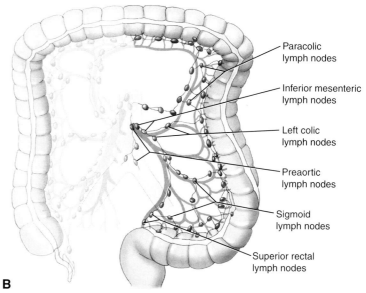

Paracolic lymph nodes

Inferior mesenteric lymph nodes

Left colic lymph nodes

Preaortic lymph nodes

Sigmoid lymph nodes

Superior rectal lymph nodes

B

- **Figure 10–13   Inferior mesenteric artery and its branches. A,** Branches of the inferior mesenteric artery. **B,** Inferior mesenteric lymph nodes

Portal System of Veins

- **Portal system of veins** – drains poorly oxygenated, but nutrient-rich, venous blood to the liver from most of the GI tract, gallbladder, pancreas, and spleen (expect variation)

- Identify the following (Figure 10–14):

  - **Portal v.** – usually 5 cm long; formed posterior to the neck of the pancreas via the union of the superior mesenteric v. and the splenic v.

  - **Superior mesenteric v.** – drains blood from the midgut; located to the right of the superior mesenteric a.

  - **Splenic v.** – courses inferior to the splenic a. and posterior to the pancreas

    - Clean and follow the splenic v. and superior mesenteric v. to their union at the portal v.

  - **Inferior mesenteric v.** – usually drains into the splenic v.; to the left of the superior mesenteric v.

  - **Right** and **left gastric vv.** – drain blood from the right and left sides of the lesser curvature of the stomach to the portal v.

  - **Gastroepiploic vv.** – drain blood from the greater curvature of the stomach to the superior mesenteric v.

  - **Superior rectal v.** – drains the rectum to the inferior mesenteric v.

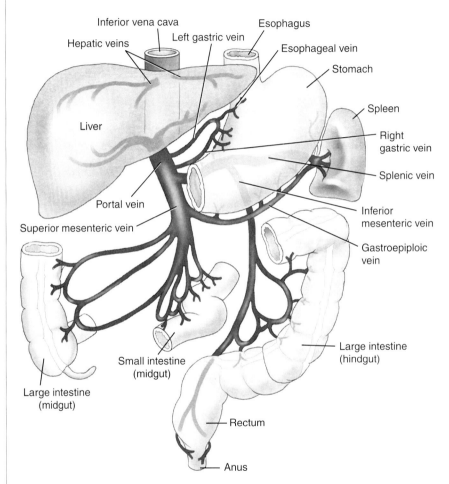

■ **Figure 10-14   Portal venous system**

Removal of the GI Tract

■ To improve studying the posterior abdominal wall, you will remove most of the GI tract; however, you will leave intact the celiac trunk and its associated organs (foregut)

■ Follow these instructions to remove the GI tract:

- Locate the **duodenojejunal junction** and the **rectum**

- Tie two pieces of string tightly around the duodenojejunal junction, about 1 inch apart (Figure 10–15A)

- Free the rectum from peritoneum

- Tie two pieces of string tightly around the rectum about 1 inch apart; this helps limit the amount of feces that will escape when you cut the rectum (Figure 10–15B)

- Cut through the GI tract between the adjacent pairs of ties

- Cut the superior mesenteric a. about 2 cm inferior to the pancreas (leave a short stump on the aorta for future reference)

- Cut the inferior mesenteric a. close to the abdominal aorta (leave a short stump on the aorta for future reference)

- Cut the superior mesenteric v. prior to its union with the splenic v.

- Bluntly dissect the peritoneal attachments while removing the GI tract

- Leave as much mesentery (and its vasculature) as possible attached to the excised small intestine

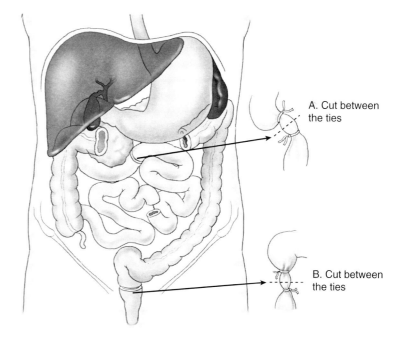

A. Cut between the ties

B. Cut between the ties

■ **Figure 10–15   Removal of the GI tract. A, Transection of the duodenojejunal junction. B, Transection of the rectum**

Posterior Abdominal Wall

Prior to dissection, you should familiarize yourself with the following structures:

MUSCLES

- Psoas major m.
- Psoas minor m.
- Iliacus m.
- Quadratus lumborum m.
  - Iliolumbar ligament
- Diaphragm
  - Right and left crura
  - Central tendon
  - Vena caval hiatus
  - Esophageal hiatus
  - Aortic hiatus
  - Medial and lateral arcuate ligaments
  - Median arcuate ligament

NERVES

- Subcostal n. (T12)
- Iliohypogastric n. (L1)
- Ilioinguinal n. (L1)
- Genitofemoral n. (L1–L2)
- Lateral femoral cutaneous n. (L2–L3)
- Obturator n. (L2–L4)

- Femoral n. (L2–L4)
- Lumbosacral trunk (L4–L5)

ARTERIES

- Abdominal aorta
  - Inferior phrenic a.
  - Celiac trunk
  - Superior mesenteric a.
  - Suprarenal aa. (superior, middle, and inferior)
  - Renal aa.
  - Gonadal (testicular/ovarian) aa.
  - Inferior mesenteric a.
  - Lumbar aa.
  - Median sacral a.
  - Common iliac aa.
    - External iliac aa.
    - Internal iliac aa.

VEINS

- Inferior vena cava
  - Renal vv.
    - Suprarenal vv.
    - Left gonadal v.
- Right gonadal v.
- Lumbar vv.

- Middle sacral v.
- Common iliac vv.
  - External iliac vv.
  - Internal iliac vv.

LYMPHATICS

- Cisterna chyli
- Lymph nodes
  - Preaortic lymph nodes
  - Common iliac lymph nodes
  - External iliac lymph nodes
  - Internal iliac lymph nodes

KIDNEY

- Perirenal fascia
- Fibrous capsule
- Renal cortex
- Renal medulla
  - Renal columns
  - Renal pyramids
  - Renal papillae
- Minor and major calyces
- Renal pelvis
- Ureter
- Hilum

TABLE 11–1 Muscles of the Posterior Abdominal Wall

Muscle	Proximal Attachment	Distal Attachment	Action	Innervation
Psoas major m.	T12–L5 vertebrae	Lesser trochanter (femur)	Flexes the thigh; flexes the vertebral column	Lumbar plexus (L2–L4 nn.)
Psoas minor m.	Vertebral bodies of T12–L1	Pecten pubis	Weak flexion of the vertebral column	First lumbar n.
Iliacus m.	Iliac fossa, ala of the sacrum	Lesser trochanter (femur)	Flexes the thigh	Femoral n. (L2–L4)
Quadratus lumborum m.	Rib 12, transverse process of L1–L5 vertebrae	Iliolumbar ligament and iliac crest	Extends and laterally flexes the vertebral column; draws rib 12 toward the pelvis	Ventral branches of T12 (subcostal) and L1–L4 nn.

Posterior Abdominal Wall – Overview

■ Identify the following (Figure 11–1):

- **Kidneys** – lateral to vertebrae T12–L3; gently remove the perirenal fat from around the kidneys

- **Ureters** – extend from the renal pelvis to the urinary bladder

- **Adrenal (suprarenal) glands** – positioned on the superior pole of each kidney

- **Abdominal aorta** – courses longitudinally to the left of the midsagittal line on the anterior surface of the vertebral bodies

- **Inferior vena cava** – courses longitudinally to the right of the midsagittal line on the anterior surface of the vertebral bodies

- **Iliac crest**

- **Diaphragm**

- **Lymphatics**
  - **Cisterna chyli**
  - **Lymph nodes** – named according to where they are situated
    - ◆ **Preaortic lymph nodes**
    - ◆ **External iliac lymph nodes**
    - ◆ **Internal iliac lymph nodes**

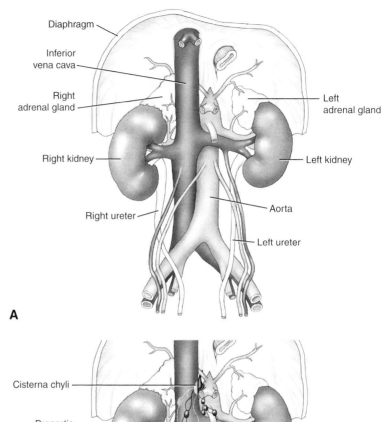

A

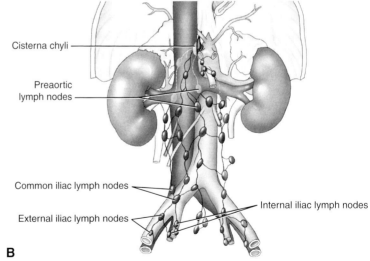

B

■ **Figure 11-1   *A*, Overview of the posterior abdominal wall. *B*, Lymphatics of the posterior abdominal wall**

Posterior Abdominal Wall – Abdominal Aorta

■ Identify the following (Figure 11–2):

- **Abdominal aorta** – courses longitudinally to the left of the midsagittal line on the anterior surface of the vertebral bodies; the bifurcation of the abdominal aorta is at L4 vertebra

  - **Inferior phrenic aa.** – paired arteries that arise immediately inferior to the diaphragm; first branches from the abdominal aorta; give superior suprarenal aa. to the suprarenal glands

  - **Celiac trunk** – T12 vertebral level; unpaired artery; supplies the foregut

  - **Superior mesenteric a.** – L1 vertebral level; unpaired artery; supplies the midgut

  - **Inferior mesenteric a.** – L3 vertebral level; unpaired artery; supplies the hindgut

  - **Middle suprarenal a.** – direct branch off the abdominal aorta, near the level of the celiac trunk

  - **Renal aa.** – L1 vertebral level; paired branches that originate between the superior and inferior mesenteric aa.; the left renal a. is shorter than the right renal a.; multiple arteries are frequently encountered

    ◆ Give rise to the **inferior suprarenal aa.**

- **Gonadal (ovarian/testicular) aa.** – L2 vertebral level; paired arteries that arise inferior to the renal aa. and superior to the inferior mesenteric a.; cross anterior to the ureters and external iliac vessels

- **Lumbar aa.** – four paired branches that pass around the sides of vertebrae L1–L4; segmental blood supply of the lumbar region

- **Median sacral a.** – arises from the abdominal aorta at its bifurcation into the common iliac aa.

- **Common iliac aa.** – the aorta bifurcates at the L4 vertebral level; each common iliac a. bifurcates into the

  - **Internal** and **external iliac aa.**

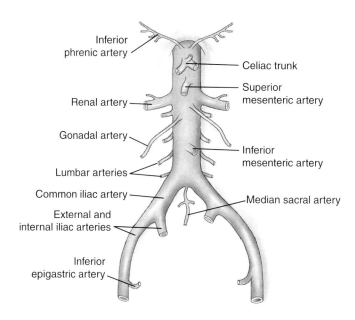

■ **Figure 11–2   Branches of the abdominal aorta**

Posterior Abdominal Wall – Inferior Vena Cava

■ Identify the following (Figure 11–3):

- **Inferior vena cava**
  - **Right renal v.** – shorter than the left renal v.
    - ◆ **Right suprarenal v.** – drains the right adrenal gland; only one suprarenal v. drains each adrenal gland compared to the three adrenal aa. that supply each gland
  - **Left renal v.** – crosses anterior to the aorta and just inferior and posterior to the superior mesenteric a.
    - ◆ **Left suprarenal v.** – one vein drains the left adrenal gland
    - ◆ **Left gonadal (testicular/ovarian) v.** – drains the left gonad into the inferior border of the left renal v.
  - **Right gonadal v.** – drains into the inferior vena cava

■ *Note:* The superior and inferior mesenteric vv. are not tributaries to the inferior vena cava but of the portal venous system

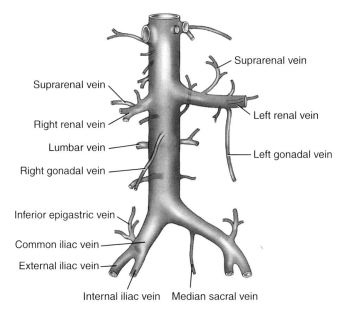

Suprarenal vein
Suprarenal vein
Right renal vein
Left renal vein
Lumbar vein
Left gonadal vein
Right gonadal vein
Inferior epigastric vein
Common iliac vein
External iliac vein
Internal iliac vein   Median sacral vein

■ **Figure 11–3   Branches of the inferior vena cava**

Kidneys and Adrenal Glands in Situ

■ Identify the following (Figure 11–4):

- **Kidneys** – left and right; observe that the right kidney is slightly lower than the left kidney, due to the liver in the upper right quadrant of the abdomen

- **Perirenal fat** – each kidney is embedded in a substantial layer of fat; with your fingers, shell out the kidneys from the perineal fat

- **Adrenal glands** – each adrenal gland is separated from the superior pole of each kidney by a layer of connective tissue; separate the adrenal glands from the kidneys *but* leave their blood vessels intact

- **Ureters** – cross the psoas major m. and course deep to the gonadal vessels en route to the true pelvis (urinary bladder)

■ Cut the right renal v. close to the inferior vena cava

■ To remove the right kidney, cut its artery and vein and then remove the kidney by cutting it free from the posterior abdominal wall

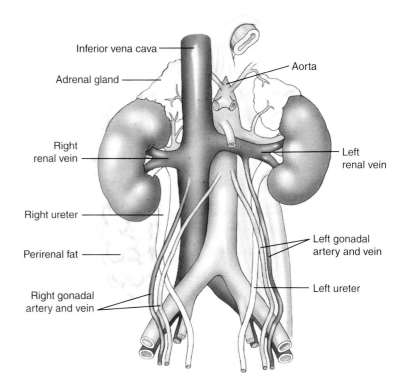

■ **Figure 11–4   Kidneys in situ**

Kidneys and Adrenal Glands in Situ—cont'd

■ Bisect the right kidney coronally to identify the following (Figure 11–5):

- **Fibrous capsule** – fibrous tissue that surrounds each kidney; easily stripped away

- **Renal cortex** – outer third of the kidney

- **Renal medulla** – consists of **renal pyramids** and **renal columns;** renal columns are extensions of the cortex between the renal pyramids

- **Renal papillae** – project into minor calyces; 1 papilla to 1 minor calyx

- **Minor calyces** – "cups" that collect urine from the renal pyramids; 2–4 minor calyces unite to form a **major calyx**

- **Renal pelvis** – formed by union of the major calyces

- **Ureter** – the continuation of the renal pelvis; connects the kidney to the urinary bladder

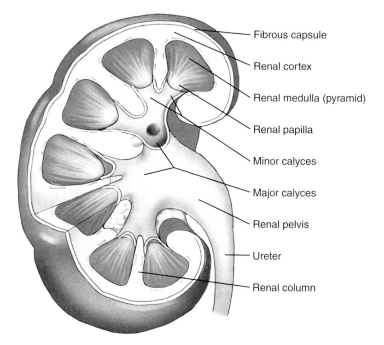

Fibrous capsule

Renal cortex

Renal medulla (pyramid)

Renal papilla

Minor calyces

Major calyces

Renal pelvis

Ureter

Renal column

■ **Figure 11-5   Coronal section of the kidney**

Adrenal Glands

■ Identify the following (Figure 11–6):

- **Adrenal glands** – superior to each kidney

- **Adrenal (suprarenal) gland vessels** – these vessels are fairly thin and may be difficult to identify

    - **Superior suprarenal aa.** – arise from the **inferior phrenic aa.**

    - **Middle suprarenal aa.** – arise from the aorta

    - **Inferior suprarenal aa.** – arise from renal aa.

    - **Suprarenal vv.** – usually drain to the renal vv.; only one vein compared to three arteries; the veins are asymmetric

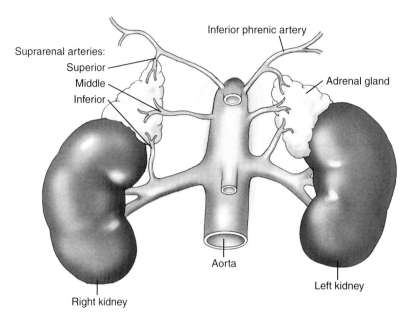

■ **Figure 11–6   Arterial supply of the adrenal glands**

Diaphragm

- A dome-shaped muscle that separates the thoracic cavity from the abdominal cavity

- Identify the following (Figure 11–7):

  - **Central tendon** – observe the opening for the inferior vena cava **(vena caval hiatus)**

  - **Muscular part** – observe the openings for the esophagus **(esophageal hiatus)** and the aorta **(aortic hiatus)**

  - **Right crus** – originates lateral to the aortic hiatus; loops around the esophagus to form the esophageal hiatus, the physiologic sphincter of the esophagus

    - Observe a muscle slip coursing inferomedially from the right crus of the diaphragm to the duodenojejunal junction – this is the **suspensory ligament of the duodenum**

  - **Left crus** – lateral to the aortic hiatus; fleshy fibers contribute to the median arcuate ligament

  - **Me*dian* arcuate ligament** – formed by the right and left crura arching over the aorta

  - **Me*dial* arcuate ligaments** – tendinous thickenings of the diaphragm; form openings for the psoas major mm.

  - **Lateral arcuate ligaments** – tendinous thickenings of the diaphragm; form openings for the quadratus lumborum mm.

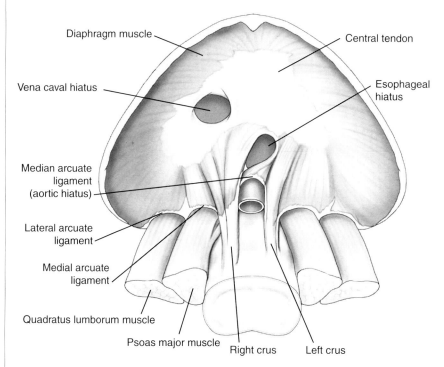

■ **Figure 11-7   Abdominal surface of the diaphragm**

Posterior Abdominal Wall – Muscles and Associated Structures

■ Identify the following (Figure 11–8):

- **Psoas major m.** – long muscle attached to the bodies and transverse processes of L1–L5 vertebrae

  - Overlaps the medial portion of the quadratus lumborum m. and crosses anterior to the sacroiliac joint; passes inferior to the inguinal ligament to insert on the lesser trochanter of the femur

  - Lumbar plexus of nerves exits the intervertebral foramina and emerges through and between the muscle fibers that arise from the transverse processes

- **Psoas minor m.** – this thin muscle courses anteriorly on the psoas major m. and inserts distally on the pecten pubis (sometimes absent)

- **Iliacus m.** – attaches in the iliac fossa and courses distally to the lesser trochanter of the femur

- **Quadratus lumborum m.** – quadrilateral-shaped muscle that is lateral and deep to the psoas major m.

  - **Iliolumbar ligament** – courses from the ilium to the lumbar vertebrae along the course of the quadratus lumborum m.

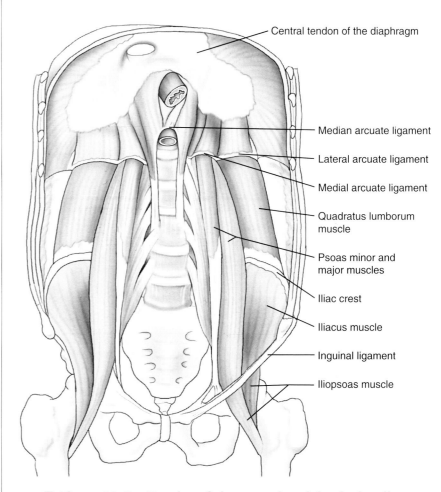

■ **Figure 11-8   Muscles of the posterior abdominal wall**

Nerves of the Posterior Abdominal Wall

■ Identify the following (Figure 11–9):

- **Subcostal n.** (T12 spinal cord segment) –
  scrape/dissect through the peritoneum and fascia inferior to rib 12

  - The subcostal n. crosses superior to the lateral
    arcuate ligament and descends on the anterior
    surface of the transversus abdominis m. before
    passing through this muscle

- **Lumbar plexus** – embedded in the muscle belly of the
  psoas major m.

  - **Iliohypogastric** and **ilioinguinal nn.** (both L1
    spinal cord segment) – diverge from their common
    trunk at the lateral, superior border of the psoas
    major m.; course across the quadratus lumborum m.
    prior to traversing the transversus abdominis m.
    above the iliac crest

    ◆ May arise as a single nerve and split within the
      layers of the abdominal wall

  - **Genitofemoral n.** (L1–L2 spinal cord segments) –
    located on the anterior surface of the psoas major m.

    ◆ Divides on the distal portion of the psoas major m.
      into the **genital branch** (exits the abdomen
      through the deep inguinal ring) and the **femoral
      branch** (courses with the external iliac a.)

  - **Lateral femoral cutaneous n.** (L2–L3 spinal cord
    segments) – located along the lateral border of the
    psoas major m., where the iliac crest touches the
    psoas major m.

    ◆ Crosses the iliacus m. before passing deep to the
      inguinal ligament

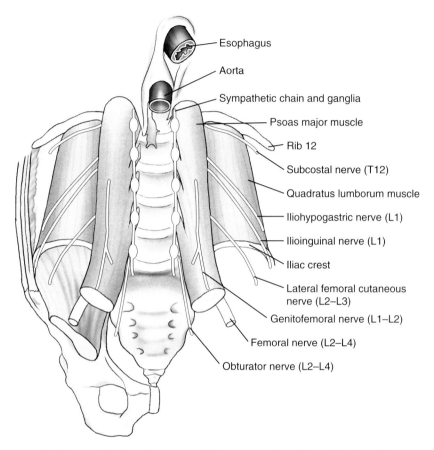

Esophagus

Aorta

Sympathetic chain and ganglia

Psoas major muscle

Rib 12

Subcostal nerve (T12)

Quadratus lumborum muscle

Iliohypogastric nerve (L1)

Ilioinguinal nerve (L1)

Iliac crest

Lateral femoral cutaneous
nerve (L2–L3)

Genitofemoral nerve (L1–L2)

Femoral nerve (L2–L4)

Obturator nerve (L2–L4)

■ **Figure 11-9   Nerves of the posterior abdominal wall**

Nerves of the Posterior Abdominal Wall—cont'd

- **Femoral n.** (L2–L4 spinal cord segments) – located lateral to the psoas major m. near its union with the iliacus m.; courses deep to the inguinal ligament en route to the thigh

- **Obturator n.** (L2–L4 spinal cord segments) – probe the medial, inferior portion of the psoas major m. to reveal the obturator nerve

- **Lumbosacral trunk** (L4–L5 spinal cord segments) – descends along the front of the sacrum to join with the ventral rami of S1–S3

Gluteal Region and Ischioanal Fossa

Prior to dissection, you should familiarize yourself with the following structures:

OSTEOLOGY

- Os coxae
  - Obturator foramen
  - Ilium
    - Iliac crest
    - Posterior superior iliac spine
  - Ischium
    - Ischial spine and tuberosity
    - Greater and lesser sciatic foramina
    - Ischiopubic (conjoined) ramus
  - Pubis
    - Pubic symphysis and arch
    - Superior pubic ramus
    - Inferior pubic ramus
    - Ischiopubic (conjoined) ramus

LIGAMENTS

- Sacrotuberous ligament
- Sacrospinous ligament
- Anococcygeal ligament

MUSCLES

- Gluteus maximus m.
- Obturator internus m.
- Piriformis m.
- External anal sphincter m.
- Pelvic diaphragm
  - Levator ani m.
    - Puborectalis m.
    - Pubococcygeus m.
    - Iliococcygeus m.
  - Coccygeus m.

ISCHIOANAL (ISCHIORECTAL) FOSSA

- Internal pudendal aa. and vv.
  - Inferior rectal n. and vessels
- Pudendal nn.
- Inferior gluteal nn., aa., and vv.
- Obturator fascia over pudendal canal
- External anal sphincter
- Sciatic nn.

Gluteal Region – Skin Incisions

■ Follow these instructions to dissect the gluteal region (avoid entering the posterior region of the thigh):

- Place the cadaver prone to make the following skin incisions (Figure 12–1):

  - *A* to the coccyx bone (*B*)

  - The coccyx bone (*B*) to *C* on the lateral side of the thigh on both sides

- Remove the skin and superficial fascia as one; this may take a while, depending on your cadaver

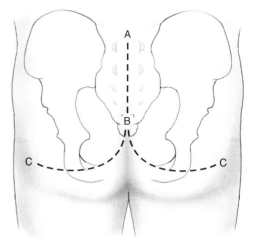

■ **Figure 12–1   Skin incisions of the gluteal region**

Gluteal Region

■ Identify the following structures on both sides (Figure 12–2):

- **Gluteus maximus m.** – study its attachments along the iliac crest, sacrum, and coccyx (it also attaches to the iliotibial tract)

  - Expose the inferior (free) border of the gluteus maximus m. by lifting up the inferior border

- **Ischial tuberosity** – attachment for the sacrotuberous ligament

- **Sacrotuberous ligament** – attached to the sacrum and ischial tuberosity

  - Cut the gluteus maximus m. from the superficial surface of this ligament, *without* cutting the ligament

  - Reflect the gluteus maximus m. laterally only as far as necessary to uncover the sciatic n. and piriformis m. (studied further in another laboratory session)

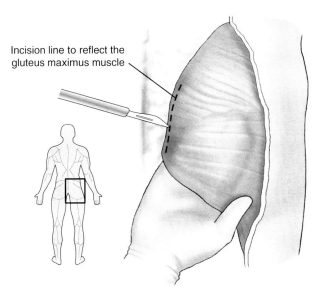

Incision line to reflect the gluteus maximus muscle

■ **Figure 12–2 Gluteus maximus muscle**

Gluteal Region—cont'd

■ Identify the following (Figure 12–3):

- **Ischial spine** – attachment for the sacrospinous ligament

- **Sacrospinous ligament** – attached to the sacrum and ischial spine

- **Piriformis m.** – attached to the sacrum and greater trochanter of the femur; the sciatic n., pudendal n., and internal pudendal a. exit the pelvis inferior to the piriformis m. via the greater sciatic foramen; studied further in another laboratory session

- **Sciatic n.** – formed from ventral primary rami of spinal cord segments L4–S3; exits the pelvis inferior to the piriformis m. via the greater sciatic foramen

- **Pudendal n.** and **internal pudendal a.** and **v.** – emerge inferior to the piriformis m., via the greater sciatic foramen

■ *Do not* reflect the gluteus maximus m. from the ilium at this time

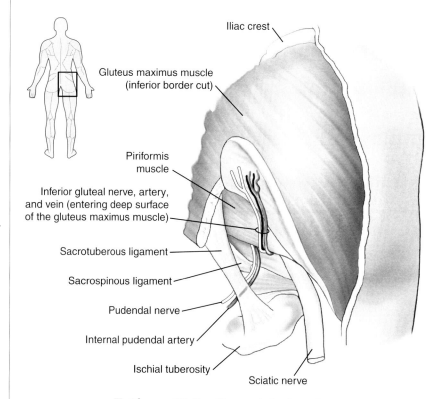

Iliac crest

Gluteus maximus muscle
(inferior border cut)

Piriformis
muscle

Inferior gluteal nerve, artery,
and vein (entering deep surface
of the gluteus maximus muscle)

Sacrotuberous ligament

Sacrospinous ligament

Pudendal nerve

Internal pudendal artery

Ischial tuberosity

Sciatic nerve

■ **Figure 12–3   Deep gluteal region**

Ischioanal (Ischiorectal) Fossa

■ Separate the legs; remove the remaining skin from the tip of the coccyx to each ischial tuberosity and across from one tuberosity to the other

■ Identify the following (Figure 12–4):

- **Pudendal n. and internal pudendal a. and v.** – they cross posterior to the ischial spine and sacrospinous ligament to enter the lesser sciatic foramen (anterior to the sacrotuberous ligament) and then the ischioanal fossa

- **Ischioanal fossa** – wedge-shaped, fat-containing space between the skin of the anal region and the pelvic diaphragm (one fossa on each side of the anus)

  - The apex (anterior recess) of each fossa is anterior and superior to the urogenital diaphragm and inferior to the pelvic diaphragm

- **Ischioanal fat pads** – the fat and connective tissue that fills the ischioanal fossae

■ Remove the fat pads in each fossa to identify the following:

- **Inferior rectal n., a., and v.** – arise from the pudendal n. and internal pudendal a. and v.; course from the lateral wall of the fossa medially to the anus

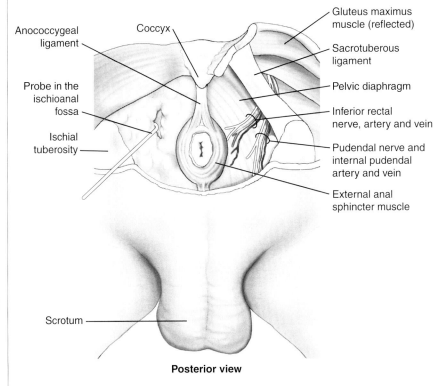

Posterior view

■ **Figure 12–4   Ischioanal fossa**

Ischioanal (Ischiorectal) Fossa—cont'd

- **Pudendal canal** – the obturator fascia splits into two layers to form the pudendal canal on the medial aspect of the obturator internus m.; follow the inferior rectal n., a., and v. laterally to the pudendal canal

  - Probe the pudendal canal by following the pudendal n. and internal pudendal vessels near the sacrospinous ligament

- **External anal sphincter** – composed of skeletal muscle; scrape the subcutaneous tissue around the anus to reveal the wisps of circumferential muscle

- **Anococcygeal ligament** – connects the anus and coccyx

- **Levator ani m.** – a broad, thin curved muscle sheet that unites with the coccygeus m. to form the pelvic diaphragm

  - Stretches between the pubis anteriorly and the coccyx posteriorly, and from one side of the pelvis to the other

- The pelvic diaphragm closes the inferior pelvic outlet somewhat like a funnel would if it were placed in the pelvic cavity

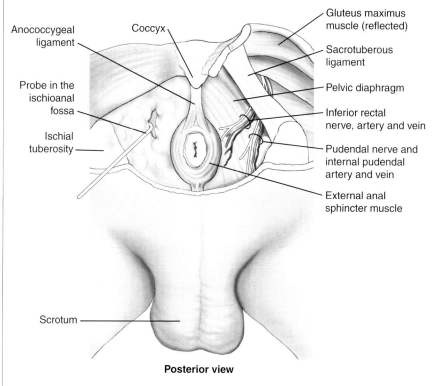

Anococcygeal ligament

Coccyx

Gluteus maximus muscle (reflected)

Sacrotuberous ligament

Probe in the ischioanal fossa

Pelvic diaphragm

Inferior rectal nerve, artery and vein

Ischial tuberosity

Pudendal nerve and internal pudendal artery and vein

External anal sphincter muscle

Scrotum

Posterior view

■ Figure 12–4   Ischioanal fossa

Urogenital Triangle

Prior to dissection, you should familiarize yourself with the following structures:

MALE
- ■ Perineum – urogenital and anal triangles
- ■ External genitalia
  - Penis – crura, body, glans, penile raphe, prepuce
  - Scrotum
  - Testis – epididymis, tunica vaginalis, tunica albuginea, seminiferous tubules, epididymal sinus, gubernaculum testis
  - Spermatic cord
    - Ductus deferens, testicular a., pampiniform plexus of vv., autonomic nn., genital branch of the genitofemoral n.
- ■ Superficial perineal space (see Table 13–1)
  - Roots of the external genitalia
    - Bulb/corpus spongiosa penis
    - Crura/corpora cavernosa penis

- Bulbospongiosus m.
- Ischiocavernosus mm.
- Spongy urethra
- Perineal nn., aa., and vv.
  - Posterior scrotal n., a., and v.
- Superficial transverse perineal mm.
- ■ Deep perineal space (see Table 13–2)
  - Perineal membrane

FEMALE
- ■ Perineum – urogenital and anal triangles
- ■ External genitalia
  - Labia majora and minora
  - Mons pubis
  - Clitoris (crura, body, and glans)
  - Urethra
  - Vagina
  - Vestibule

- ■ Superficial perineal space (see Table 13–1)
  - Roots of the external genitalia
    - Bulb of the vestibule
    - Crura/corpora cavernosa clitoris
  - Bulbospongiosus m.
  - Ischiocavernosus mm.
  - Spongy urethra
  - Perineal nn., aa., and vv.
    - Posterior labial n., a., and v.
  - Superficial transverse perineal mm.
  - Greater vestibular glands
  - Vagina
- ■ Deep perineal space (see Table 13–2)
  - Perineal membrane (UG diaphragm)

- ■ Male cadavers go to page 150
- ■ Female cadavers go to page 161

TABLE 13-1   Contents of the Superficial Perineal Space

Male	Female
1. Roots of external genitalia a. Bulb*/corpus spongiosum penis b. Crura†/corpora cavernosa penis 2. Bulbospongiosus m. – covers the bulb of the penis 3. Ischiocavernosus m. – covers the crura 4. Spongy urethra 5. Perineal nerves, arteries, and veins a. Posterior scrotal n., a., and v. b. Muscular branches 6. Superficial transverse perineal mm. 7. — 8. —	1. Roots of external genitalia a. Bulb of the vestibule b. Crura†/corpora cavernosa clitoris 2. Bulbospongiosus m. – covers the vestibular bulbs 3. Ischiocavernosus m. – covers the crura 4. Spongy urethra 5. Perineal nerves, arteries, and veins a. Posterior labial n., a., and v. b. Muscular branches 6. Superficial transverse perineal mm. 7. Greater vestibular glands and ducts 8. Vagina

*Changes name distal to the suspensory ligament to the corpus spongiosum penis
†Change names distal to suspensory ligament to corpora cavernosa penis/clitoris

TABLE 13-2   Contents of the Deep Perineal Space

Male	Female
1. Membranous urethra 2. Deep transverse perineal mm. 3. Sphincter urethrae m. 4. Internal pudendal a. a. Artery to the bulb b. Deep a. of the penis – no paired v. c. Dorsal a. of the penis d. Muscular branches 5. Internal pudendal v. and branches 6. Branches of the pudendal n. – dorsal n. and muscular branches of the perineal n. 7. Bulbourethral glands and ducts 8. –	1. Membranous urethra 2. Deep transverse perineal mm. 3. Sphincter urethrovaginalis m. 4. Internal pudendal a. a. Artery to the bulb b. Deep a. of the clitoris – no paired v. c. Dorsal a. of the clitoris d. Muscular branches 5. Internal pudendal v. and branches 6. Branches of the pudendal n. – dorsal n. and muscular branches of the perineal n. 7. – 8. Vagina

TABLE 13–3 Structure of the Abdominal Wall

Layers of the Abdominal Wall	Corresponding Layers in the Scrotum and Spermatic Cord
1. Skin	1. Skin
2. Fatty (Camper's)	2.
3. Membranous (Scarpa's)	3. } Dartos muscle and fascia
4. External abdominal oblique m. and fascia	4. External spermatic fascia
5. Internal abdominal oblique m. and fascia	5. Cremaster muscle and fascia
6. Transversus abdominis m. and fascia	6. Cremaster muscle and fascia
7. Transversalis fascia	7. Internal spermatic fascia
8. Extraperitoneal fatty tissue	8. Areolar tissue and fat
9. Peritoneum	9. Tunica vaginalis

MALE

Spermatic Cord and Scrotum Overview

- Both the spermatic cord and scrotum are surrounded by fascial coverings that are continuous and derived from the abdominal wall (Table 13–3)

- To study the spermatic cord, the scrotum has to be cut (Figure 13–1):

  - Make cutaneous, vertical incisions along the left and right sides of the scrotum to reveal the spermatic cord

- Identify the following:

  - **Spermatic cord** – begins at the deep inguinal ring, lateral to the inferior epigastric vessels, and ends in the scrotum at the posterior border of the testis

  - **Scrotal sac** – hangs inferior to the pubis and the root of the penis

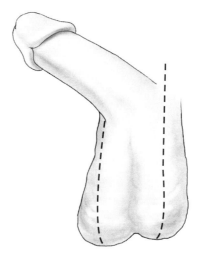

■ **Figure 13–1   Scrotal incisions**

MALE

Scrotum – Tissue Layers

■ Identify the following layers of the scrotum (Figure 13–2):

- **Superficial scrotal (Dartos) fascia** – very thin layer that is continuous with Scarpa's fascia; devoid of fat; forms the scrotal septum; probe the fascia to find the following:

  - **Posterior scrotal nn. and vessels** – originate from branches of the perineal branch of the pudendal n. and perineal branches of the internal pudendal vessels

    ◆ These structures may not be evident

  - **Anterior scrotal nn. and vessels** – branches from the ilioinguinal n. and external pudendal vessels from the femoral vessels

■ Layers of the anterior abdominal wall cover the spermatic cord because the testis carried the layers during its descent from the posterior abdominal wall to the scrotum; of the layers listed below, you probably will only be able to identify the cremasteric fibers (middle layer)

- **External spermatic fascia** – outer layer of spermatic fascia continuous with the external oblique aponeurosis; deep to the Dartos fascia

- **Cremasteric fibers/fascia** – middle layer of spermatic fascia; composed of loose connective tissue and thin fibers of cremasteric m. derived from the internal oblique m.

- **Internal spermatic fascia** – innermost layer of spermatic fascia; continuous with the transversalis fascia

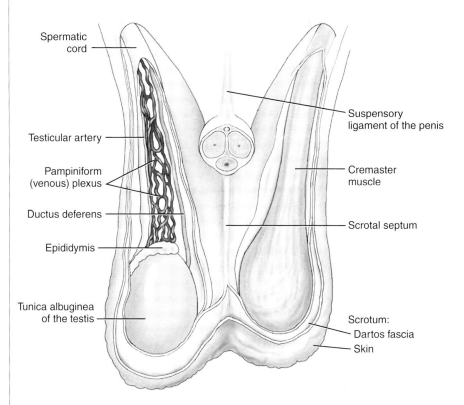

■ **Figure 13-2   Spermatic cord and scrotum**

MALE

Spermatic Cord – Contents

■ Identify the following structures within the spermatic cord (Figure 13–3):

- **Ductus deferens** – muscular tube that conveys sperm from the epididymis to the ejaculatory duct

  - Palpate this firm, tubular structure within the spermatic cord (the ductus deferens is the hardest structure in the spermatic cord)

- **Testicular a.** – arises from the aorta; in the spermatic cord, the artery courses with the pampiniform plexus of veins

- **Pampiniform plexus of veins** – venous network that drains into the right and left testicular vv.

- **Sympathetic n. fibers** – on the wall of the testicular a.

- **Genital branch of the genitofemoral n.** – motor supply to the cremaster m.; this nerve may not be evident

- **Testis** – primary sex organ in the male

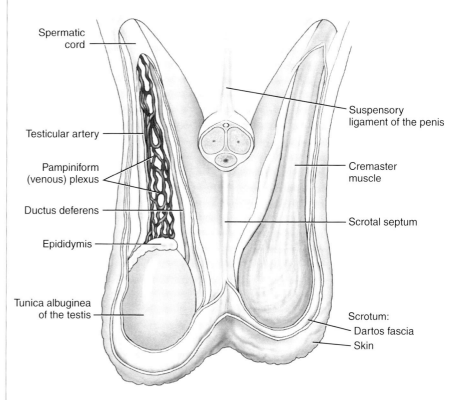

■ **Figure 13–3   Coronal section through the anterior wall of the scrotal sac; penis cut in cross-section for reference**

MALE

Testis

■ Identify the following (Figures 13–4 and 13–5):

- **Gubernaculum testis** – ligament that attached the inferior pole of the testis to the inferior surface of the scrotal sac
- **Tunica vaginalis** – derived from the parietal peritoneum; it covers most of the testis; it has a parietal layer (fused with the internal spermatic fascia) and visceral layer (bound to the anterolateral surface of the testis and epididymis)
  - Covers the anterior, medial, and lateral surfaces of the testis, but not the posterior surface

■ Incise the tunica vaginalis to inspect the interior of the serous sac

- **Sinus of the epididymis** – separates the testis from the body of the epididymis
- **Epididymis** – attached to the posterior surface of the testis; observe its **head, body,** and **tail regions;** the ductus deferens begins at the tail of the epididymis

■ Cut the testis longitudinally into left and right portions but leave the epididymis intact

- **Tunic albuginea** – deep to the visceral layer of the tunica vaginalis; tough layer of connective tissue that is the capsule of the testis; observe the numerous pyramidal **lobules** of the testis
- **Seminiferous tubules** – fine, threadlike tubes that produce sperm

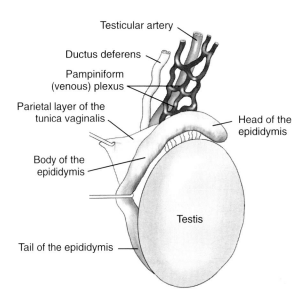

■ **Figure 13-4   Lateral view of the testis and epididymis**

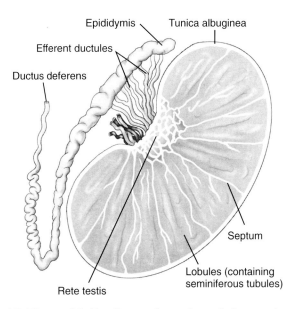

■ **Figure 13-5   Coronal section of the testis**

MALE

Overview of the Urogenital (UG) Triangle

■ The perineum is a diamond-shaped area between the thighs; it is divided into two triangular areas (Figure 13–6):

- **Anal triangle** – triangular region bounded by the two ischial tuberosities and the coccyx bone

- **Urogenital triangle** – triangular region bounded by the ischial tuberosities, ischiopubic (conjoined) rami, and pubic symphysis

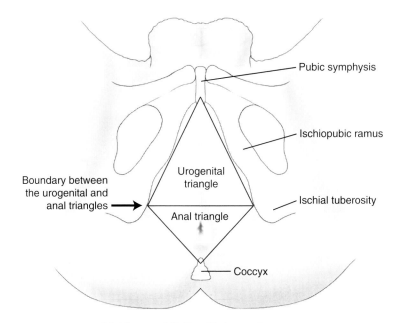

■ **Figure 13–6   Male perineum**

MALE

Dissection of the Urogenital (UG) Triangle

- Turn the cadaver supine and abduct the legs

- Identify the following structures (Figure 13–7):

  - **Shaft of the penis** – the body of the penis

  - **Glans penis** – the head or tip of the penis

  - **Penile raphe** – located on the ventral surface of the penis (erect in the anatomic position)

- To remove the skin and fat from both sides of the UG triangle and scrotum, follow these instructions (see Figure 13–7):

  - Separate both testicles from the scrotum

  - Cut off the scrotum where it is attached posteriorly and laterally over the urogenital triangle, and from the base of the penis but *do not* cut the penis

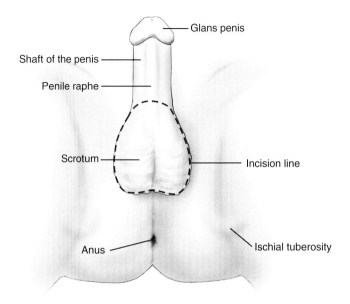

■ **Figure 13–7   Skin incision in the male perineum**

MALE

Superficial Perineal Space – Contents

■ To observe the structures in the superficial perineal space, follow these instructions (Figure 13–8):

● Cut the penis in cross-section, about 2 cm from its root

■ Observe the following structures on the cross-section of the penis (homologous to the female) (Figure 13–9):

● **Superficial penile fascia** – continuous with Scarpa's and Dartos fasciae; devoid of fat tissue

  ● **Dorsal superficial v.** – drains to the external pudendal v.

● **Deep penile fascia** – also called Buck's fascia

● **Tunica albuginea** – connective tissue sheath surrounding the erectile tissues; forms the septum penis that separates the corpora cavernosa

● **Dorsal deep v.** – courses deep to the symphysis pubis to drain into the prostatic plexus of veins

● **Dorsal a.** – originates from the internal pudendal a.

● **Dorsal n.** – originates from the pudendal n.

● **Corpora cavernosa (two)** – paired dorsal erectile tissue bodies

● **Corpus spongiosum** – ventral erectile tissue that contains the **spongy urethra**

● **Deep a.** – in the center of each of the two corpora cavernosa; branch of the internal pudendal a.; no paired veins

■ Retain the detached penile cross-section for review

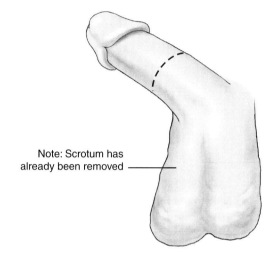

Note: Scrotum has already been removed

■ **Figure 13–8   Transection of the penile shaft**

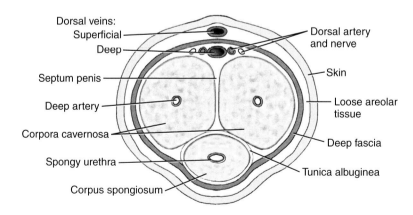

Dorsal veins:
Superficial
Deep
Septum penis
Deep artery
Corpora cavernosa
Spongy urethra
Corpus spongiosum
Dorsal artery and nerve
Skin
Loose areolar tissue
Deep fascia
Tunica albuginea

■ **Figure 13–9   Cross-section of the penis**

MALE

Superficial Perineal Space

- The superficial perineal space is the region between the membranous layer of the superficial fascia (continuous with Scarpa's and Dartos fasciae) and the inferior fascia of the urogenital (UG) diaphragm called the perineal membrane; this space extends into the scrotal sac (Figure 13–10)

- Palpate the area between the bulb and crura of the penis; your finger is stopped by the inferior fascia of the UG diaphragm

- **UG diaphragm**

  - Located between the ischial tuberosities, ischiopubic (conjoined) rami, and inferior arch of the pubic symphysis

  - Consists of two muscles and their superior and inferior fasciae

- The contents of the superficial perineal space include three paired muscles and penile erectile tissue (see Table 13–1)

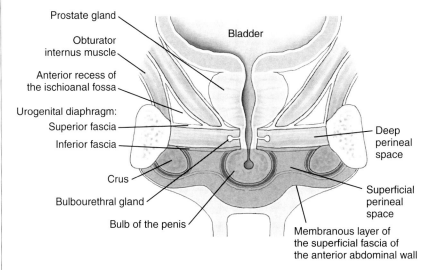

Prostate gland
Bladder
Obturator internus muscle
Anterior recess of the ischioanal fossa
Urogenital diaphragm:
Superior fascia
Inferior fascia
Deep perineal space
Crus
Superficial perineal space
Bulbourethral gland
Bulb of the penis
Membranous layer of the superficial fascia of the anterior abdominal wall

■ **Figure 13–10 Coronal section of the male perineum**

MALE

Superficial Perineal Space – Contents— cont'd

- Follow these instructions to reveal the muscles of the superficial perineal space

  - Reflect remaining skin and superficial fascia that extend inferiorly from the anterior abdominal wall to the penis; note the fundiform ligament on the dorsal surface of the penis while you dissect the deeper fibrous **suspensory ligament of the penis** (not shown)

- Identify the following muscles of the superficial perineal space (Figure 13–11):

  - **Ischiocavernosus m.** – covers each crus of the **corpora cavernosa**

    - Follow the paired corpora cavernosa to their respective attachment on each ischiopubic (conjoined) ramus

    - These fixed (nonpendulous) parts are called **crura** (legs); both crura are covered superficially (inferiorly) by an ischiocavernosus m.

  - **Bulbospongiosus m.** – covers the **bulb of the penis**

- Follow the corpus spongiosum posteriorly to find the perineal body and the inferior fascia of the UG diaphragm

  - **Superficial transverse perineal m.** – crosses transversely between the medial aspects of the ischial tuberosities and the perineal body (discussed later)

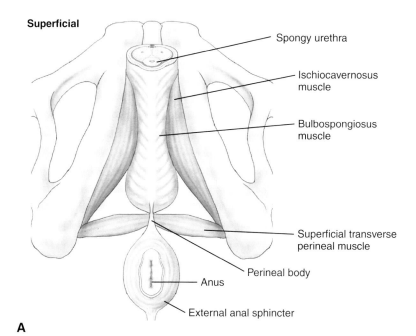

Superficial

- Spongy urethra
- Ischiocavernosus muscle
- Bulbospongiosus muscle
- Superficial transverse perineal muscle
- Perineal body
- Anus
- External anal sphincter

A

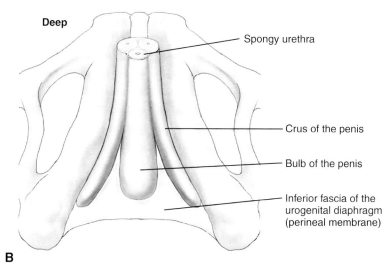

Deep

- Spongy urethra
- Crus of the penis
- Bulb of the penis
- Inferior fascia of the urogenital diaphragm (perineal membrane)

B

- **Figure 13-11   Superficial perineal space of the male.**
  A, Muscles – overlay the erectile tissue. B, Erectile tissue – muscle removed from the erectile tissue

MALE

Deep Perineal Space

■ To identify some of the structures in the deep perineal space (see Figure 13–10), follow these instructions (Figure 13–12):

- Make an incision through the inferior fascia of the UG diaphragm, between the crura and bulb

- Spread the cut inferior fascia to reveal the deep transverse perineal m.

■ Identify the following:

- **Perineal body** – the central attachment point of the muscles in the perineum; identify the converging muscles:

  - **Deep transverse perineal m.**

  - **Superficial transverse perineal m.**

  - **External anal sphincter m.**

  - **Levator ani m.**

■ *Note:* The anterior portion of the deep transverse perineal m. is also called the external urethral sphincter; this muscle surrounds the membranous portion of the urethra and acts as a voluntary sphincter for urination; the urethral sphincter is difficult to dissect without destroying everything in the superficial perineal space

■ See Table 13–2 for contents of the deep perineal space

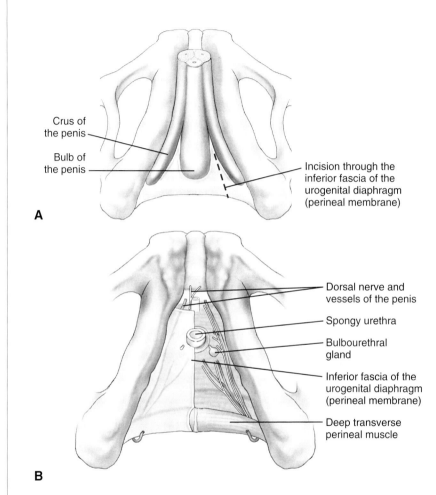

Crus of the penis

Bulb of the penis

Incision through the inferior fascia of the urogenital diaphragm (perineal membrane)

A

Dorsal nerve and vessels of the penis

Spongy urethra

Bulbourethral gland

Inferior fascia of the urogenital diaphragm (perineal membrane)

Deep transverse perineal muscle

B

■ **Figure 13-12   Deep perineal space. *A,* Incision in the inferior fascia of the urogenital diaphragm (UG). *B,* Structures associated with the deep perineal space**

MALE

Review – Branches of the Pudendal Artery and Nerve

- Please review this page and note the compartmental nerve, artery, and vein distribution (Figure 13–13)

- Blood supply to the urogenital region:

  - **Internal pudendal a.**
    - **Inferior rectal a.**
    - **Perineal a.**
      - ◆ **Posterior scrotal a.**
    - **Deep a. of the penis** – no paired veins
    - **Dorsal a. of the penis**
  - **External pudendal a.**
    - **Anterior scrotal a.**

- Innervation to the urogenital region:

  - **Pudendal n.**
    - **Inferior rectal n.**
    - **Deep perineal n.**
    - **Dorsal n. of the penis**
    - **Superficial perineal n.**
    - **Posterior scrotal nn.**
  - **Ilioinguinal n. (L1)**
    - **Anterior scrotal nn.**

- *Note:* Most arteries have an accompanying vein; remember the presence of the portal-caval anastomosis between the superior rectal v. and the middle/inferior rectal vv.

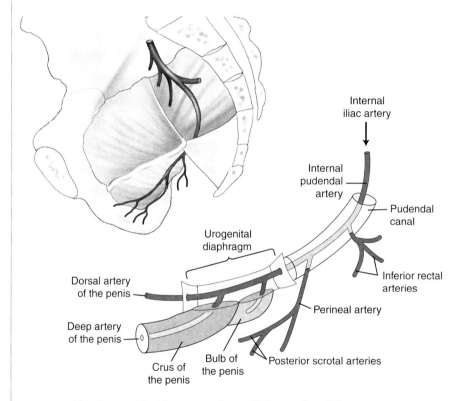

■ **Figure 13-13** **Branches of the pudendal artery**

FEMALE

Overview of the Urogenital (UG) Triangle

■ The perineum is a diamond-shaped area between the thighs; it is divided into two triangular areas (Figure 13–14):

- **Anal triangle** – triangular region bounded by the two ischial tuberosities and the coccyx bone

- **Urogenital triangle** – triangular region bounded by the ischial tuberosities, ischiopubic (conjoined) rami, and pubic symphysis

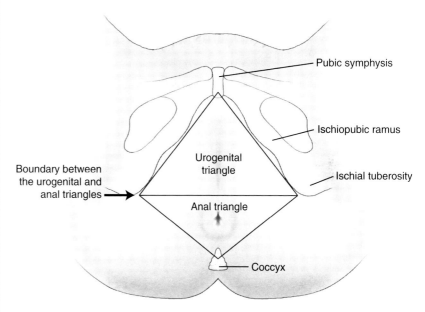

■ **Figure 13–14   Female perineum**

FEMALE

External Genitalia – Perineum

■ Turn the cadaver prone to identify the following (Figure 13–15):

- **Mons pubis** – skin and adipose tissue superficial to the pubic symphysis; point of attachment for the round ligament of the uterus

- **Labia majora**
  - Two large folds of skin filled with adipose tissue
  - Joined anteriorly by the anterior labial commissure
  - Pubic hair lines this region of the perineum

- **Labia minora**
  - Two thin delicate folds of adipose free, hairless skin
  - Lie between the labia majora and enclose the vestibule of the vagina
  - Lie on each side of the urethra and vagina

- **Vestibule** – space between the labia minora

- **Clitoris** – located between the anterior ends of the labia minora; composed of erectile tissue (and its muscle), like the penis, and is capable of erection

- **External urethral orifice** – anterior to the vaginal orifice

- **Vaginal orifice** – opening for the vagina; positioned between the external urethral orifice and the anus

- **Anus** – posterior to the vaginal orifice

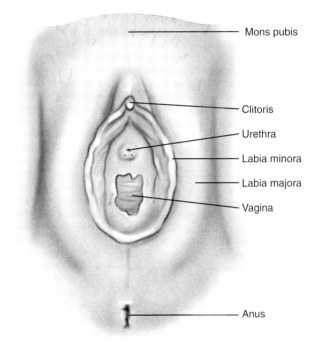

■ **Figure 13-15   Female external genitalia**

FEMALE

Dissection of the Urogenital (UG) Triangle

■ Turn the cadaver supine and abduct the legs

■ Make the following incisions (Figure 13–16):

- Cut circumferentially around the lateral surface of the labia minora
  - Remove the skin overlying the labia majora and minora
- Reflect the skin and fat to reveal the underlying muscles in the superficial perineal space

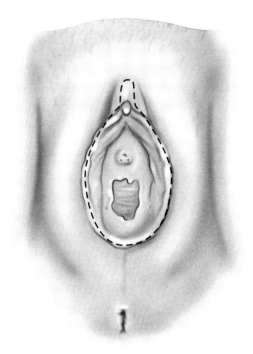

■ **Figure 13–16   Dissection of the labia majora and minora**

FEMALE

Superficial Perineal Space

- The superficial perineal space is the region between the membranous layer of the superficial fascia (continuous with Scarpa's fascia) and the inferior fascia of the urogenital (UG) diaphragm called the perineal membrane (Figure 13–17)

- Palpate the area between the bulb and crura of the clitoris; your finger is stopped by the inferior fascia of the UG diaphragm

- The contents of the superficial perineal space include three paired muscles and clitoral erectile tissue and greater vestibular glands (see Table 13–1)

- **UG diaphragm**

  - Located between the ischial tuberosities, ischiopubic (conjoined) rami, and inferior arch of the pubic symphysis

  - Consists of two muscles and their superior and inferior fasciae

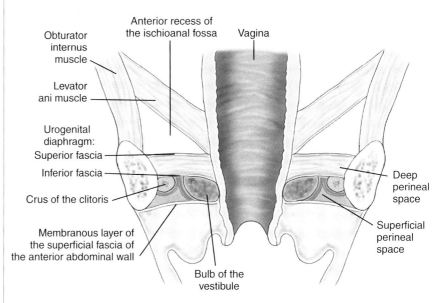

■ Figure 13–17   Coronal section of the female perineum

FEMALE

Superficial Perineal Space – Contents

■ To observe the structures in the superficial perineal space, follow these instructions:

- Expose the clitoris and cut it in cross-section

■ Observe the following gross structures on the cross-section of the cut clitoris (Figure 13–18):

- **Tunica albuginea** – connective tissue sheath surrounding the erectile tissues; forms the septum clitoris that separates the corpora cavernosa tissue

- **Corpora cavernosa (two)** – paired erectile tissue bodies

- **Deep a.** – in the center of each of the two corpora cavernosa; branches of the internal pudendal a.

■ Attempt to locate the smaller structures (homologous to the male):

- **Superficial clitoral fascia** – continuous with Scarpa's fascia

  - **Dorsal superficial v.** – drains to the external pudendal v.

- **Deep clitoral fascia**

- **Dorsal deep v.** – courses deep to the pubic symphysis

- **Dorsal a.** – originates from the internal pudendal a.; no paired vv.

- **Dorsal n.** – originates from the pudendal n.

- **Greater vestibular glands** – located posterior to each labia majora; do not try to locate these glands

■ Retain the detached clitoris cross-section for review

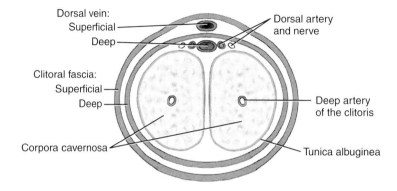

Dorsal vein:
Superficial
Deep

Clitoral fascia:
Superficial
Deep

Corpora cavernosa

Dorsal artery and nerve

Deep artery of the clitoris

Tunica albuginea

■ **Figure 13–18   Cross-section of the clitoris**

FEMALE

Superficial Perineal Space – Contents—cont'd

■ Identify the following muscles of the superficial perineal space (Figure 13–19):

• **Ischiocavernosus m.** – covers each crus of the **corpora cavernosa**

  ● Follow the paired corpora cavernosa to their respective attachment on each ischiopubic (conjoined) ramus

  ● These fixed (nonpendulous) parts are called **crura** (legs); both crura are covered superficially (inferiorly) by a thin muscle called the ischiocavernosus m.

• **Bulbospongiosus m.** – covers the **bulb of the vestibule**

  ● Clean the area between the crura and vaginal orifice to locate the bulbospongiosus m. fibers on the lateral surface of the labia minora

  ● Follow the bulbospongiosus m. posteriorly to its attachment to the perineal body and the inferior fascia of the UG diaphragm

• **Superficial transverse perineal m.** – crosses transversely between the medial aspect of the ischial tuberosities and the perineal body (discussed later)

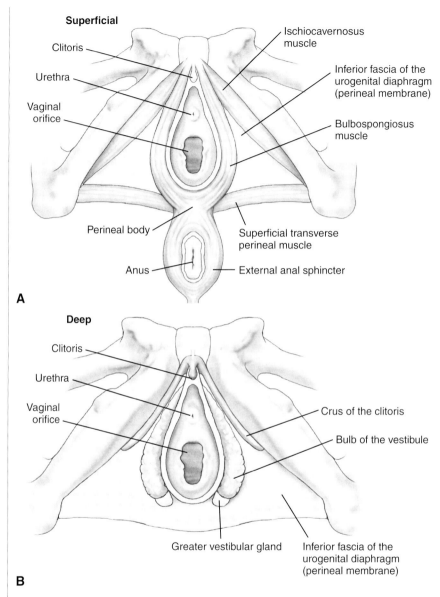

■ Figure 13-19 **Superficial perineal space of the female. A, Muscles—overlay the erectile tissue. B, Structures associated with the deep perineal space**

FEMALE

Deep Perineal Space

■ To identify some of the structures in the deep perineal space, follow these instructions (Figure 13–20):

- Make an incision through the inferior fascia of the UG diaphragm, between the crura and bulb

- Spread the cut inferior fascia to reveal the deep transverse perineal muscle

■ Identify the following:

- Perineal body – the central attachment point of the muscles of the perineum; identify the converging muscles:

  - **Deep transverse perineal m.**

    - ◆ **Sphincter urethrovaginalis**

  - **Superficial transverse perineal m.**

  - **External anal sphincter m.**

  - **Levator ani m.**

■ *Note:* The anterior portion of the deep transverse perineal m. contains both the sphincter urethra and sphincter vagina mm.; the sphincter urethrovaginalis is difficult to dissect without destroying everything in the superficial perineal space

■ See Table 13–2 for contents of the deep perineal space

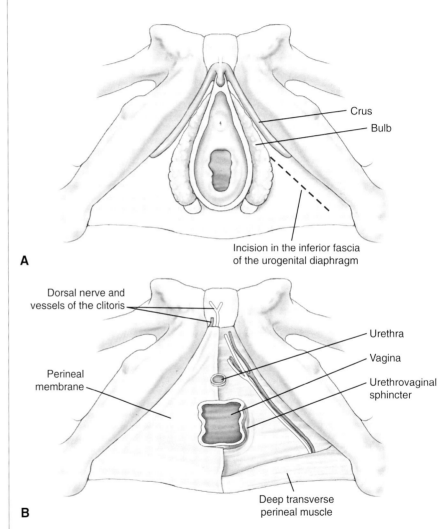

A

Crus
Bulb
Incision in the inferior fascia of the urogenital diaphragm

B

Dorsal nerve and vessels of the clitoris
Perineal membrane
Urethra
Vagina
Urethrovaginal sphincter
Deep transverse perineal muscle

■ **Figure 13–20   Deep perineal space. *A,* Incision in the inferior fascia of the urogenital diaphragm (UG). *B,* Structures associated with the deep perineal space**

FEMALE

Review – Branches of the Pudendal Artery and Nerve

- Please review this page and note the compartmental nerve, artery, and vein distribution (Figure 13–21)

- Blood supply to the urogenital region:

  - **Internal pudendal a.**
    - **Inferior rectal a.**
    - **Perineal a.**
      - ◆ **Posterior labial a.**
    - **Deep a. of the clitoris** – no paired vein
    - **Dorsal a. of the clitoris**
  - **External pudendal a.**
    - **Anterior labial a.**

- Review the innervation to the urogenital region:

  - **Pudendal n.**
    - **Dorsal n. of the clitoris**
    - **Inferior rectal n.**
    - **Deep perineal n.**
    - **Superficial perineal n.**
    - **Posterior labial nn.**
  - **Ilioinguinal n.**
    - **Anterior labial n.**

- *Note*: Most arteries have an accompanying vein; remember the presence of the portal-caval anastomosis between the superior rectal v. and the middle/inferior rectal vv.

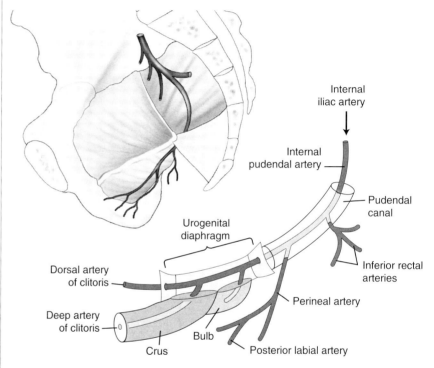

■ **Figure 13-21   Branches of the pudendal artery**

Pelvis and Reproductive Systems

Lab 14

Prior to dissection, you should familiarize yourself with the following structures:

OSTEOLOGY

- Pelvic brim
  - False pelvis
  - True pelvis
- Pelvic inlet and outlet

FEMALE PELVIC STRUCTURES

- Transverse cervical (cardinal) ligament
- Broad ligament
  - Mesometrium
  - Mesovarium
  - Mesosalpinx
- Ligament of the ovary
- Suspensory ligament of the ovary
- Round ligament of the uterus
- Vesicouterine pouch
- Rectouterine pouch

FEMALE PELVIC ORGANS

- Bladder
  - Trigone
  - Obliterated urachus
- Uterus – fundus, body, uterine horn, uterine cavity, internal os, cervix, isthmus, external os
- Vagina – rugae; anterior, two lateral, and posterior fornices; hymen
- Uterine tube – fimbriae, isthmus, ampulla
- Ovary

MALE PELVIC STRUCTURE

- Rectovesical pouch

MALE PELVIC ORGANS

- Bladder
  - Trigone
  - Obliterated urachus
- Prostate gland
- Seminal vesicle
- Ductus deferens

MUSCLES

- UG diaphragm
- Pelvic diaphragm
  - Levator ani m.
    - Puborectalis m.
    - Pubococcygeus m.
    - Iliococcygeus m.
  - Coccygeus m.
- Piriformis m.
- Obturator internus m.

NERVES

- Lumbosacral trunk (L4, L5)
- Sacral plexus (ventral rami of S1–S4)
  - Sciatic n.
  - Pudendal n.
- Superior and inferior gluteal nn.

ARTERIES AND VEINS

- Common iliac a. and v.
  - External iliac a. and v.
    - Inferior epigastric a. and v.
  - Internal iliac a. and v.
    - Iliolumbar aa. and vv.
    - Lateral sacral a. and v.
    - Obturator a. and v.
    - Superior gluteal a. and v.
    - Inferior gluteal a. and v.
    - Obliterated umbilical a. and v.
      - Branch to the ductus deferens (male)
    - Inferior vesical aa. and vv. (male)
    - Uterine and vaginal aa. and vv. (female)
    - Superior vesical a.
    - Vaginal aa. (female)
    - Middle rectal a. and vv.
    - Internal pudendal a. and v.
      - Inferior rectal a. and v.

Pelvis – Overview (Both Sexes)

■ Identify the following (Figures 14–1 and 14–2):

- **Common iliac a.** – bifurcates at the sacroiliac joint into the internal and external iliac aa.

- **Ureters** – paired tubes that drain urine from the kidneys; course inferiorly along both sides of the pelvis; cross anterior to the common/external iliac vessels on their way to the urinary bladder

- **Urinary bladder** – a reservoir for urine; located in the true pelvis, inferior to the peritoneum, where it rests on the pelvic floor posterior to the pubic symphysis

- **Ductus deferens (male)** – follow the ductus deferens from the deep inguinal ring to the posterior surface of the bladder and then to the prostate gland

- **Round ligament of the uterus (female)** – courses from the uterus at the junction of the uterine tube, between the layers of the broad ligament, and across the pelvic wall to the deep inguinal ring

- **Gonadal aa. (testicular/ovarian)** – observe their course from the aorta to the deep inguinal ring to the abdominal aorta (male) and from the abdominal aorta to the ovaries (female)

- **Left gonadal v. (testicular/ovarian)** – observe its course from the deep inguinal ring to the left renal v.

- **Right gonadal v. (testicular/ovarian)** – observe its course from the deep inguinal ring to the inferior vena cava

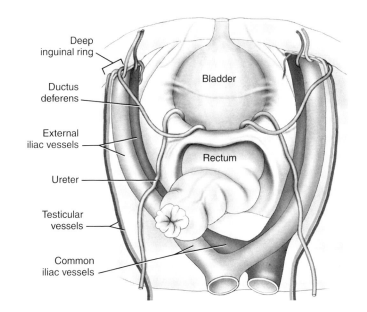

■ **Figure 14-1   Superior view of the male pelvis**

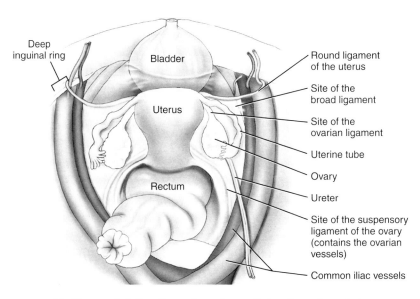

■ **Figure 14-2   Superior view of the female pelvis**

FEMALE

Pelvic Peritoneum – Broad Ligament

■ Identify the following components of the broad ligament (Figure 14–3):

- **Broad ligament** – a drape of pelvic peritoneum over the female reproductive organs

- **Mesosalpinx** – the mesentery of the uterine (fallopian) tube; most superior part of the broad ligament, between the uterine tube and ovary

- **Mesovarium** – the mesentery of the ovary; posterior (perpendicular) extension of the broad ligament; to reveal the mesovarium, pull the ovary posteriorly 90 degrees to the plane of the mesosalpinx and broad ligament

  ● The remainder of the broad ligament is inferior to the ovary

- **Round ligament of the uterus** – courses from the attachment of the ligament of the ovary with the uterine wall to the deep inguinal ring and ends by attaching to the labia majora

- **Ovarian ligament** – within the broad ligament; formed by the broad ligament where it drapes the ovarian branches of the uterine a. and v.

- **Suspensory ligament of the ovary** – formed by the peritoneum where it drapes the ovarian vessels

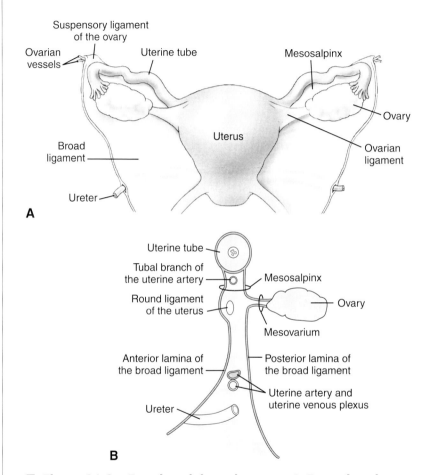

■ **Figure 14–3   Female pelvic peritoneum. *A*, Posterior view of the uterus. *B*, Sagittal section through the uterine tube**

BOTH SEXES

Pelvic Diaphragm

■ Identify the following (Figure 14–4):

- **Obturator canal** – a small opening on the internal, superior aspect of the **obturator internus m.;** traversed by the obturator a., v., and n.

- **Tendinous arch** – inferior to the opening of the obturator canal, the obturator internus fascia thickens and forms a tendinous arch between the ischial spine and pubic bone; this tendinous arch is the lateral attachment for the **iliococcygeus** part of the **levator ani m.**

- **Pelvic diaphragm** – a funnel-shaped muscle that forms the floor of the pelvic outlet; the pelvic organs are anchored to the middle of the pelvic diaphragm; composed of two paired muscles that fuse at the midline:

  - **Levator ani m.** – consists of three contiguous parts that are named by their attachments:

    ◆ **Puborectalis m.** – medial part

    ◆ **Pubococcygeus m.** – intermediate part

    ◆ **Iliococcygeus m.** – lateral part (arises from the tendinous arch of the obturator internus fascia)

  - **Coccygeus (ischiococcygeus) m.** – arises from the ischial spines and from the posterior part of the pelvic diaphragm; attaches to the coccyx

- Closure of the pelvic outlet posterior to the pelvic diaphragm is completed by the **piriformis m.**

  - Arises from the sacrum and attaches to the femur

  - The gap between the piriformis m. and pelvic diaphragm is the primary passageway for nerves and vessels from the pelvis to the gluteal region and perineum

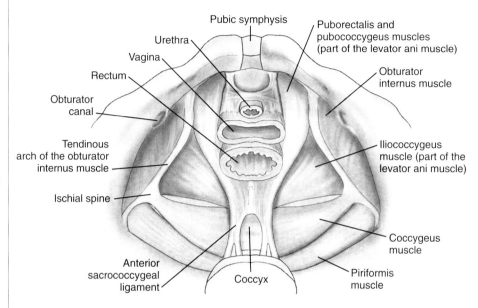

■ **Figure 14–4   Superior view of the female pelvic diaphragm**

BOTH SEXES

Hemisections

- After this part of the dissection, your cadaver will be divided into three segments; these steps are taken to improve dissection and understanding of the pelvic wall

- *Note:* Not all courses bisect the cadaver; follow the instructions from your course director

- To prepare the cadaver to be bisected, follow these instructions (Figure 14–5):

  - Cut the coronary and triangular ligaments of the liver

  - Cut the inferior vena cava between the liver and diaphragm

  - Reflect the liver to the left

  - With a scalpel, on both sides of the vertebral bodies, cut horizontally through intercostal space 9 superior to the diaphragm (*A*)

  - Use a hand bone saw to cut through the T10 intervertebral disk across the plane marked *A*

  - Cut any soft tissue necessary to divide your cadaver into superior and inferior segments

- Saw

  - Cut (with a handsaw or band saw) through the pubic symphysis in a parasagittal plane (*B*) up to the horizontal cut (*A*)

    - If you are using a handsaw, use a scalpel to cut through the soft tissues; use the saw for cutting through bone

    - **Male pelvis** – go to page 174

    - **Female pelvis** – go to page 175

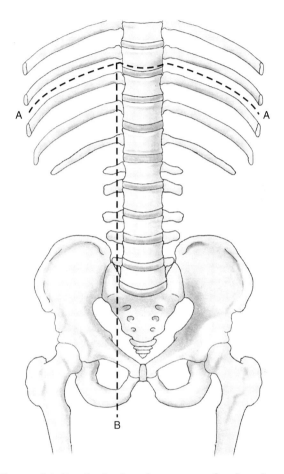

■ **Figure 14–5   Scalpel and saw cuts for hemisection**

MALE

Pelvis

■ Identify the following structures (Figure 14–6):

- **Bladder** – superior and posterior to the pubic symphysis

  - **Ureters** – enter the lateral sides of the bladder

  - **Trigone** – triangular space between the two ureteral orifices and the internal urethral orifice (you will only see half)

- **Seminal vesicle** – inferior and posterior to the bladder

- **Prostate gland** – dark organ between the bladder and urogenital diaphragm; normally about the size of a walnut

  - **Ejaculatory ducts** – you may see the ejaculatory ducts leading to the prostatic urethra; formed by the union of the ductus deferens and seminal vesical

- **Urethra** – divided into three regions based on location:

  - **Prostatic urethra** – travels through the prostate gland

  - **Membranous urethra** – travels through the UG diaphragm

  - **Spongy urethra** – travels through the corpus spongiosum penis

- **Ductus deferens** – traverses the deep inguinal ring, crosses the external iliac vessels and obturator n. and vessels to join the duct of the seminal vesicle to become the ejaculatory duct

- **UG diaphragm** (perineal structure, not a pelvic one) – inferior to the prostate gland; contains the **bulbourethral glands, membranous urethra,** and **sphincter urethrae m.**

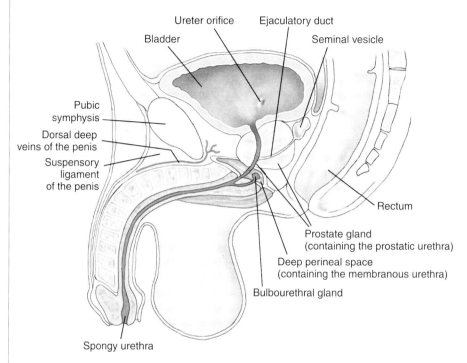

■ **Figure 14-6   Sagittal section of the male pelvis**

FEMALE

Pelvis

■ Identify the following structures (Figure 14–7):

- **Bladder** – superior and posterior to the pubic symphysis

- **Ureters** – enter the lateral sides of the bladder

- **Trigone** – triangular space between the two ureteral orifices and the internal urethral orifice (you will only see half)

- **Uterus** – posterior to the bladder; identify the fundus, body, and cervix and the following:
  - **Uterine tubes**
  - **Fimbria** – distal ends of the uterine tubes
  - **Ovaries**

- **Cervix** – protrudes into the vaginal canal; **external os** opens into the vagina
  - **Transverse cervical (cardinal) ligament** – thickening of the pelvic fascia around the uterine a. and v.

- **Vagina** – has anterior, posterior, and two lateral fornices

- **Urethra** – between the clitoris and vagina

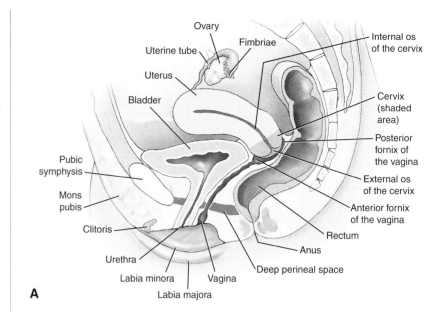

A

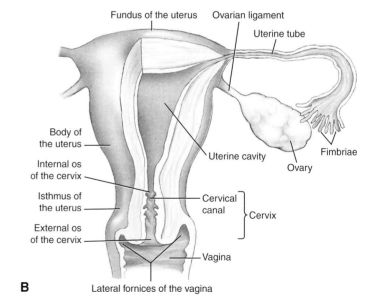

B

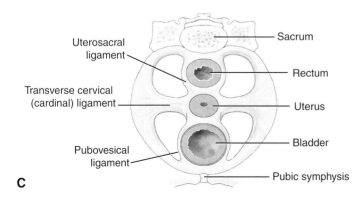

C

■ **Figure 14–7   Female pelvis. *A*, Sagittal section. *B*, Coronal section through the uterus. *C*, Cross-section of the female pelvis**

BOTH SEXES

Pelvic Vasculature and Innervation

- Remove the peritoneum and extraperitoneal fat covering the iliac vessels and nerves from the half of the pelvis that does not have the bulk of the pelvic organs attached

- Dissect the pelvic vasculature, following these instructions:

  - Remove branches of the internal iliac vv. to more clearly demonstrate the arterial distribution

  - Identify the following (Figure 14–8):

    - **Common iliac a.**

      - ◆ **Internal iliac a.** – has anterior and posterior divisions

      - ◆ **External iliac a.** – destined for the lower extremity

    - **Lymph nodes**

- Dissection of the male pelvis: Go to page 177

- Dissection of the female pelvis: Go to page 179

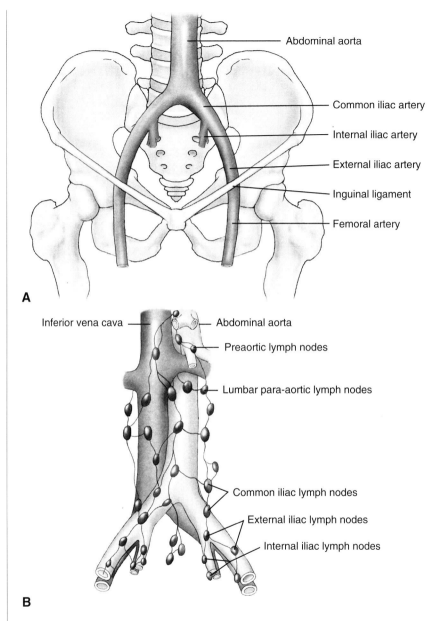

A

B

■ **Figure 14–8   Common iliac vessels. *A*, Common iliac artery. *B*, Lymphatics**

MALE

Pelvic Vasculature – Internal Iliac Artery (Anterior Division)

■ Identify the following branches (Figure 14–9):

- **Anterior division of the internal iliac a.**

  - **Umbilical a.** – continuation of the anterior division to the anterior abdominal wall; becomes the **medial umbilical ligament (obliterated umbilical a.);** may give off 3–4 **superior vesical aa.**

  - **Obturator a.** – courses through the obturator canal

  - **Inferior vesical a.** – trace to the posteroinferior part of the bladder, prostate gland, and seminal vesicles (the inferior vesical a. is not present in females)

  - **Middle rectal a.** – to the rectum

  - **Internal pudendal a.** – exits the pelvis through the greater sciatic foramen to enter the gluteal region, inferior to the piriformis m.; enters the perineum by traveling through the lesser sciatic foramen

  - **Inferior gluteal a.** – exits the pelvis, between S2 and S3 nn. usually; courses through the greater sciatic foramen to enter the gluteal region, inferior to the piriformis m.

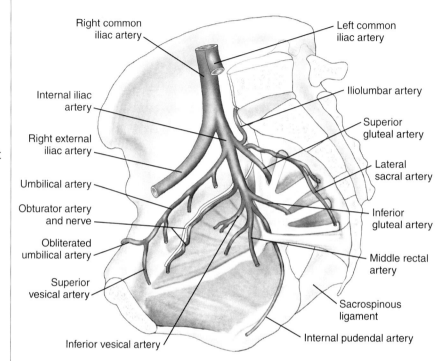

■ **Figure 14-9   Branches of the internal iliac artery of the male**

MALE

Pelvic Vasculature – Internal Iliac Artery (Posterior Division)

■ Identify the following branches (Figure 14–10):

- **Posterior division of the internal iliac a.**

  - **Superior gluteal a.** – exits the pelvis between the lumbosacral trunk (L4, L5) and S1 nn. usually; courses through the greater sciatic foramen to enter the gluteal region, anterior to the piriformis m.

  - **Iliolumbar a.** – located near the lumbosacral trunk, along the vertebrae

  - **Lateral sacral aa.** – located along the lateral borders of the sacrum, with branches exiting through the sacral foramina

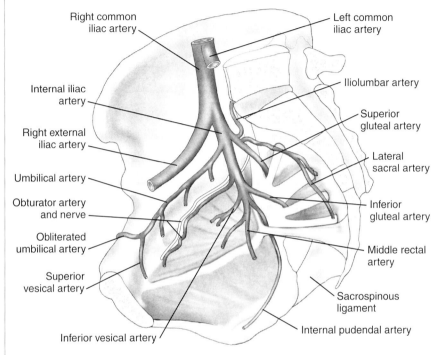

■ **Figure 14-10   Branches of the internal iliac artery of the male**

FEMALE

Pelvic Vasculature – Internal Iliac Artery (Anterior Division)

■ Identify the following branches (Figure 14–11):

- **Anterior division of the internal iliac a.**

  - **Umbilical a.** – continues to the anterior abdominal wall, becomes the **medial umbilical ligament (obliterated umbilical a.);** may give off 3–4 superior vesical aa.

  - **Obturator a.** – courses through the obturator canal

  - **Uterine a.** – trace it to the isthmus of the uterus; this artery is within the cardinal ligament; divides into a large superior branch to the body and fundus of the uterus and a smaller branch to the cervix, vagina, and inferior region of the bladder; anastomoses with the ovarian and vaginal aa.; the uterine a. crosses superior to the ureter, near the lateral fornix of the vagina

    - ◆ **Vaginal a.** – usually is a branch of the uterine a.; trace it to the vagina and posteroinferior surface of the urinary bladder

  - **Internal pudendal a.** – exits the pelvis through the greater sciatic foramen, inferior to the piriformis m.

  - **Middle rectal a.** – to the lateral wall of the rectum

  - **Inferior gluteal a.** – exits the pelvis between S2 and S3 nn.; courses through the inferior part of the greater sciatic foramen to enter the gluteal region inferior to the piriformis m.

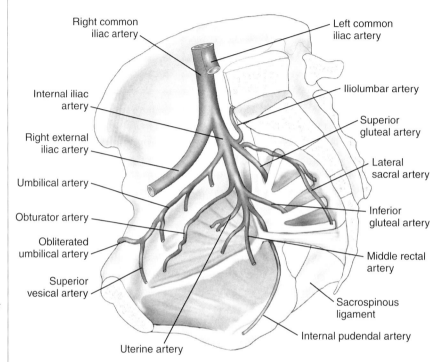

■ **Figure 14–11   Branches of the internal iliac artery of the female**

FEMALE

Pelvic Vasculature – Internal Iliac Artery (Posterior Division)

■ Identify the following branches (Figure 14–12):

- **Posterior division of the internal iliac a.**

  - **Superior gluteal a.** – exits the pelvis between the lumbosacral trunk (L4–L5) and S1 superiorly through the greater sciatic foramen and enters the gluteal region above the piriformis m.

  - **Iliolumbar a.** – located near the lumbosacral trunk along the vertebrae

  - **Lateral sacral aa.** – located along the lateral borders of the sacrum, with branches exiting through the sacral foramina

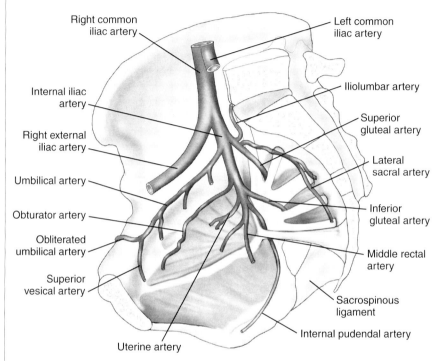

Labels on figure:
- Right common iliac artery
- Internal iliac artery
- Right external iliac artery
- Umbilical artery
- Obturator artery
- Obliterated umbilical artery
- Superior vesical artery
- Uterine artery
- Left common iliac artery
- Iliolumbar artery
- Superior gluteal artery
- Lateral sacral artery
- Inferior gluteal artery
- Middle rectal artery
- Sacrospinous ligament
- Internal pudendal artery

■ **Figure 14–12   Branches of the internal iliac artery of the female**

BOTH SEXES

Pelvic Innervation

■ Identify the following (Figure 14–13):

- **Lumbosacral trunk** (L4, L5 spinal cord segments) – contributes to the sacral plexus

- **Anterior sacral nerves** (ventral rami of S1–S4) – enter the pelvis through the anterior sacral foramina

- **Sciatic n.** (L4–S3 spinal cord segments) – exits the pelvis through the greater sciatic foramen and enters the gluteal region anterior to the piriformis m.

- **Pudendal n.** (S2, S3, S4 spinal cord segments)

- **Superior gluteal n.** – courses with the superior gluteal a.; exits the pelvis between the lumbosacral trunk and S1 n. usually

- **Inferior gluteal n.** – courses with the inferior gluteal a.; exits the pelvis between S1 and S2 nn. usually

- **Obturator n.** – courses with the obturator a. and v.

- **Sympathetic trunk** – the sympathetic chain and its ganglia are medial to the sacral foramina

■ *Note:* The arteries, veins, and nerves exit the pelvis en route to the gluteal region (and beyond) by passing between the coccygeus and piriformis mm.

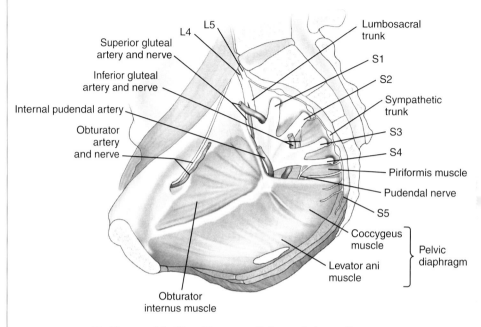

■ **Figure 14–13   Nerves of the pelvic wall**

Pelvic Autonomics

■ Identify the following (Figure 14–14):

- **Superior hypogastric nn.** – lie inferior to the bifurcation of the aorta; branches from the superior hypogastric plexus descend into the pelvis along the common iliac aa., as large nerve trunks known as the **left and right hypogastric nn.**

- **Inferior hypogastric nn.** – each inferior hypogastric n. surrounds the corresponding internal iliac a.

- **Pelvic splanchnic nn.** – preganglionic parasympathetic fibers derived from S2–S4 sacral nn.; contribute to the inferior hypogastric nerve plexus

- **Lumbar and sacral splanchnic nn.**

- **Visceral plexus** – extension of the right and left inferior hypogastric plexuses in the walls of the pelvic viscera

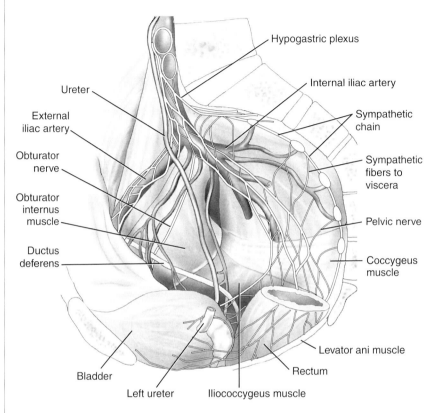

■ **Figure 14–14   Pelvic autonomics**

Unit 2 Abdomen, Pelvis, and Perineum Overview

At the end of Unit 2, you should be able to identify the following structures on cadavers, skeletons, and/or radiographs:

Osteology

- Sternum
  - Xiphoid process
- Ribs (12 pairs)
  - Costal margin and costal cartilage
- Os coxae
  - Acetabulum
  - Ischiopubic (conjoint) ramus
  - Obturator foramen
  - Greater sciatic notch/foramen
  - Pelvic brim, inlet, and outlet
  - False and true pelvis
  - Ilium
    - Iliac crest and fossa
    - Anterior superior iliac spine
    - Posterior superior/inferior iliac spine
  - Ischium
    - Ischial ramus (ischiopubic ramus)
    - Ischial tuberosity and spine
    - Lesser sciatic notch/foramen
  - Pubis
    - Pubic tubercle
    - Superior pubic ramus
      - ◆ Pecten pubis/pectineal line
    - Pubic symphysis and arch
    - Inferior pubic ramus (ischiopubic ramus)

Muscles and Associated Structures

- Rectus abdominis m.
  - Tendinous intersections
  - Rectus sheath
  - Arcuate line
- Pyramidalis m.
- External oblique m.
  - Inguinal ligament
    - Lacunar ligament
    - Pectineal ligament
    - Superficial inguinal ring
      - ◆ Medial and lateral crura
      - ◆ Intercrural fibers
- Internal oblique m.
  - Cremaster m.
- Transversus abdominis m.
  - Conjoint tendon
  - Deep inguinal ring
- Linea alba
  - Umbilicus
- Linea semilunaris
- Inguinal canal
  - Spermatic cord (male)
  - Round ligament of the uterus (female)
- Inguinal triangle

- Quadratus lumborum m.
- Pelvic diaphragm
  - Levator ani m.
    - Puborectalis m.
    - Pubococcygeus m.
    - Iliococcygeus m.
  - Coccygeus m.
- Piriformis m.
- Gluteus maximus and medius mm.
- Obturator internus m.
- External anal sphincter
- Diaphragm
  - Right and left crura
  - Central tendon
  - Vena caval hiatus
  - Esophageal hiatus
  - Aortic hiatus
  - Medial and lateral arcuate ligaments
  - Median arcuate ligament
- Psoas major and minor mm.
- Iliacus m.

Arteries

- Aortic arch
  - Subclavian a.
  - Internal thoracic a.

- Superior epigastric a.
- Musculophrenic a.
- Aorta
  - Inferior phrenic a.
    - Superior suprarenal aa.
  - Lumbar aa.
  - Median sacral a.
  - Celiac trunk
    - Left gastric a.
      - Esophageal a.
    - Splenic a.
      - Left gastroepiploic a.
      - Short gastric aa.
    - Common hepatic a.
      - Hepatic a. proper
        - Right gastric a.
        - Left and right hepatic aa.
        - Cystic a.
      - Gastroduodenal a.
        - Supraduodenal a.
        - Anterior superior pancreaticoduodenal a.
        - Posterior superior pancreaticoduodenal a.
        - Right gastroepiploic a.
  - Superior mesenteric a.
    - Anterior and posterior inferior pancreaticoduodenal aa.
    - Jejunal aa., vasa recta
    - Ileal aa., arcades
    - Ileocolic a.
      - Appendicular a.
    - Right colic a.
    - Middle colic a.
    - Marginal a.
- Aorta
  - Inferior mesenteric a.
    - Left colic a.
    - Sigmoid aa.
    - Superior rectal a.

- Marginal a.
- Middle suprarenal a.
- Renal a.
  - Inferior suprarenal a.
- Gonadal aa. (ovarian/testicular)
- Common iliac a.
  - Internal iliac a.
    - Iliolumbar a.
    - Lateral sacral a.
    - Obturator a.
    - Superior and inferior gluteal aa.
    - Umbilical a.
      - Branch to the ductus deferens (Male)
    - Inferior vesical a. (male)
    - Uterine and vaginal aa. (female)
    - Superior vesical a.
    - Middle rectal a.
    - Internal pudendal a.
      - Inferior rectal a.
      - Perineal a.
      - Posterior labial a. (female) or scrotal a. (male)
      - Artery to the bulb of the vestibule (female) or penis (male)
      - Dorsal a. of the clitoris (female) or penis (male)
      - Deep a. of the clitoris (female) or penis (male)
  - External iliac a.
    - Inferior epigastric a.
    - Deep circumflex iliac a.
- External iliac a.
  - Femoral a.
    - Superficial epigastric a.
    - Superficial circumflex iliac a.

Veins
- Inferior vena cava
  - Inferior phrenic vv.

- Lumbar vv.
- Right and left hepatic vv.
- Renal v.
  - Left suprarenal v.
  - Left ovarian v. (female) or testicular v. (female)
- Right suprarenal v.
- Right ovarian v. (female) or testicular v. (male)
- Common iliac v.
  - Median sacral v.
  - Iliolumbar v.
  - Internal iliac v.
    - Superior gluteal vv.
    - Inferior gluteal vv.
    - Obturator vv.
    - Lateral sacral vv.
    - Vesical vv.
    - Deep dorsal v. of the clitoris (female) or penis (male)
    - Uterine vv. (female)
    - Superior vesical v.
    - Middle rectal v.
    - Internal pudendal v.
      - Deep vv. of the clitoris (female) or penis (male)
      - Inferior rectal v.
    - Posterior labial vv. (female) or scrotal vv. (male)
    - Vein of the bulb of the vestibule (female) or penis (male)
- Common iliac v.
  - External iliac v.
    - Inferior epigastric v.
    - Deep circumflex iliac v.

- **Portal venous system**
  - **Portal v.**
    - Umbilical part
      - Ligamentum venosum
        - Round ligament of the liver

- Cystic v.
- Paraumbilical vv.
- Left gastric v.
- Right gastric v.
- **Superior mesenteric v.**
  - ◆ Jejunal and ileal vv.
  - ◆ Right gastroepiploic (omental) vv.
  - ◆ Pancreatic vv.
  - ◆ Ileocolic vv.
  - ◆ Right and middle colic vv.
- **Splenic v.**
  - ◆ Pancreatic vv.
  - ◆ Short gastric vv.
  - ◆ Left gastroepiploic (omental) vv.
  - ◆ Inferior mesenteric v.
    - ◆ Left colic v.
    - ◆ Sigmoid vv.
    - ◆ Superior rectal v.
- Lateral thoracic v.
  - Thoracoepigastric v.

Lymphatics

- Thoracic duct
- Cisterna chyli
- Lumbar and intestinal lymphatics
- Preaortic lymph nodes
- Lumbar lymph nodes
- Superior mesenteric lymph nodes
- Inferior mesenteric lymph nodes
- Common iliac lymph nodes
- External iliac lymph nodes
- Internal iliac lymph nodes

Autonomic Nerves

- Sympathetic part
  - Thoracic ganglia
    - Greater splanchnic n.
    - Lesser splanchnic n.

- Least splanchnic n.
  - Lumbar ganglia
  - Sacral ganglia
- Parasympathetic part
  - Pelvic ganglia
    - Pelvic splanchnic nn.
- Peripheral autonomic plexuses/ganglia
  - Celiac ganglia and plexuses
  - Aorticorenal ganglia and plexus
  - Superior mesenteric ganglia and plexuses
  - Renal plexus
  - Inferior mesenteric plexus
  - Superior hypogastric plexus
    - Hypogastric n.
  - Inferior hypogastric plexus
  - Uterovaginal plexus (female)
  - Prostatic plexus (male)

Spinal Nerves

- Intercostal nn. (ventral rami)
  - Lateral cutaneous nn.
  - Anterior cutaneous nn.
- Subcostal n.
- Iliohypogastric and ilioinguinal nn.
- Genitofemoral n.
  - Genital and femoral branches
- Lateral femoral cutaneous n.
- Obturator n.
- Femoral n.
- Lumbosacral trunk
- Sacral plexus
  - Superior and inferior gluteal nn.
  - Posterior femoral cutaneous n.
  - Pudendal n.
    - Inferior rectal nn.
    - Dorsal n. of the clitoris or penis
    - Perineal nn.
      - ◆ Posterior labial and scrotal nn.

- Coccygeal n.
  - Anococcygeal n.
- Sciatic n.

Anterior Abdominal Wall (internally)

- Median fold (obliterated urachus)
- Medial folds (obliterated umbilical aa.)
- Lateral folds (inferior epigastric vessels)
- Medial and lateral inguinal fossae
- Superior epigastric vessels
- Obliterated umbilical v.
  - Ligamentum teres
- Falciform ligament
- Umbilicus

Layers of the Anterolateral Abdominal Wall

- Skin
- Superficial fascia
  - Fatty (Camper's fascia) layer
  - Membranous (Scarpa's fascia) layer
- External oblique m./aponeurosis
- Internal oblique m./aponeurosis
- Transversus abdominis m./aponeurosis
- Transversalis fascia
- Extraperitoneal fat
- Parietal peritoneum

Ligaments

- Sacrotuberous ligament
- Sacrospinous ligament
- Iliolumbar ligament
- Anococcygeal ligament

Peritoneum

- ■ Mesentery
  - ● Root of the mesentery
- ■ Mesocolon
  - ● Transverse mesocolon
  - ● Sigmoid mesocolon
  - ● Mesoappendix
- ■ Lesser omentum
  - ● Hepatophrenic ligament
  - ● Hepatogastric ligament
  - ● Hepatoduodenal ligament
- ■ Greater omentum
  - ● Gastrosplenic ligament
  - ● Splenorenal ligament
- ■ Peritoneal attachments of the liver
  - ● Coronary ligament
    - ▪ Falciform ligament
    - ▪ Triangular ligaments
    - ▪ Hepatorenal ligament
- ■ Recesses, fossae, and folds
- ■ Peritoneal cavity
  - ● Greater sac
  - ● Lesser sac (omental bursa)
    - ▪ Omental (epiploic) foramen
  - ● Paravesical fossa
  - ● Rectovesical pouch (male)
  - ● Vesicouterine pouch (female)
  - ● Rectouterine pouch (female)

Stomach

- ■ Greater and lesser curvatures
- ■ Pylorus and pyloric sphincter
- ■ Body, cardia, and fundus
- ■ Gastric rugae

Small Intestine

- ■ Duodenum (four parts)
  - ● Major and minor duodenal papillae
  - ● Duodenojejunal junction
  - ● Suspensory ligament of the duodenum
  - ● Pancreaticoduodenal ampulla
  - ● Duodenal sphincter
  - ● Circular folds
- ■ Jejunum: circular folds
- ■ Ileum: circular folds
  - ● Ileocecal junction
  - ● Peyer's patches of lymphoid tissue

Large Intestine

- ■ Cecum
  - ● Vermiform appendix
- ■ Ascending colon – right colic (hepatic) flexure
- ■ Transverse colon
- ■ Descending colon – left colic (splenic) flexure
- ■ Sigmoid colon: semilunar folds
- ■ Rectum and anus
- ■ Colon
  - ● Omental (epiploic) appendices
  - ● Haustra
  - ● Taenia coli

Abdominal Organs

- ■ Pancreas
  - ● Head, neck, body, and tail
  - ● Main and accessory pancreatic ducts
  - ● Uncinate process
- ■ Spleen
- ■ Liver
  - ● Diaphragmatic and visceral surfaces
  - ● Left and right lobes
  - ● Quadrate and caudate lobes
  - ● Falciform and round ligaments
  - ● Porta hepatis
  - ● Subphrenic recess
  - ● Bare area
  - ● Common hepatic duct
    - ▪ Cystic duct
    - ▪ Common bile duct
- ■ Gallbladder
- ■ Kidney
  - ● Perirenal fascia and fibrous capsule
  - ● Hilum, renal sinus, renal pelvis, and ureter
  - ● Major and minor calyces
  - ● Renal cortex: renal columns
  - ● Renal medulla
    - ▪ Renal pyramids and papillae
- ■ Bladder
  - ● Apex, body, and fundus
  - ● Trigone
  - ● Internal urethral orifice
  - ● Medial umbilical ligament

Female Internal Genitalia

- ■ Ovary
  - ● Ligament of the ovary
  - ● Suspensory ligament of the ovary
- ■ Uterine tube
  - ● Infundibulum
  - ● Fimbria
  - ● Ampulla and isthmus
- ■ Uterus
  - ● Fundus and body
  - ● Uterine horn
  - ● Uterine cavity
  - ● Internal os
  - ● Cervix
  - ● Isthmus
  - ● External os
  - ● Round ligament of the uterus
  - ● Pubocervical ligament
  - ● Cardinal (transverse cervical) ligament
  - ● Uterosacral (rectouterine) ligament
- ■ Vagina
  - ● Vaginal fornices
    - ▪ Anterior, posterior, and two lateral

- Hymen
- Vaginal rugae

Female External Genitalia

- Mons pubis
- Labia majora
- Labia minora
- Vestibule
  - Bulb of the vestibule
  - Vaginal orifice
  - Greater and lesser vestibular glands
- Clitoris
  - Crus, body, and glans
  - Corpus cavernosum of the clitoris
  - Suspensory ligament of the clitoris
- Female urethra

Male Internal Genitalia

- Testis
  - Tunica vaginalis
    - Parietal and visceral layers
    - Sinus of the epididymis
  - Tunica albuginea
  - Mediastinum of the testis
  - Lobules of the testis
  - Seminiferous tubules
  - Gubernaculum testis
- Epididymis
  - Head, body, and tail
- Spermatic cord
  - External spermatic fascia
  - Cremaster m.
  - Cremasteric fascia
  - Internal spermatic fascia
  - Processus vaginalis
  - Pampiniform plexus of vv.

- Ductus deferens
- Seminal vesicle
- Ejaculatory duct
- Prostate gland
- Bulbourethral gland

Male External Genitalia

- Penis
  - Root, body, dorsum, and crus
  - Glans penis
  - Prepuce/foreskin
    - Frenulum
  - Raphe of the penis
  - Corpus cavernosum penis (crus)
  - Corpus spongiosum penis
    - Bulb of the penis
  - Tunica albuginea of corpora cavernosa and corpus spongiosum
  - Septum penis
  - Suspensory ligament of the penis
  - Fundiform ligament of the penis
- Male urethra
  - Prostatic urethra
  - Membranous urethra
  - Spongy urethra
  - External urethral orifice
- Scrotum
  - Raphe of the scrotum
  - Dartos fascia
    - Dartos m.

Perineum

- Triangles
  - Urogenital triangle
  - Anal triangle
- External anal sphincter

- Perineal body (Anococcygeal body/ ligament)
- Superficial perineal space
  - Superficial perineal fascia
  - Superficial transverse perineal m.
  - Ischiocavernosus m.
  - Bulbospongiosus m.
- Deep perineal space
  - Perineal membrane
  - Deep transverse perineal m.
  - External urethral sphincter
  - Sphincter urethrovaginalis m. (female)
- Ischioanal fossa
  - Fat
  - Pudendal canal
    - Pudendal n.
    - Internal pudendal a. and v.

Urogenital Peritoneum

- Paravesical fossa
- Transverse vesical fold
- Vesicouterine pouch (female)
- Broad ligament of the uterus (female)
  - Mesometrium
  - Mesosalpinx
  - Mesovarium
- Suspensory ligament of the ovary (female)
- Rectouterine fold (female)
- Rectouterine pouch (female)
- Rectovesical pouch (male)
- Pararectal fossa

Miscellaneous

- Pudendal canal
  - Obturator fascia over the pudendal canal

Neck and Head

Posterior Triangle of the Neck

Prior to dissection, you should familiarize yourself with the following structures:

OSTEOLOGY

- Manubrium
  - Jugular notch
- Clavicle
- Scapula
  - Acromion
- Hyoid bone
- Temporal bone
  - Mastoid process
- Mandible
  - Mental symphysis
  - Mental tubercles
  - Ramus of the mandible

MUSCLES

- Platysma m.
- Sternocleidomastoid m.
- Trapezius m.
- Splenius capitis m.
- Levator scapulae m.
- Scalene mm.
  - Anterior
  - Middle
  - Posterior
- Infrahyoid mm.
  - Omohyoid m.

NERVES

- Spinal accessory n. (CN XI)
- Cervical plexus
  - Great auricular n.
  - Lesser occipital n.
  - Transverse cervical n.
  - Supraclavicular nn.
  - Phrenic n.
- Brachial plexus
  - Suprascapular n.

VEINS

- Brachiocephalic v.
  - Internal jugular v.
    - External jugular v.
      - Anterior jugular v.
        - Jugular venous arch

ARTERIES

- Aorta
  - Subclavian a.
    - Thyrocervical trunk
      - Transverse cervical a.
      - Suprascapular a.
  - Common carotid a.
    - External carotid a.
      - Occipital a.

CERVICAL FASCIA

- Superficial cervical fascia
- Deep cervical fascia
  - Investing fascia
  - Pretracheal fascia
  - Prevertebral fascia
  - Carotid sheath

MISCELLANEOUS

- Subtriangles of the posterior triangle of the neck
  - Supraclavicular triangle
  - Suboccipital triangle
    - Interscalene triangle

TABLE 15–1   Muscles of the Posterior Triangle of the Neck

Muscle	Attachments		Action	Innervation
Platysma	Inferior border of the mandible and subcutaneum	Fascia over the pectoralis major and deltoid mm.	Grimace expression, tightens skin of the neck	Cervical branch of the facial n.
Trapezius	Occipital bone, nuchal ligament, C7–T12 vertebrae	Lateral third of the clavicle, acromion, and spine	Elevates, adducts, and depresses the scapula	Spinal accessory n. and cervical n.
Sternocleidomastoid	Clavicle and manubrium	Mastoid process	Laterally flexes and rotates the neck	Spinal accessory n.
Omohyoid	Hyoid	Superior angle of the scapula	Depresses the hyoid bone	Cervical nn. C1–C3 via ansa cervicalis
Scalene, anterior	Transverse processes of C4–C6 vertebrae	Rib 1	Laterally flex the neck; bilaterally stabilize the neck	Cervical nn. C4–C6
Scalene, middle	Posterior tubercle of the transverse processes of C4–C6 vertebrae			Ventral rami of cervical spinal nn. C3–C8
Scalene, posterior		Rib 2		Ventral rami of cervical spinal nn. C6–C8
Levator scapulae	Posterior tubercle of the transverse processes of C1–C4 vertebrae	Superior angle of the scapula	Elevates the scapula	Dorsal scapular n.
Splenius capitis	Spinous process of T1–T6 vertebrae	Mastoid process and superior nuchal line	Extends the head and neck; laterally flexes the neck	Dorsal rami of cervical spinal nn.

TABLE 15–2   Schematic of the Cervical Plexus

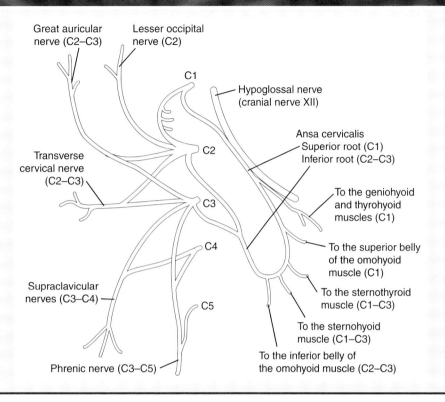

Great auricular nerve (C2–C3)

Lesser occipital nerve (C2)

C1

Hypoglossal nerve (cranial nerve XII)

Ansa cervicalis
Superior root (C1)
Inferior root (C2–C3)

C2

Transverse cervical nerve (C2–C3)

To the geniohyoid and thyrohyoid muscles (C1)

C3

To the superior belly of the omohyoid muscle (C1)

C4

To the sternothyroid muscle (C1–C3)

Supraclavicular nerves (C3–C4)

C5

To the sternohyoid muscle (C1–C3)

To the inferior belly of the omohyoid muscle (C2–C3)

Phrenic nerve (C3–C5)

Fascial Planes – Overview

■ Throughout this dissection, you will observe the following fascial planes in the neck (Figure 15–1):

- **Skin**

- **Superficial cervical fascia** (subcutaneum) – encircles the neck and surrounds the structures in the neck; envelopes the platysma m.

- **Deep cervical fascia**

  - **Investing fascia** – envelops the sternocleidomastoid and trapezius mm.

  - **Pretracheal fascia** – covers the anterior aspect of the neck; encloses the thyroid gland, trachea, and esophagus

  - **Prevertebral fascia** – covers the prevertebral muscles and is continuous with the deep fascia covering the muscular floor of the posterior triangle

  - **Carotid sheath** – contains the common carotid a., internal jugular v., and the vagus n.; part of the ansa cervicalis courses within the connective tissue that comprises the carotid sheath

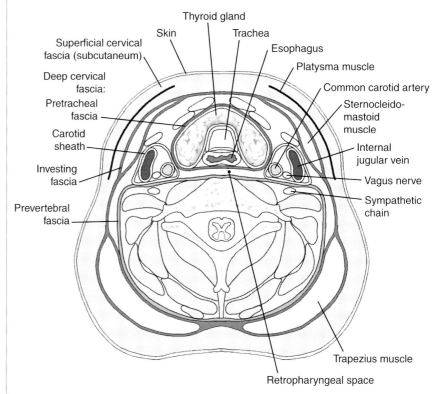

■ **Figure 15–1   Overview of the fascial planes of the neck**

Surface Anatomy of the Neck

■ Identify the following surface landmarks (Figure 15–2):

- **Clavicle** – palpate the clavicle where it articulates with the manubrium and the acromion

- **Acromion of the scapula** – the lateral tip of the spine of the scapula; attached to the lateral aspect of the clavicle

- **Jugular notch of the manubrium** – also called the suprasternal notch

- **Mastoid process of the temporal bone** – a palpable bump deep to the skin, posterior to the external ear

- **Inferior border of the mandible** – the inferior edge of the jaw

- **Laryngeal prominence** – formed by the thyroid cartilage ("Adam's apple")

■ You will expose both the posterior and anterior triangles of the neck at this time, enabling you to study structures found in both triangles

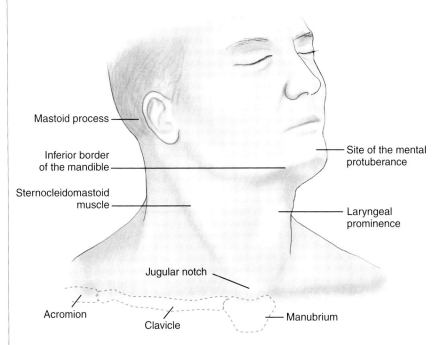

Mastoid process

Inferior border of the mandible

Sternocleidomastoid muscle

Site of the mental protuberance

Laryngeal prominence

Jugular notch

Acromion

Clavicle

Manubrium

■ **Figure 15–2   Surface anatomy of the neck**

Superficial Aspect of the Neck

- Make the following skin incisions on both sides of the neck; note the thinness of the skin on the neck (Figure 15–3A):

  - Mandible (A) to the jugular notch (B)

  - Acromion (C) to the jugular notch (B) along the superior border of the clavicle

  - Mental protuberance (A) of the mandible along the inferior margin of the mandible; continue the cut far enough posteriorly (inferior to the ear) to expose the anterior 2 cm of the trapezius m. (D)

- Grasp an edge of the skin with forceps to cut the skin from the superficial fascia; *use a sharp scalpel!*

- *Note:* The platysma m. is attached to the skin; try not to remove the platysma m. or cutaneous branches of the cervical plexus as you remove the skin (stay superficial to the platysma m.)

- Identify the following (Figure 15–3B):

  - **Platysma m.** – located in the superficial fascia and attached to the skin; covers superficial regions of both the posterior and anterior triangles of the neck

  - **Supraclavicular nn.** – medial, intermediate, and lateral branches; formed by the ventral rami of C3–C4 (cervical plexus); the nerves emerge through the platysma m. 1–2 cm above the clavicle

- Carefully reflect the platysma m. from the clavicle toward the face but do not cut the platysma from the mandible

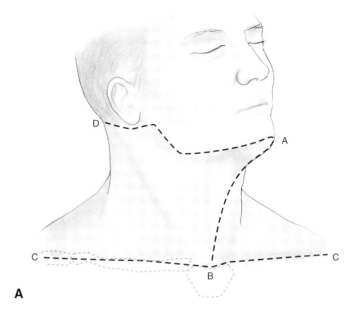

A

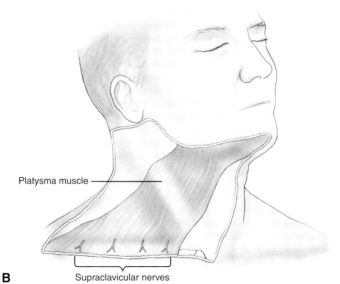

Platysma muscle

Supraclavicular nerves

B

■ **Figure 15–3   Superficial aspect of the neck. A, Skin incisions of the neck. B, Platysma muscle**

Posterior Triangle of the Neck

- Identify the following (Figure 15–4):

  - Borders of the posterior triangle:
    - Posterior border of the sternocleidomastoid m.
    - Anterior border of the trapezius m.
    - Middle third of the clavicle
    - "Roof" – investing fascia
    - "Floor" – deep fascia over the deep muscles of the neck (prevertebral fascia)

  - **Omohyoid m. (inferior belly)** – passes obliquely through the posterior triangle, dividing it into two smaller triangles:
    - Occipital triangle
    - Supraclavicular triangle

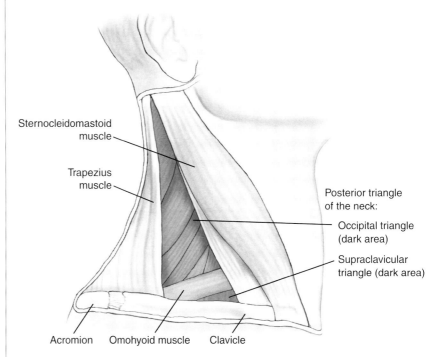

Sternocleidomastoid muscle

Trapezius muscle

Posterior triangle of the neck:

Occipital triangle (dark area)

Supraclavicular triangle (dark area)

Acromion   Omohyoid muscle   Clavicle

■ **Figure 15–4   Borders of the posterior triangle of the neck**

Posterior Triangle of the Neck – Contents

- Identify the following structures (see Figure 15–5):

  - **Occipital triangle** – the larger, superior triangle bordered by the trapezius, sternocleidomastoid, and omohyoid mm.; named because the occipital a. appears at the apex of this triangle (Figure 15–5)

  - **Supraclavicular triangle** – the smaller, inferior triangle bounded by the omohyoid m., sternocleidomastoid m., and clavicle

  - **External jugular v.** – emerges posterior to the ramus of the mandible and crosses superficial to the sternocleidomastoid m.; the vein pierces the fascial "roof" of the posterior triangle

  - **Investing fascia** – forms the "roof" of the posterior triangle; this fascia spans from the sternocleidomastoid m. to the trapezius m.; the investing fascia is tough so be patient as you dissect through it

- Identify the following nerves:

  - **Spinal accessory n. (CN XI)** – emerges from the posterior border of the sternocleidomastoid m., approximately 5 cm above the clavicle, and courses to the deep surface of the trapezius m.

  - **Great auricular n.** – usually posterior and parallel to the external jugular v.; ascends on the superolateral surface of the sternocleidomastoid m. toward the ramus of the mandible

  - **Lesser occipital n.** – emerges from the posterior border of the sternocleidomastoid m.; trace the lesser occipital n. superiorly along the posterior border of the sternocleidomastoid m.

  - **Transverse cervical n.** – traverses anteriorly across the middle of the sternocleidomastoid m. toward the anterior triangle of the neck

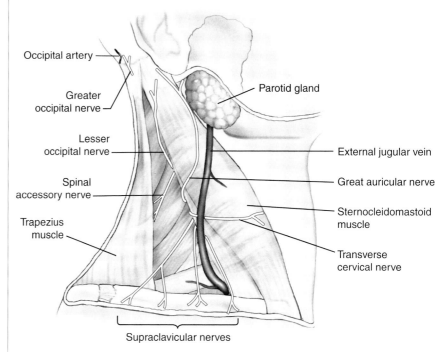

■ **Figure 15–5   Contents of the superficial layer of the posterior triangle of the neck**

Posterior Triangle of the Neck – Contents—cont'd

- Trace the great auricular, lesser occipital, and transverse cervical nn. to a common trunk (C2–C3) immediately deep to the posterior border of the sternocleidomastoid m.; *do not cut* the sternocleidomastoid m. at this time

- Identify the following structures (Figure 15–6):

  - **Omohyoid m.** – the inferior belly passes through the posterior triangle, deep to the clavicular portion of the sternocleidomastoid m. and trapezius m.

  - **Transverse cervical a.** (expect variation) – branch of the **thyrocervical trunk** (branch of the subclavian a.) that is deep to the sternocleidomastoid m. and superficial to the anterior scalene m.

    - Courses laterally and posteriorly across to the posterior triangle to supply the trapezius m. **(superficial branch),** and the rhomboid mm., and levator scapulae m. **(deep branch)**

  - **Occipital a.** – branch of the external carotid a.; enters the posterior triangle at its apex

- Remove the prevertebral fascia from the "floor" of the posterior triangle to identify the following muscles:

  - **Splenius capitis m.**

  - **Levator scapulae m.**

  - **Scalene mm.** – there are 3 scalene muscles, named the **anterior, middle,** and **posterior scalenes**

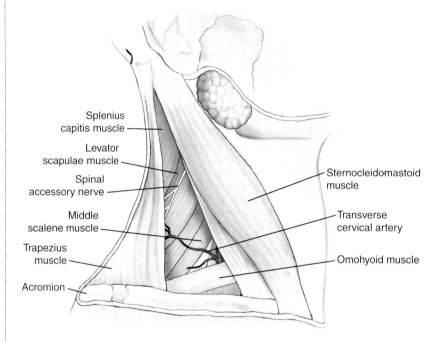

Splenius capitis muscle
Levator scapulae muscle
Spinal accessory nerve
Middle scalene muscle
Trapezius muscle
Acromion
Sternocleidomastoid muscle
Transverse cervical artery
Omohyoid muscle

■ **Figure 15–6   Contents of the intermediate layer of the posterior triangle of the neck**

Posterior Triangle of the Neck – Contents—cont'd

■ Identify the following structures (Figure 15–7):

- **Suprascapular a.** – branch of the thyrocervical trunk; the artery passes inferolaterally across the anterior scalene m., near the clavicle

  - Crosses the subclavian a. and the cords of the brachial plexus as it courses toward the scapula (do *not* dissect the scapular area at this time)

- **Interscalene triangle**

  - Borders – anterior scalene m., the first rib, and middle scalene m.

  - Contents – brachial plexus and subclavian a.

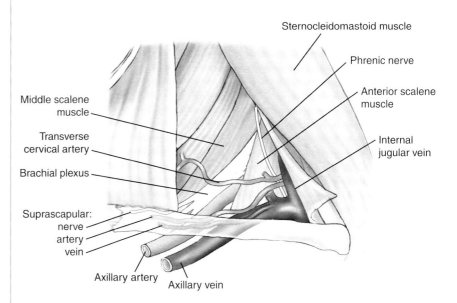

■ **Figure 15–7   Contents of the deep layer of the posterior triangle of the neck**

Anterior Triangle of the Neck

Prior to dissection, you should familiarize yourself with the following structures:

CAROTID TRIANGLE

- Carotid sheath
  - Common carotid a.
  - Internal jugular v.
  - Vagus n.
- Cervical sympathetic chain
- External carotid a.
  - Ascending pharyngeal a.
  - Superior thyroid a.
    - Superior laryngeal a.
  - Lingual a.
  - Facial a.
  - Occipital a.
  - Posterior auricular a.
- Internal carotid a.
- Carotid body and sinus
- Nerves
  - Hypoglossal n.
  - Ansa cervicalis
    - Superior root
    - Inferior root
  - Spinal accessory n.
  - Superior laryngeal n.
    - Internal laryngeal n.
    - External laryngeal n.
- Deep cervical lymph nodes

MUSCULAR TRIANGLE

- Infrahyoid mm.
  - Sternohyoid m.
  - Omohyoid m.
  - Sternothyroid m.
  - Thyrohyoid m.
- Thyroid gland
- Parathyroid glands

SUBMANDIBULAR TRIANGLE

- Suprahyoid mm.
  - Digastric m.
    - Intermediate tendon
    - Fibrous sling
  - Mylohyoid m.
  - Stylohyoid m.
  - Geniohyoid m.
- Submandibular gland
- Submandibular lymph nodes
- Hypoglossal n.
- Mylohyoid n.
- Facial a. and v.
- Submental a. and v.

SUBMENTAL TRIANGLE

- Mylohyoid m. (fibrous raphe)
- Submental lymph nodes
- Small vv. from the anterior jugular v.

VEINS

- External jugular v.
- Internal jugular v.
- Retromandibular v.
- Facial v.
- Common facial v.
- Anterior jugular v.
- Posterior auricular v.

LYMPHATICS

- Superficial cervical lymph nodes
- Deep cervical lymph nodes
- Submental lymph nodes
- Submandibular lymph nodes

MISCELLANEOUS

- Sternocleidomastoid m.
- Hyoid bone
- Thyrohyoid membrane
- Thyroid cartilage
- Cricothyroid membrane
- Cricoid cartilage
- First tracheal ring
- Laryngeal prominence
- Larynx
- Branches of the cervical plexus

TABLE 16–1   Muscles Associated with the Anterior Triangle of the Neck

Muscles	Attachments		Action	Innervation
Sternocleidomastoid	Manubrium and medial part of the clavicle	Mastoid process	Rotates the head and neck, laterally flexes the neck	Spinal accessory n. (CN XI)
Suprahyoid mm.				
• Mylohyoid	Mylohyoid line of the mandible	Hyoid bone	Elevates the hyoid bone and tongue while swallowing and speaking	Mylohyoid n.: inferior alveolar n. (CN V-3)
• Geniohyoid	Inferior genial tubercle of the mandible	Body of the hyoid bone	Moves the hyoid bone anterosuperiorly	C1 and C2 coursing with the hypoglossal n. (CN XII)
• Stylohyoid	Styloid process (temporal bone)	Body of the hyoid bone	Elevates the hyoid bone	Cervical branch of the facial n. (CN VII)
• Digastric	Anterior belly: digastric fossa of the mandible Posterior belly: mastoid notch	Body and greater horn of the hyoid bone	Depresses the mandible; elevates the hyoid bone	Anterior belly: mylohyoid n. (inferior alveolar n., CN V-3) Posterior belly: facial n.
Infrahyoid mm				
• Sternohyoid	Manubrium and medial part of the clavicle	Body of the hyoid bone	Depresses the hyoid bone	C1–C3 (ansa cervicalis)
• Omohyoid	Superior border of the scapula	Inferior border of the hyoid bone	Depresses and stabilizes the hyoid bone	C1–C3 (ansa cervicalis)
• Sternothyroid	Dorsal region of the manubrium	Thyroid cartilage	Depresses the hyoid bone and larynx	C2–C3 (ansa cervicalis)
• Thyrohyoid	Thyroid cartilage	Greater horn of the hyoid bone	Depresses the hyoid bone and elevates the thyroid cartilage	C1 (via hypoglossal n.)

TABLE 16–2 Schematic of the Cervical Plexus

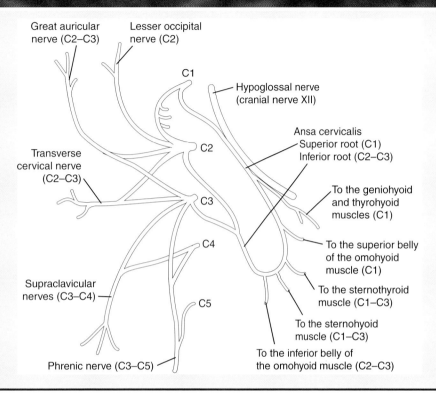

Great auricular nerve (C2–C3)

Lesser occipital nerve (C2)

C1

Hypoglossal nerve (cranial nerve XII)

Ansa cervicalis
Superior root (C1)
Inferior root (C2–C3)

C2

Transverse cervical nerve (C2–C3)

To the geniohyoid and thyrohyoid muscles (C1)

C3

To the superior belly of the omohyoid muscle (C1)

C4

To the sternothyroid muscle (C1–C3)

Supraclavicular nerves (C3–C4)

C5

To the sternohyoid muscle (C1–C3)

To the inferior belly of the omohyoid muscle (C2–C3)

Phrenic nerve (C3–C5)

Anterior Triangle of the Neck

■ Identify the following:

- **Borders of the anterior triangle of the neck** (Figure 16–1*A*)

  - Midline of the neck

  - Anterior border of the sternocleidomastoid m.

  - Inferior border of the mandible

- The four smaller triangles into which the anterior triangle is subdivided include the following (Figure 16–1*B*):

  1. Carotid

  2. Muscular

  3. Submental

  4. Submandibular

■ During this dissection, you will identify anatomic structures within each smaller triangle

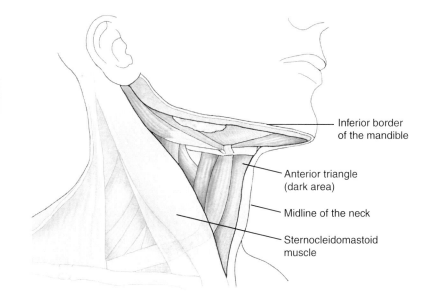

Inferior border of the mandible

Anterior triangle (dark area)

Midline of the neck

Sternocleidomastoid muscle

A

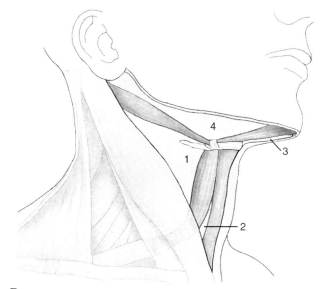

B

■ **Figure 16–1   Anterior triangle of the neck. *A*, Lateral view of the borders. *B*, Lateral view of the subtriangles**

Superficial Structures of the Anterior Triangle of the Neck

■ Identify and palpate the following surface landmarks along the anterior midline of the anterior triangle of the neck on your cadaver (Figure 16–2):

- **Hyoid bone**

- **Thyroid cartilage**

- **Cricoid cartilage**

- **Thyroid gland**

- **Trachea**

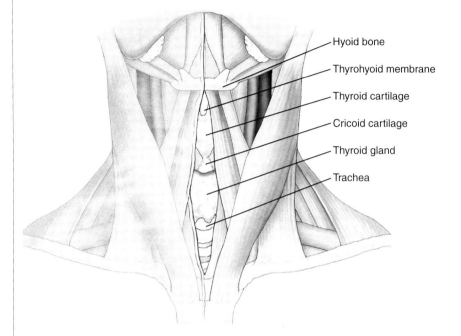

Hyoid bone

Thyrohyoid membrane

Thyroid cartilage

Cricoid cartilage

Thyroid gland

Trachea

■ **Figure 16–2   Superficial structures in the anterior triangle**

Anterior Triangle of the Neck – Superficial Structures

■ Identify the following (expect variation) (Figure 16–3*A, B*):

- **External jugular v.** – descends superficially, posterior to the mandible (superficial to the sternocleidomastoid m.), and pierces the investing fascia about 2–3 cm superior to the clavicle before joining the subclavian v.

- **Superficial cervical lymph nodes** – parallel the external jugular v.

- **Retromandibular v.** – formed by union of the superficial temporal and maxillary vv. (studied further in another laboratory session)

  - The posterior branch of the retromandibular v. unites with the posterior auricular v. (deep to the parotid gland) to form the external jugular v.

- **Posterior auricular v.** – courses posterior to the ear

- **Facial v.** – courses along the inferior border of the mandible, where the facial v. is joined by the anterior branch of the retromandibular v. to form the common facial v.

  - **Common facial v.** – joins the internal jugular v. deep to the sternocleidomastoid m.

- **Anterior jugular v.** – located in the superficial fascia of the anterior triangle of the neck; occasionally, this vein joins the facial v. via a communicating branch

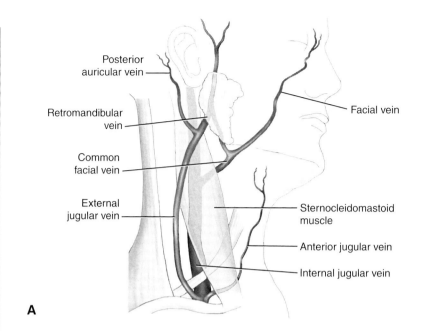

A

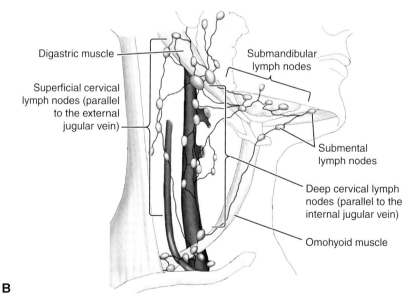

B

■ **Figure 16–3 Superficial structures of the neck.** *A*, **Lateral view of veins.** *B*, **Lateral view of the superficial lymph nodes**

Carotid Triangle

- Identify the following (Figure 16–4):

  - **Borders of the carotid triangle**

    - Posterior belly of the digastric m.

    - Superior belly of the omohyoid m.

    - Anterior border of the sternocleidomastoid m.

- On one side of the neck only, make the following transections to allow access to deeper structures (Figure 16–5):

  - **Sternocleidomastoid m.** – cut horizontally across the center of its belly; free and clean the superior portion of this muscle

    - *Note:* Try not to damage the cervical plexus of nerves that radiate from the posterior border of this muscle

  - **External jugular v.** – may already be cut; if not, cut horizontally across the center of the vein as it crosses superficially over the sternocleidomastoid m.

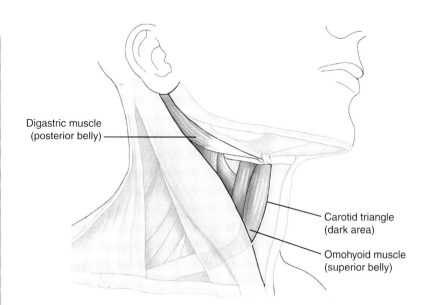

Digastric muscle (posterior belly)

Carotid triangle (dark area)

Omohyoid muscle (superior belly)

■ Figure 16–4 Boundaries of the carotid triangle

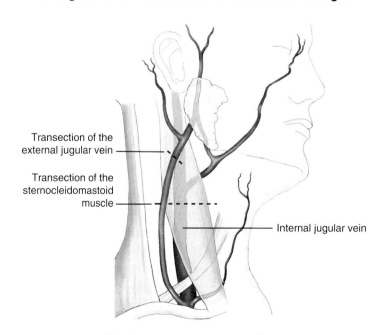

Transection of the external jugular vein

Transection of the sternocleidomastoid muscle

Internal jugular vein

■ Figure 16–5 Transections

Carotid Triangle—cont'd

- Identify the following structures (Figure 16–6):

  - **Spinal accessory n.** (CN XI) – enters the deep surface of the sternocleidomastoid m., about 4 cm inferior to the mastoid process

  - **Hypoglossal n.** (CN XII) – courses immediately superior to the tip of the greater horn of the hyoid bone

    - Trace the hypoglossal n. posteriorly to locate where it passes deep to the posterior belly of the digastric m.

    - Retract the posterior belly of the digastric m. to demonstrate the course of the hypoglossal n. from the hypoglossal canal

    - **Carotid sheath** – a tubular fascial investment that extends from the base of the skull to the root of the neck; contains the following (Figure 16–7):

      - **Common carotid a., internal carotid a., internal jugular v.,** and the **vagus n.**

- **Ansa cervicalis** (C1–C3 spinal cord segments) – embedded in the connective tissue that is the carotid sheath (ansa means loop)

  - **Inferior root of the ansa cervicalis** and the **superior root of the ansa cervicalis** – form a loop superficial to the internal jugular v.

  - Attempt to follow the terminal branches of the ansa cervicalis to the infrahyoid muscles that they supply (Table 16–2)

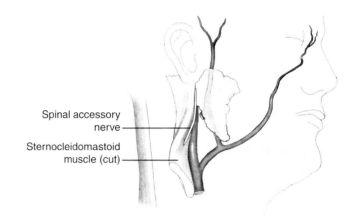

Spinal accessory nerve

Sternocleidomastoid muscle (cut)

■ **Figure 16–6   Spinal accessory nerve**

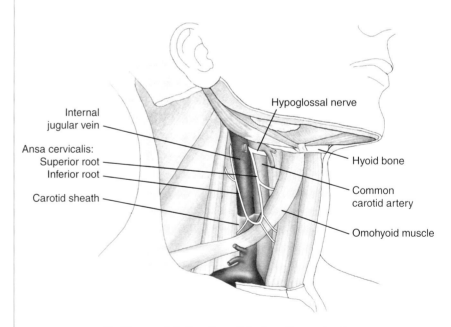

Internal jugular vein

Ansa cervicalis:
Superior root
Inferior root

Carotid sheath

Hypoglossal nerve

Hyoid bone

Common carotid artery

Omohyoid muscle

■ **Figure 16–7   Carotid sheath contents**

Carotid Triangle—cont'd

■ Identify the following:

- **Deep cervical lymph nodes** along the carotid sheath, near the internal jugular v. (Figure 16–8A)

- **Vagus n.** – deep, between the common carotid a. and the internal jugular v. (Figure 16–8B)

- **Common carotid a.** – medial position within the sheath

  - Bifurcates to become the **internal** and **external carotid aa.;** the proximal portion of the internal carotid a. is dilated and is called the **carotid sinus** (contains blood pressure receptors)

  - **Carotid body** – small, dark fibrous mass of tissue at the carotid bifurcation; contains chemoreceptors for oxygen; difficult to identify (not illustrated)

- **Internal carotid a.** – has no branches in the neck; the artery enters the skull via the carotid canal to supply the brain and orbital structures

- **External carotid a.** – anteromedial to the internal carotid a.; provides arterial branches to the neck and face

  - Terminates deep to the parotid gland, where it divides into the maxillary and superficial temporal aa.

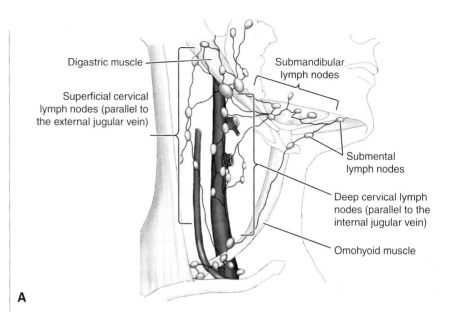

Digastric muscle

Superficial cervical lymph nodes (parallel to the external jugular vein)

Submandibular lymph nodes

Submental lymph nodes

Deep cervical lymph nodes (parallel to the internal jugular vein)

Omohyoid muscle

A

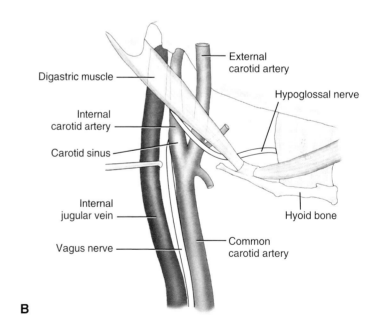

Digastric muscle

Internal carotid artery

Carotid sinus

Internal jugular vein

Vagus nerve

External carotid artery

Hypoglossal nerve

Hyoid bone

Common carotid artery

B

■ **Figure 16–8   Carotid region. *A*, lateral view of the deep lymph nodes. *B*, Lateral view**

Carotid Triangle—cont'd

- Identify the following branches of the external carotid a. (Figure 16–9):

  - **Superior thyroid a.** – most inferior branch of the external carotid a.; inferior and posterior to the greater horn of the hyoid bone; descends anteriorly and inferiorly, deep to the infrahyoid mm., to reach the thyroid gland (superior pole)

    - **Superior laryngeal a.** – arises from the superior thyroid a.; pierces the thyrohyoid membrane with the **internal laryngeal n.**

  - **Lingual a.** – arises from the external carotid a., immediately posterior to the greater horn of the hyoid bone and deep to the posterior belly of the digastric m. and stylohyoid m.

  - **Facial a.** – arises superior to the lingual a. (sometimes the lingual and facial aa. branch from a common trunk); enters a groove under the mandible and emerges from the superficial surface of the mandible, in the region of the angle (seen later)

  - **Ascending pharyngeal a.** – arises near the carotid bifurcation; ascends to the pharynx, deep and medial to the internal carotid a.

  - **Occipital a.** – arises near the posterior surface of the external carotid a.; courses parallel and deep to the posterior belly of the digastric m. to reach the occipital region

  - **Posterior auricular a.** – arises from the posterior surface of the external carotid a.; small in size; ascends between the ear and mastoid process

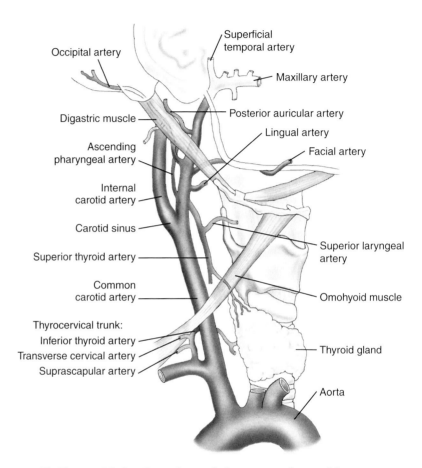

- **Figure 16–9**   **Branches of the external carotid artery**

Carotid Triangle—cont'd

■ On one side of the neck only, cut the omohyoid and sternohyoid mm. from their attachments to the hyoid bone

■ Reflect the muscles to identify the following (Figure 16–10):

- **Thyrohyoid membrane** – spans between the thyroid cartilage and hyoid bone

- **Superior laryngeal n.** – arises from the vagus n. (CN X) and passes deep to the carotid a.

  - **Internal laryngeal n.** – arises from the superior laryngeal n.; pierces the thyrohyoid membrane inferior to the greater horn of the hyoid bone

    ◆ *Do not* pursue this nerve into the larynx; this will be studied in another laboratory session

- **External laryngeal n.** – also arises from the superior laryngeal n.; motor to the cricothyroid m.

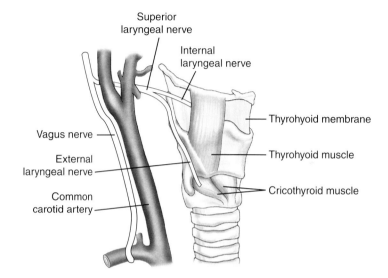

■ **Figure 16–10   Branches of the superior laryngeal nerve**

Muscular Triangle and Infrahyoid Muscles

■ The muscular triangle contains the infrahyoid ("strap") mm. and thyroid and parathyroid glands

■ Identify the following:

- **Borders of the muscular triangle** (Figure 16–11):
  - Superior belly of the omohyoid m.
  - Anterior border of the sternocleidomastoid m.
  - Midsagittal plane in the neck
- **Infrahyoid mm.** – named by their muscular attachments (Figure 16–12):
  - **Sternohyoid mm.** – superficial and parallel to the midline of the neck
  - **Omohyoid mm.** – lateral to the sternohyoid m.
  - **Sternothyroid mm.** – deep to the sternohyoid m.
  - **Thyrohyoid mm.** – deep to the sternohyoid m.

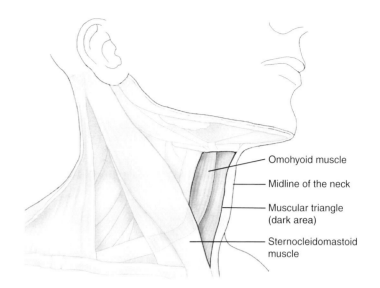

■ **Figure 16–11   Borders of the muscular triangle**

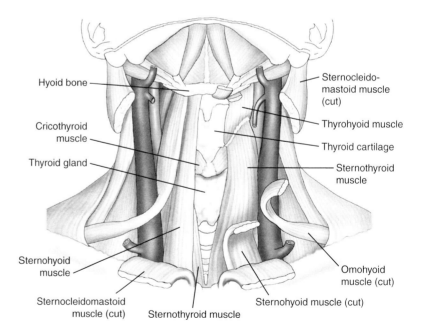

■ **Figure 16–12   Chin tilted upward to show the infrahyoid muscles**

Submental Triangle

■ Identify the following (Figure 16–13):

- **Borders of the submental triangle**
  - Hyoid bone
  - Anterior bellies of the two digastric mm.
- **Mylohyoid m.** – forms the floor of the triangle; the two mylohyoid mm. fuse together to form a median fibrous raphe
- **Submental lymph nodes**
- **Submental v.**

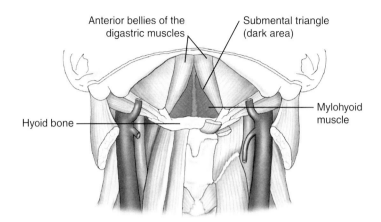

Anterior bellies of the digastric muscles

Submental triangle (dark area)

Hyoid bone

Mylohyoid muscle

■ **Figure 16–13   Submental triangle (chin lifted up)**

Submandibular Triangle

- Identify the following (Figures 16–14 and 16–15):

  - **Borders of the submandibular triangle**

    - Inferior border of the mandible

    - **Anterior belly of the digastric m.**

    - **Posterior belly of the same digastric m.**

  - **Submandibular gland** – almost fills the triangle; this gland has superficial and deep parts that form a U shape superficial and deep to the mylohyoid m.

  - **Submandibular lymph nodes**

  - **Facial a.** and **v.** – identify these vessels as they exit the submandibular triangle en route to the face

    - **Facial a.** – usually takes a course *deep* to the submandibular gland

    - **Facial v.** – usually takes a course *superficial* to the submandibular gland

- On one side of the neck, remove the superficial part of the submandibular gland to locate the deep contents of the submandibular triangle

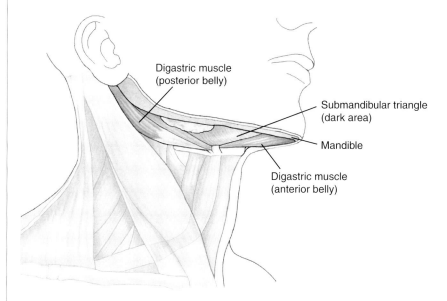

■ Figure 16–14   Borders of the submandibular triangle

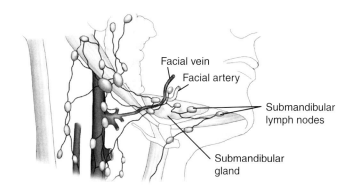

■ Figure 16–15   Contents of the submandibular triangle

Contents of the Submandibular Triangle

- Identify the following (Figure 16–16):

  - **Digastric m.** (anterior and posterior bellies) – identify the **intermediate tendon** attached to the greater horn of the hyoid bone via a **fibrous sling**

  - **Stylohyoid m.** – anterior to the digastric m.; the intermediate tendon of the digastric m. splits the stylohyoid m. near the greater horn of the hyoid bone

  - **Mylohyoid n.** – pull the anterior belly of the digastric m. medially to expose the mylohyoid n. (branch of CN V-3)

  - **Facial a.** and **v.** – deep to the stylohyoid m. and posterior belly of the digastric m.

  - **Hypoglossal n.** (CN XII) – courses through the submandibular triangle; this nerve is posterior and deep to the intermediate tendon of the digastric mm. and disappears under the mylohyoid m.

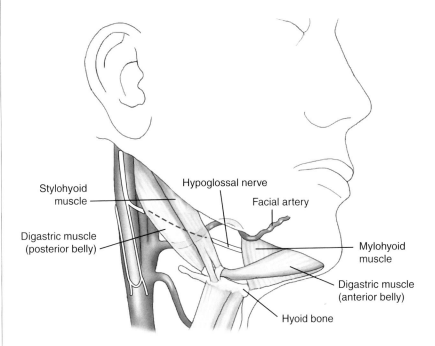

Stylohyoid muscle

Hypoglossal nerve

Facial artery

Digastric muscle (posterior belly)

Mylohyoid muscle

Digastric muscle (anterior belly)

Hyoid bone

■ **Figure 16–16   Contents of the submandibular triangle after removal of the submandibular gland**

Visceral Compartment of the Neck, and the Brain

Lab 17

Prior to dissection, you should familiarize yourself with the following structures:

ARTERIES

- Subclavian a.
  - Vertebral aa.
    - Posterior inferior cerebellar a.
    - Basilar a.
      - Anterior inferior cerebellar aa.
        - Labyrinthine a.
      - Pontine aa.
      - Superior cerebellar aa.
      - Posterior cerebral aa.
      - Posterior communicating a.
  - Thyrocervical trunk
    - Inferior thyroid a.
- Common carotid a.
  - External carotid a.
    - Superior thyroid a.
    - Maxillary a.
      - Middle meningeal a.
  - Internal carotid a.
    - Ophthalmic a.
    - Anterior cerebral a.
      - Anterior communicating a.
    - Middle cerebral a.

VEINS

- Brachiocephalic v.

- Inferior thyroid v.
- Internal jugular v.
  - Superior bulb of the jugular v.
  - Common facial v.
  - Facial v.
  - Retromandibular v.
    - Superficial temporal v.
    - Transverse facial v.
    - Maxillary v.
    - Pterygoid plexus of vv.
      - Middle meningeal v.
      - Deep temporal v.
  - Superior thyroid v.
  - Middle thyroid v.
  - Superior laryngeal v.
- Subclavian v.
  - External jugular v.
    - Posterior auricular v.
      - Occipital v.
    - Anterior jugular v.

LYMPHATICS

- Thoracic (lymphatic) duct
- Deep cervical lymph nodes
- Superficial cervical lymph nodes

MUSCLES

- Occipitofrontalis m.
- Longus coli m.
- Anterior scalene m.

GLANDS

- Thyroid gland
  - Right lobe, left lobe, and isthmus
- Parathyroid glands

OSTEOLOGY

- Calvarium
- Sutures
  - Sagittal suture
  - Coronal suture
  - Lambdoid suture
- Bregma and lambda (cranial landmarks)
- Frontal bone
- Parietal bones
- Occipital bone
  - External occipital protuberance
  - Internal occipital protuberance
  - Grooves for the transverse and confluence of sinuses

- Temporal bone
  - Mastoid process
- Ethmoid bone
  - Cribriform plate and crista galli
- Sphenoid bone
- Cranial fossae
  - Anterior cranial fossa
  - Middle cranial fossa
  - Posterior cranial fossa
    - Foramen magnum

SCALP LAYERS

- Skin
- Superficial fascia
- Aponeurosis of the muscular layer
- Loose connective tissue
- Pericranium

DURAL VENOUS SINUSES

- Superior sagittal sinus
- Inferior sagittal sinus
- Straight sinus
- Transverse sinuses
- Occipital sinus
- Confluence of sinuses
- Sigmoid sinuses
- Inferior petrosal sinuses
- Superior petrosal sinuses

- Basilar plexus
- Cavernous sinuses
- Sphenoparietal sinus

MENINGES

- Dura mater
  - Falx cerebri
  - Tentorium cerebelli
  - Falx cerebelli
- Arachnoid mater
  - Arachnoid granulations
  - Subarachnoid space
- Pia mater

BRAIN

- Cerebral gyri (ridges)
- Cerebral sulci (valleys)
  - Lateral sulcus
  - Central sulcus
- Cerebrum
- Poles
  - Frontal pole
  - Temporal pole
  - Occipital pole
- Lobes
  - Frontal lobe
  - Temporal lobe
  - Parietal lobe
  - Occipital lobe
- Cerebellum

VENTRICULAR SYSTEM

- Lateral ventricles
  - Interventricular foramen
- Third ventricle
  - Cerebral aqueduct
- Fourth ventricle
  - Median aperture
  - Lateral apertures
  - Central canal of the spinal cord

NERVES

- Phrenic n.
- Cranial nn.
  - Olfactory n. (CN I)
  - Optic n. (CN II)
  - Oculomotor n. (CN III)
  - Trochlear n. (CN IV)
  - Trigeminal n. (CN V)
  - Abducens n. (CN VI)
  - Facial n. (CN VII)
  - Vestibulocochlear n. (CN VIII)
  - Glossopharyngeal n. (CN IX)
  - Vagus n. (CN X)
    - Superior laryngeal n.
      - Internal laryngeal n.
      - External laryngeal n.
    - Recurrent laryngeal n.
  - Spinal accessory n. (CN XI)
  - Hypoglossal n. (CN XII)

MISCELLANEOUS

- Posterior atlanto-occipital membrane

Thyroid Gland and Associated Structures

■ Identify the following (Figure 17–1):

- **Thyroid gland** – reflect the infrahyoid mm.
  - **Left** and **right lobes** of the thyroid gland
  - **Isthmus** – inferior to the cricoid cartilage, on the anterior surface of the trachea; connection between the left and right lobes of the thyroid gland
- **Superior thyroid a.** – branch of the external carotid a.
- **Inferior thyroid a.** – pull a lobe of the thyroid gland anteriorly; the inferior thyroid a. is deep to the inferior poles of the thyroid gland; arises from the thyrocervical trunk
- **Superior, middle,** and **inferior thyroid vv.**
- **Laryngeal prominence** – the anterior projection of thyroid cartilage ("Adam's apple")
- **Thyrohyoid membrane** – deep to the thyrohyoid m.; spans the space between the thyroid cartilage and hyoid bone
- **Cricothyroid membrane** – deep to the cricothyroid m.; spans the space between the cricoid and thyroid cartilages
- **Cricoid cartilage** – the only complete ring (shaped like a signet ring) of cartilage in the respiratory tract; inferior to the thyroid cartilage
- **First tracheal ring** – C-shaped band of hyaline cartilage; inferior to the cricoid cartilage

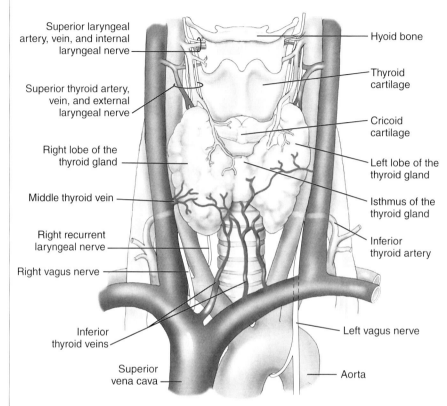

■ **Figure 17–1   Thyroid gland and associated structures**

Recurrent Laryngeal Nerves and the Parathyroid Glands

- Follow these dissection instructions:

  - Cut the thyroid gland through its isthmus
  - Reflect the lobes laterally

- Identify the following structures associated with the posterior surface of the lobes of the thyroid gland (Figures 17–2 and 17–3):

  - **Recurrent laryngeal nn.** – ascend bilaterally between the thyroid gland and trachea

  - **Parathyroid glands** – usually located lateral to the recurrent laryngeal nn.; small, brownish bodies about 1–2 mm in diameter; 2–5 glands on each side

    - *Note:* You may not be able to locate the parathyroid glands because of their tiny size and because they are usually embedded in the thyroid gland

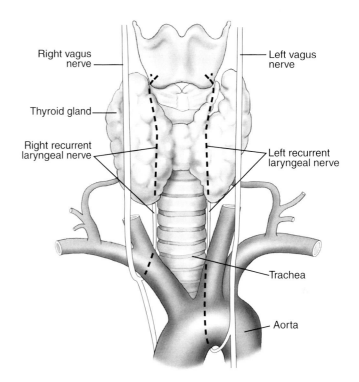

- **Figure 17–2   Anterior view of the thyroid gland**

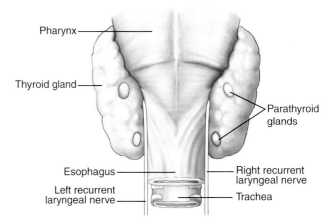

- **Figure 17–3   Posterior view of the pharynx**

Root of the Neck

- Follow these dissection instructions (Figure 17–4A):

  - On the left side, if not already done, cut and reflect the sternohyoid m., cut the sternothyroid m. from the manubrium; cut and reflect the superior belly of omohyoid m.

  - Cut the middle third of the clavicle to expand the dissection field

- Identify the following (Figure 17–4B):

  - **Borders of the root of the neck**
    - Manubrium, rib 1, and the body of T1 vertebra

  - **Thoracic duct**
    - ◆ Find the junction of the left subclavian v. and the left internal jugular v.

    - ◆ Usually at the V where these two veins join is the entry of the thoracic duct into the venous system

    - ◆ The thoracic duct is usually located on the posterior surface of the venous junction; it often has a dilated segment (due to valves) at its junction with the veins

    - An alternative approach is to locate the thoracic duct in the posterior mediastinum, where you may follow the thoracic duct superiorly into the root of the neck

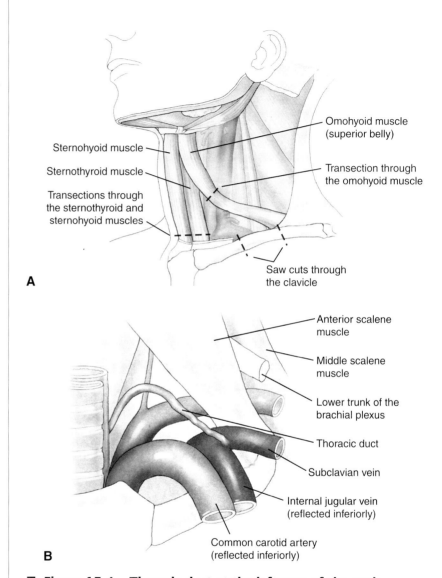

■ **Figure 17–4 Thoracic duct at the left root of the neck.
A, Lateral view. *B*, Anterior view**

Root of the Neck—cont'd

- Continue dissection on the left side of the neck only

- Identify the following (Figure 17–5):

  - **Phrenic n.** – courses on the anterior surface of the anterior scalene m.

  - **Thyrocervical trunk** – arises from the subclavian a. near the medial border of the **anterior scalene m.**

    - **Inferior thyroid a.** – a branch of the thyrocervical trunk; located posterior to the carotid sheath and enters the inferior region of the thyroid gland

  - **Triangle of the vertebral a.**

    - **Borders – longus coli m.** and **anterior scalene m.** form two sides; the apex of the triangle is the transverse process of C6 vertebra (the common carotid a. passes anterior to C6 vertebra)

    - **Vertebral a.** – enters the transverse foramen of C6 vertebra at the apex of this triangle

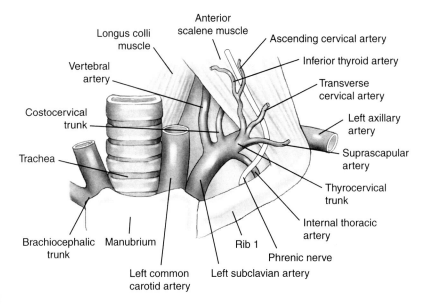

- **Figure 17–5   Triangle of the vertebral artery on the left side of the neck**

Removal of the Calvarium to Expose the Meninges and Brain – Overview

■ Because of the difficulty of the following dissection, we offer this visual overview of the dissection sequence. The ensuing pages detail each of the dissection steps.

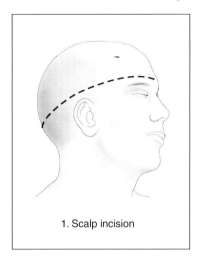

1. Scalp incision

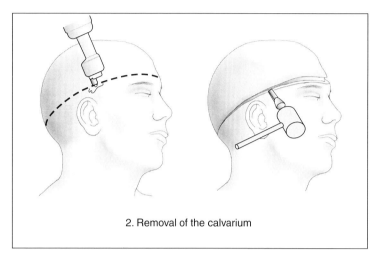

2. Removal of the calvarium

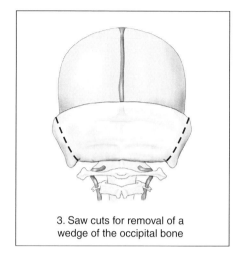

3. Saw cuts for removal of a wedge of the occipital bone

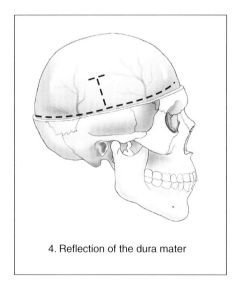

4. Reflection of the dura mater

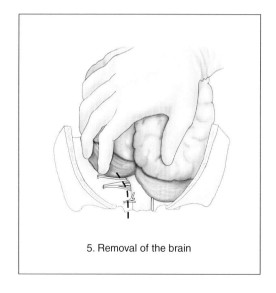

5. Removal of the brain

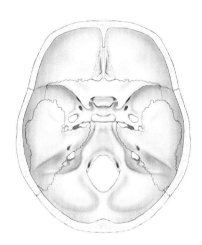

Note: The base of the skull will be studied during the next laboratory session

Scalp

■ **Incision** – using a scalpel, make the following continuous incision through the scalp (Figure 17–6):

• Around the circumference of the skull, approximately 2 cm above the supraorbital margin and 2 cm above the external occipital protuberance

■ Identify the following (Figure 17–7):

• **Scalp layers** – five layers, from superficial to deep:

1. **(S) Skin** – external surface covered with hair

2. **(C) Superficial Connective tissue (fascia)** – consists of dense connective tissue; many nerves and vessels are contained within it

3. **(A) Aponeurosis** – of the **frontalis m.** anteriorly and the **occipitalis m.** posteriorly, connected by a broad aponeurosis called the **galea aponeurotica**

4. **(L) Loose connective tissue** – deep to the aponeurosis

5. **(P) Pericranium** – the deepest layer is the periosteum that ensheaths the cranial bones

■ The bones of the skull are flat bones that have an external lamina, a central spongy bone (diploë), and an internal lamina

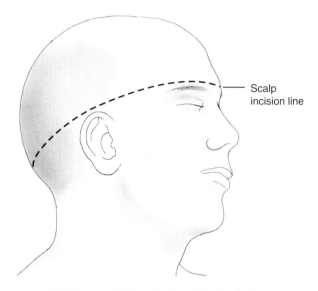

■ **Figure 17-6   Scalp skin incision**

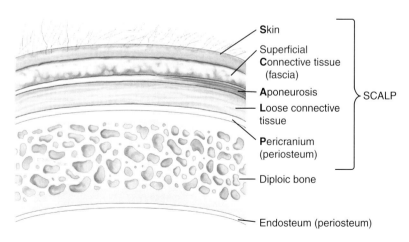

■ **Figure 17-7   Scalp layers**

Removal of the Calvarium

- You will use a bone saw, chisel, mallet, and finally your fingers to remove the calvarium from the rest of the skull

- *Note:* Make sure the bone saw is turned off before plugging it in

- Follow these instructions to remove the calvarium (Figure 17–8):

  - **Bone saw (a)** – cut through the skull where the skin incision was made

    - The key to sawing through the skull **(b)** is to only saw into the diploë region and leave the internal lamina intact; this approach minimizes damage (cuts) to the underlying dura mater and brain

  - **Chisel and mallet (c)** – once you have completed sawing, break through the internal lamina of the skull by repeatedly inserting a chisel along the saw cut and striking the chisel with a mallet

    - Continue this procedure until the calvarium is loose

  - **Separate the calvarium from the dura mater (d)** – this step sounds easier than it is; it takes patience and strength

    - Using a probe and your fingers, separate the dura mater from the internal surface of the calvarium; they are strongly bonded

    - *Be careful* of sharp edges of the bone; they will pierce rubber gloves and puncture your skin

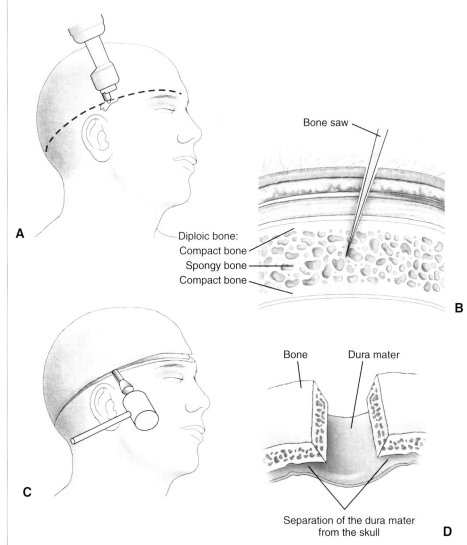

- **Figure 17–8   Cutting the skull with a bone saw. *A*, Saw cut along the scalpel incision line. *B*, Desired depth of penetration of the saw blade. *C*, Hammer and chisel use along the saw cut line. *D*, Separation of the dura mater from the inside of the skull**

Calvarium – Internal Surface

■ Once the calvarium is removed, examine its internal surface

■ Identify the following (Figure 17–9):

- **Grooves of the middle meningeal a.**
- **Groove for the superior sagittal sinus**
  - **Indentations (impressions) of the arachnoid villi (granulations)**
- **Frontal bone**
- **Parietal bones**
- **Occipital bone**
- **Coronal suture**
- **Sagittal suture**
- **Lambdoid suture**
- **Sutural bones** – extra bones within a suture; sutural bones are commonly found in the lambdoid region (not shown)

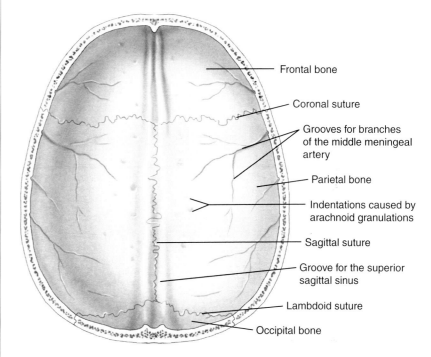

Frontal bone

Coronal suture

Grooves for branches of the middle meningeal artery

Parietal bone

Indentations caused by arachnoid granulations

Sagittal suture

Groove for the superior sagittal sinus

Lambdoid suture

Occipital bone

■ **Figure 17-9   Internal view of the calvarium**

Removal of a Wedge of the Occipital Bone to Better Expose the Brain

■ Place the cadaver prone and follow these instructions:

● Scrape away skin, muscles, and soft tissue that attach to the occipital bone and surround the **foramen magnum**

● Locate the space between the occipital bone and C1 vertebra (atlas); preserve the vertebral aa. (Figure 17–10A):

  ● With a scalpel, carefully cut through the **posterior atlanto-occipital membrane** from one vertebral a. to the other

  ● Ensure that all muscle and connective tissue have been removed from the area

  ● Insert a probe between the dura mater and occipital bone; sweep the probe from right to left; peel the dura mater from the occipital bone

● **Make the following bone saw cuts** (Figure 17–10A, B):

  ● Along the lines indicated, from point A to point B, on both sides of the occipital bone; as when you sawed through the calvarium, saw only into the spongy bone and use a chisel and mallet to finish the job

● Examine the removed bony wedge to identify the following (see Figure 17–10B):

  ● **Grooves for the transverse dural sinuses**

  ● **Groove for the confluence of dural venous sinuses**

  ● **Groove for the occipital dural sinus**

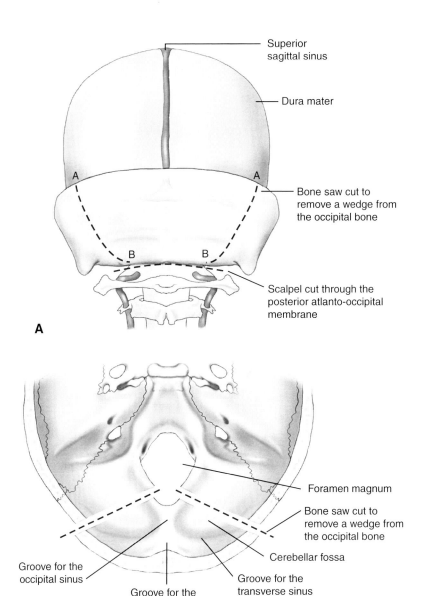

A

B

■ **Figure 17–10**   **Occipital wedge removal. A, Posterior view. B, Superior view**

Examination of the Cranial Dura Mater and Dural Venous Sinuses

■ Identify the following structures (Figure 17–11):

- **Dura mater** – surrounds the cerebrum and cerebellum

- **Branches of the middle meningeal a.** and **v.** – observe the anterior and posterior branches along the dura mater; observe the corresponding grooves on the internal surface of the calvarium

- **Dural venous sinuses** – venous sinuses are not true veins but are endothelial-lined venous structures that drain venous blood from the brain; in several locations inside the skull; the dural venous sinuses are formed by separation of the two layers of the dura mater (endosteal and meningeal layers)

  - **Superior sagittal sinus** – extends from anterior to posterior in the midline, deep to the calvarium

  - **Arachnoid granulations** – tufts that provide absorption of cerebrospinal fluid (CSF) into the dural venous sinuses (Figure 17–11B):

  - Cut open the superior sagittal sinus to observe these numerous cauliflower-like masses; you can usually see their outline through the dura mater

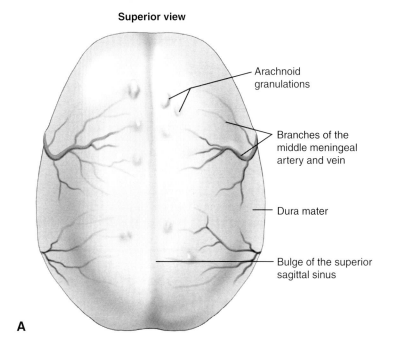

Superior view

Arachnoid granulations

Branches of the middle meningeal artery and vein

Dura mater

Bulge of the superior sagittal sinus

A

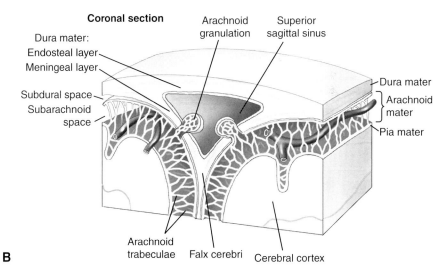

Coronal section

Dura mater:
Endosteal layer
Meningeal layer

Subdural space
Subarachnoid space

Arachnoid granulation

Superior sagittal sinus

Dura mater

Arachnoid mater

Pia mater

Arachnoid trabeculae   Falx cerebri   Cerebral cortex

B

■ **Figure 17–11   Dura mater. *A*, Superior view. *B*, Coronal section**

Cranial Meninges

■ Identify the following (Figure 17–12):

- **Dura mater** – the most superficial meninx; in the cranial cavity, the dura mater consists of two layers:
  - **Endosteal layer** – the external layer; endosteum (periosteum on the inside of the cranial bones)
  - **Meningeal layer** – the internal layer; continuous with the spinal cord's dura mater
- Cut the dura mater along the circumference of the edge of the saw cut through the skull; make a vertical incision with a scalpel, as illustrated in Figure 17-12, to reflect the dura mater
- **Arachnoid mater** – the second meninx; smoothly drapes over the surface of the brain; does not drop into the **sulci** (valleys)
  - **Subarachnoid space** – space deep to the arachnoid mater in which CSF is located
- **Pia mater** – the third, and deepest meninx; cannot be distinguished from the surface of the brain; the pia mater is intimately applied to every **gyrus** (ridge) and sulcus

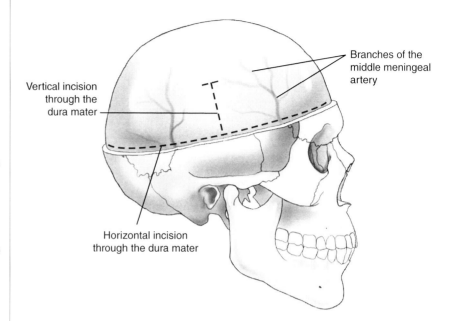

Vertical incision through the dura mater

Branches of the middle meningeal artery

Horizontal incision through the dura mater

■ **Figure 17-12   Lateral view of the skull to demonstrate removal of the dura mater**

Cranial Dural Folds

- Identify the following (Figure 17–13):

  - **Dural folds** – projections of the dura mater that compartmentalize the cranial cavity and thereby stabilize the brain; contain the dural venous sinuses

    - **Falx cerebri** – biggest dural projection; sickle-shaped septum between the two cerebral hemispheres; attached from the crista galli (ethmoid bone) anteriorly to the tentorium cerebelli posteriorly

    - **Tentorium cerebelli** – transverse projection between the cerebellum inferiorly and occipital lobes of the cerebrum superiorly; attached to the superior border of the petrous portion of the temporal bones

    - **Falx cerebelli** – short dural projection that separates the cerebellar hemispheres; attached to the internal occipital crest and tentorium cerebelli

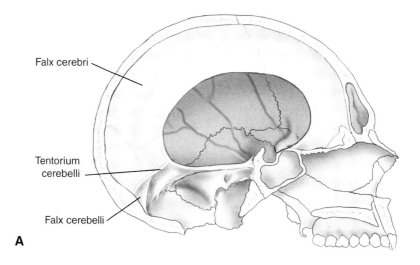

A

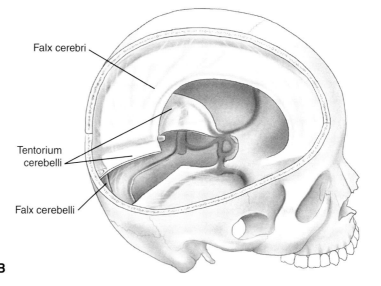

B

- **Figure 17–13** **Brain removed to demonstrate the dural folds.** *A,* Medial view. *B,* Superior and oblique view

Cranial Dural Venous Sinuses

■ Identify the following (Figure 17–14):

- **Dural venous sinuses** – drain venous blood from the brain; the two layers of the dura mater split to enclose the dural venous sinuses; veins draining the brain enter the dural venous sinuses, which ultimately drain to the great veins of the neck

  - **Superior sagittal sinus** – located along the superior border of the falx cerebri

  - **Inferior sagittal sinus** – located along the inferior border of the falx cerebri

  - **Straight sinus** – located along the line of fusion between the falx cerebri and tentorium cerebelli

  - **Great cerebral v.** – drains into the junction of the inferior sagittal and straight sinuses

  - **Transverse sinuses** – located along the lateral edges of the tentorium cerebelli, from the internal occipital protuberance to the petrous portion of the temporal bones

  - **Sigmoid sinuses** – S-shaped dural venous sinuses that end at the jugular foramen where the jugular bulb (dilation) is located (beginning of the internal jugular v.)

  - **Superior and inferior petrosal sinuses** – located along the petrous portion of the temporal bone

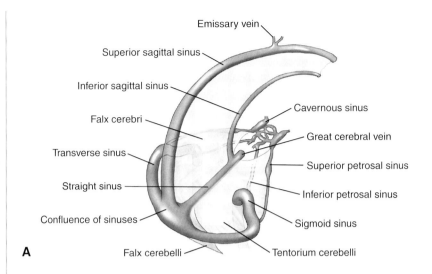

A

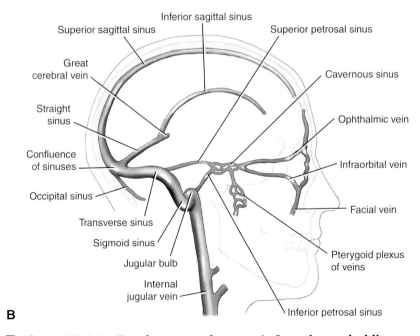

B

■ **Figure 17–14   Dural venous sinuses. A, Superior and oblique view. B, Medial view**

Removal of the Brain

- To remove the brain, all of its attachments will be cut

  - Attachments include nerves, vessels, and dura mater at the base of the brain and skull

- To remove the brain, place the cadaver prone and complete the following instructions:

  - Transect the spinal cord and both vertebral aa. inferior to the foramen magnum of the occipital bone (Figure 17–15A)

  - Cut the dura mater along the edge of the occipital bone (where the occipital wedge was removed) from A to B on both sides

  - Reflect the dural flap superiorly to expose the brain

  - Gently lift up the occipital lobes and cerebellum (Figure 17–15B)

  - Gradually extend a scalpel incision along the tentorium cerebelli from its internal cranial attachments along the occipital bones and the rim of the petrous portion of the temporal bones

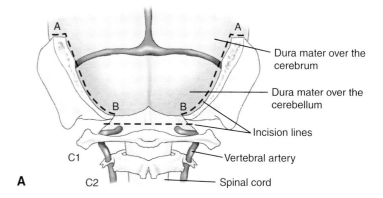

Dura mater over the cerebrum

Dura mater over the cerebellum

Incision lines

Vertebral artery

Spinal cord

A

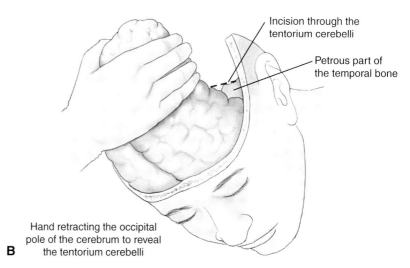

Incision through the tentorium cerebelli

Petrous part of the temporal bone

Hand retracting the occipital pole of the cerebrum to reveal the tentorium cerebelli

B

■ **Figure 17–15   Removal of the brain. *A,* Posterior view of the skull. *B,* Superior and oblique view**

Removal of the Brain—cont'd

■ To finish removal of the brain, complete the following instructions, bilaterally:

● Posterior region of the brain and skull (cadaver prone)

  ● Using a scalpel or scissors, transect **cranial nerves XII, XI, X, IX, VIII,** and **VII** near the base of the skull (Figure 17–16A)

    ◆ Leave one of the cut stubs attached to the brain and the other stub attached to the skull

● Anterior region of the brain and skull (cadaver supine)

  ● Gently lift both frontal poles of the cerebrum (Figure 17–16B)

  ● Identify the **olfactory bulbs** and **tracts** (CN I) on the **cribriform plates of the ethmoid bone;** using a probe, pry the bulbs from the cribriform plates to keep the bulbs and tracts attached to the brain

  ● Identify and transect the following structures:

    ◆ **Optic n.** (CN II) – cut just anterior to the optic chiasma

    ◆ **Internal carotid a.** – cut it close to the bone to keep the cerebral arterial circle intact on the brain

    ◆ **Cut the following nerves sequentially:**

      ◆ **Oculomotor nn. (CN III), abducens nn. (CN VI), trochlear nn. (CN IV),** and **trigeminal nn. (CN V)**

    ◆ You will also need to cut the falx cerebri and remaining dura mater from the crista galli of the ethmoid bone

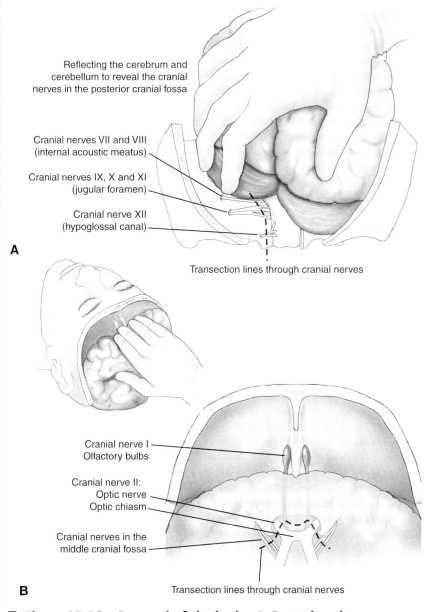

Reflecting the cerebrum and cerebellum to reveal the cranial nerves in the posterior cranial fossa

Cranial nerves VII and VIII (internal acoustic meatus)

Cranial nerves IX, X and XI (jugular foramen)

Cranial nerve XII (hypoglossal canal)

Transection lines through cranial nerves

A

Cranial nerve I
Olfactory bulbs

Cranial nerve II:
Optic nerve
Optic chiasm

Cranial nerves in the middle cranial fossa

Transection lines through cranial nerves

B

■ **Figure 17–16   Removal of the brain. *A,* Posterior view. *B,* Superior view**

Examination of the Removed Brain

■ Gently pull the brain out of the cranial vault

■ Identify the following (Figure 17–17):

- *Poles* of the cerebrum
  - **Frontal**
  - **Temporal**
  - **Occipital**
- *Lobes* of the cerebrum
  - **Frontal** – anterior cranial fossae
  - **Temporal** – middle cranial fossae
  - **Parietal** – superior to the temporal lobes
  - **Occipital** – posterior cranial fossae
- **Main sulci**
  - **Lateral sulcus** – separates the temporal lobe from the parietal and frontal lobes
  - **Central sulcus** – separates the primary motor cortex (located in the precentral gyrus of the frontal lobes) from the primary sensory cortex (located in the post-central gyrus of the parietal lobes)

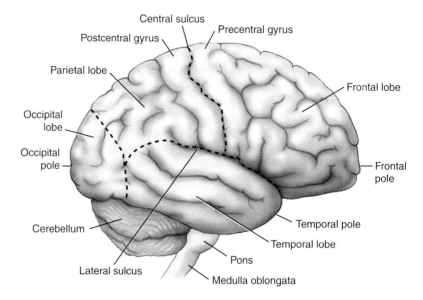

■ **Figure 17–17   Lateral view of the brain**

Arterial Supply of the Brain

- Peel away the arachnoid mater to reveal arteries at the brain's base

  - Identify the following arteries (Figure 17–18):

    - **Vertebral aa.** – located along the ventral surface of the brainstem

      - ◆ **Posterior inferior cerebellar aa.** – arise near the inferior end of the olive of the medulla oblongata

      - ◆ **Anterior spinal a.** – prior to the joining of the vertebral aa., each gives off a root of the anterior spinal a. that runs down the anterior median fissure of the spinal cord

    - **Basilar a.** – formed by the union of the vertebral aa.; courses on the inferior surface of the pons

      - ◆ **Anterior inferior cerebellar aa.** – branches from the inferior part of the basilar a.; course ventral to CN VI, VII, and VIII

      - ◆ **Labyrinthine a.** – may branch from the basilar a. but more often from the anterior inferior cerebellar a.; accompanies the facial (CN VII) and vestibulo-cochlear nn. (CN VIII) into the internal acoustic meatus; separated from the anterior inferior cere-bellar a. by the abducens n. (CN VI)

      - ◆ **Pontine aa.** – branches from the basilar a. to supply the pons

      - ◆ **Superior cerebellar aa.** – arise near the superior end of the basilar a.

      - ◆ **Posterior cerebral aa.** – also arise from the superior end of the basilar a.; separated from the superior cerebral a. by the oculomotor n. (CN III)

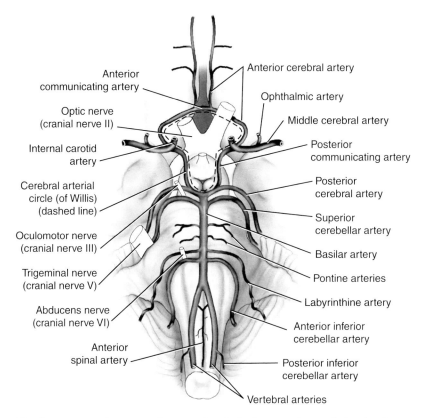

■ **Figure 17–18   Arteries on the inferior surface of the brain**

Arterial Supply of the Brain—cont'd

- **Internal carotid a.** – enters the cranial base via the carotid canal and turns to travel anteriorly through the cavernous sinus (on the side of the sphenoid bone); divides into the following branches:

  - **Middle cerebral a.** – courses laterally, deep to the temporal lobe, in the lateral sulcus; largest of the branches of the internal carotid a.

  - **Ophthalmic a.** – enters the cavernous sinus medial to the anterior clinoid process of the sphenoid bone; enters the orbit via the optic canal, inferolateral to the optic n.

  - **Anterior cerebral a.** – courses anteriomedially superior to the optic n. to reach the longitudinal cerebral fissure

    - ◆ **Anterior communicating a.** – connects the right and left anterior cerebral aa.; about 4 mm long

  - **Posterior communicating aa.** – connects the right and left middle cerebral aa. to the right and left posterior cerebral aa., respectively

- **Cerebral arterial circle (of Willis)** – a hexagonal vascular ring that is formed by union of the internal carotid and vertebral aa. via the anterior and posterior communicating aa.; surrounds the optic chiasma and stalk of the pituitary gland

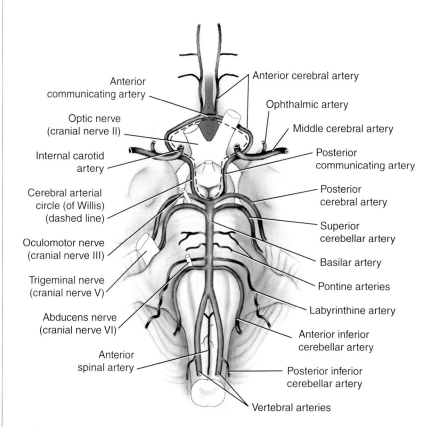

■ **Figure 17–18   Arteries on the inferior surface of the brain**

Ventricles of the Brain

■ Unless the brain is sliced in the gross anatomy course, you will not identify the ventricular system of the brain, other than radiographically

■ The following ventricular structures are identifiable on radiographs (Figure 17–19):

- **Lateral ventricles (paired)**
  - **Interventricular foramen**
- **Third ventricle**
  - **Cerebral aqueduct**
- **Fourth ventricle**
  - **Lateral apertures (paired)**
  - **Median aperture**
    - ◆ Both apertures allow CSF to flow into the **cisterna magna,** which communicates with the subarachnoid space
- **Central canal of the spinal cord**

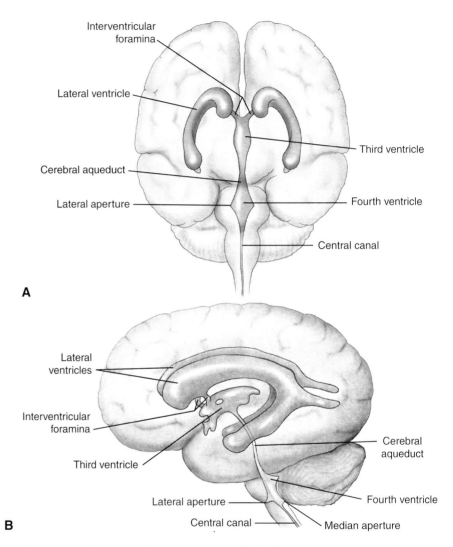

A

B

■ **Figure 17–19   Ventricular system of the brain. *A,* Anterior view. *B,* Lateral view**

Base of the Skull and Cranial Nerves

Lab 18

Prior to dissection, you should familiarize yourself with the following structures:

ANTERIOR CRANIAL FOSSA

■ **Frontal bone**
- Orbital plates
- Foramen cecum
  - Emissary vv. to the superior sagittal sinus

■ **Ethmoid bone**
- Crista galli
- Cribriform plate

■ **Sphenoid bone**
- Lesser wings
- Anterior clinoid processes

MIDDLE CRANIAL FOSSA

■ **Sphenoid bone**
- Greater wings
  - Grooves for the middle meningeal a.
- Chiasmatic groove (optic chiasma for CN II)
- Sella turcica and pituitary gland
  - Dorsum sellae
  - Posterior clinoid processes
- Groove for the internal carotid a.
- Foramen spinosum
  - Middle meningeal a. and v., meningeal branch of the mandibular n.

- Foramen ovale
  - Mandibular division of the trigeminal nerve (CN V-3)
- Foramen rotundum
  - Maxillary division of the trigeminal nerve (CN V-2)
- Superior orbital fissures
  - Oculomotor n. (CN III), trochlear n. (CN IV), ophthalmic division of the trigeminal nerve (CN V-1), abducens n. (VI), and superior ophthalmic v.
- Optic canal
  - Optic n. (CN II) and ophthalmic a.

■ **Temporal bone**
- Greater petrosal canal with accompanying nerves
- Foramen lacerum

POSTERIOR CRANIAL FOSSA

■ **Temporal bone**
- Internal acoustic meatus
  - Facial n. (CN VII), vestibulocochlear n. (CN VIII), and labyrinthine a.
- Groove for the superior petrosal sinus
- Groove for the sigmoid sinus

■ **Parietal bone**
- Groove for the middle meningeal a.

■ **Occipital bone**
- Clivus
- Groove for the inferior petrosal sinus
- Groove for the transverse and occipital sinuses
- Internal occipital crest and protuberance
- Groove for the superior sagittal sinus
- Foramen magnum
  - Medulla oblongata
  - Spinal cord
- Hypoglossal canal
  - Hypoglossal n. (CN XII)
- Jugular foramen
  - Inferior petrosal sinus, glossopharyngeal n. (CN IX), vagus n. (CN X), spinal accessory n. (CN XI), sigmoid sinus, and posterior meningeal a.

DURAL VENOUS SINUSES

■ Superior sagittal sinus

■ Inferior sagittal sinus

■ Straight sinus

Continued

- Transverse sinus
- Sigmoid sinus
- Confluence of sinuses
- Occipital sinus
- Basilar plexus
- Cavernous sinus
- Inferior petrosal sinus
- Superior petrosal sinus
- Sphenoparietal sinus

CRANIAL NERVES (CN)

- Olfactory n. (I)
- Optic n. (II)
- Oculomotor n. (III)
- Trochlear n. (IV)
- Trigeminal n. (V)
  - Trigeminal ganglion
    - V-1 – ophthalmic division
    - V-2 – maxillary division
    - V-3 – mandibular division

- Abducens n. (VI)
- Facial n. (VII)
  - Nerve to the pterygoid canal
- Vestibulocochlear n. (VIII)
- Glossopharyngeal n. (IX)
- Vagus n. (X)
- Spinal accessory n. (XI)
- Hypoglossal n. (XII)

Cranial Fossae

■ The base of the skull (internal surface of the cranium) is divided into three depressions (fossae) (Figure 18–1*A*)

■ Identify the three cranial fossae (Figure 18–1*B*):

- **Anterior cranial fossa** – separated from the middle cranial fossa by the sharp posterior crests of the lesser wings of the sphenoid bone and the anterior margin of the optic (chiasmatic) groove

- **Middle cranial fossa** – butterfly-shaped fossa formed mainly by the greater wings of the sphenoid bone and the temporal bones

- **Posterior cranial fossa** – largest of the cranial fossae; separated from the middle cranial fossa by the **dorsum sellae** and the posterior superior margins of the **petrous parts of the right and left temporal bones**

■ If a dry skull is available, use it to study the anatomic features of the cranial fossae

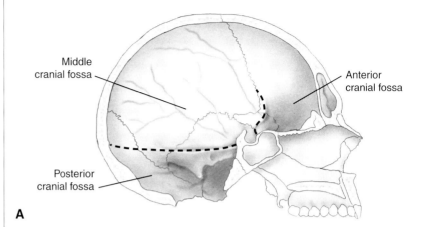

A

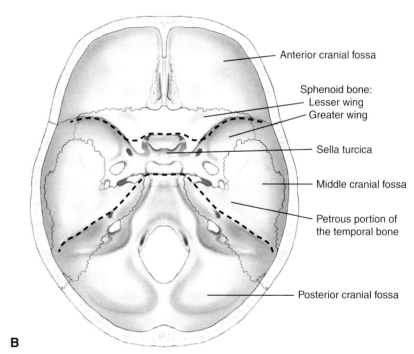

B

■ **Figure 18–1   Cranial fossae. *A*, Medial view in sagittal section. *B*, Superior view**

Anterior Cranial Fossa

■ Identify the following (Figure 18–2):

- **Frontal bone**
  - ● **Orbital plates**
  - ● **Foramen cecum** – transmits an emissary v. en route to the superior sagittal sinus
- **Ethmoid bone**
  - ● **Crista galli** – attachment for the dura mater (falx cerebri)
  - ● **Cribriform plate** – location of the olfactory bulbs from CN I
- **Sphenoid bone**
  - ● **Lesser wings**
  - ● **Anterior clinoid processes**

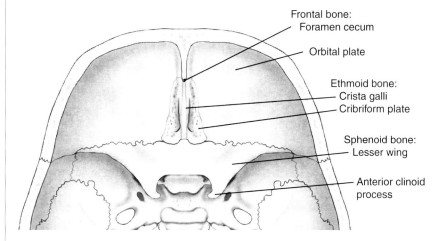

Frontal bone:
  Foramen cecum

Orbital plate

Ethmoid bone:
  Crista galli
  Cribriform plate

Sphenoid bone:
  Lesser wing

Anterior clinoid process

■ **Figure 18-2   Internal view of the anterior cranial fossa**

Middle Cranial Fossa

■ Identify the following (Figure 18–3):

- **Sphenoid bone**
  - **Sella turcica** – houses the pituitary gland
  - **Foramen rotundum** – round foramen; transmits the **maxillary division of the trigeminal n.** (CN V-2) to the pterygopalatine fossa
  - **Foramen ovale** – oval foramen; posterolateral to the foramen rotundum; transmits the **mandibular division of the trigeminal n.** (CN V-3) to the infratemporal fossa
  - **Foramen lacerum** – jagged gap superior to the carotid canal; covered by cartilage
  - **Foramen spinosum** – small round foramen posterolateral to the foramen ovale; transmits the **middle meningeal vessels** and the **accessory meningeal branch of the mandibular division of the trigeminal n. (CN V-3)**
  - **Superior orbital fissure** – inferior to the lesser wings of the sphenoid bone; transmits CN III, IV, V-1, and VI, and the **ophthalmic vv.** to the orbit
  - **Optic canal** – transmits the optic n. and the ophthalmic a.

- **Temporal bone**
  - **Hiatus for the greater petrosal n.**
  - **Petrous portion** – the densest bone in the skull, which is necessary for sound conduction

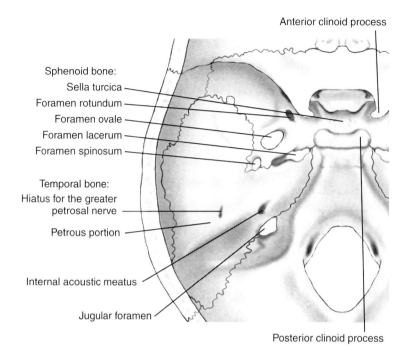

■ **Figure 18-3   Internal view of the middle cranial fossa**

Middle Cranial Fossa and Dural Venous Sinuses

■ Identify the following structures:

- **Superior petrosal sinuses** – paired dural venous sinuses located along the superior border of the petrous portions of the temporal bones; located at the lateral attachment of the **tentorium cerebelli** (Figure 18–4A)

- **Inferior petrosal sinuses** – paired dural venous sinuses located inferior to the superior petrosal sinuses; course from the cavernous sinuses along the petro-occipital sutures to the anterior compartment of the jugular foramen to terminate at the bulb of the internal jugular vv.

- **Basilar venous plexus** – located in the posterior cranial fossa

  - The superior petrosal sinuses, inferior petrosal sinuses, and basilar venous plexus are located in the posterior cranial fossa but are associated with the cavernous sinus

- **Cavernous sinuses** – paired, irregularly shaped dural venous sinuses located on the lateral walls of the sella turcica; the dura mater in the lateral wall of the cavernous sinus contains the following, from superior to inferior (Figure 18–4B), when the cavernous sinus is cut coronally:

  - **Oculomotor n. (CN III)**

  - **Trochlear n. (CN IV)**

  - **Trigeminal n. (CN V) branches V-1 and V-2**

- **Lumen of the cavernous sinus** – the following course through the lumen of the cavernous sinus, surrounded by venous blood:

  - **Internal carotid a.**

  - **Abducens n. (CN VI)**

- **Sphenoparietal sinus** – drains to the cavernous sinus

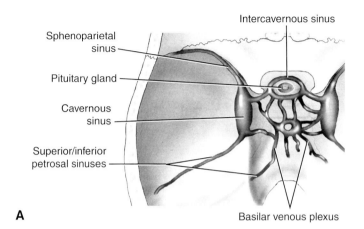

A

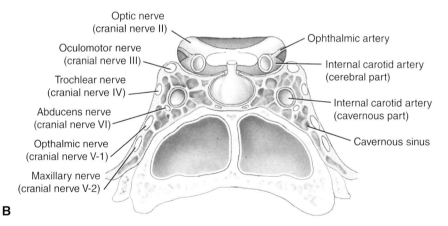

B

■ **Figure 18–4   Cavernous sinuses. A, Internal view. B, Coronal section through the sphenoid bone**

Middle Cranial Fossa and Dural Venous Sinuses—cont'd

■ Identify the following, after you observe the structures associated with the cavernous sinuses (Figure 18–5):

● On one side of the petrous portion of the temporal bone, peel away the dura mater to reveal the divisions of the trigeminal n. and the trigeminal ganglion

  ● **Trace V-1 (ophthalmic division)** – exits through the superior orbital fissure

  ● **Trace V-2 (maxillary division)** – exits through the foramen rotundum

  ● **Trace V-3 (mandibular division)** – exits through the foramen ovale

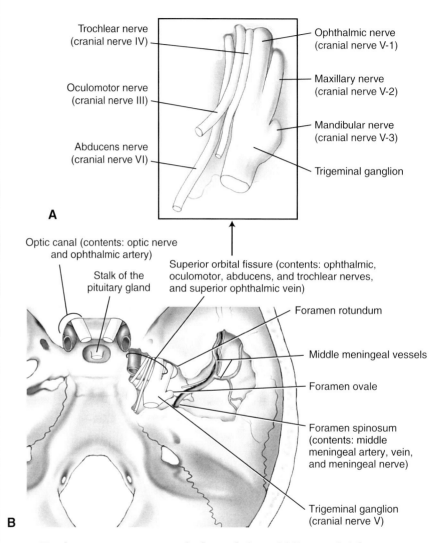

Trochlear nerve (cranial nerve IV)

Ophthalmic nerve (cranial nerve V-1)

Oculomotor nerve (cranial nerve III)

Maxillary nerve (cranial nerve V-2)

Mandibular nerve (cranial nerve V-3)

Abducens nerve (cranial nerve VI)

Trigeminal ganglion

A

Optic canal (contents: optic nerve and ophthalmic artery)

Stalk of the pituitary gland

Superior orbital fissure (contents: ophthalmic, oculomotor, abducens, and trochlear nerves, and superior ophthalmic vein)

Foramen rotundum

Middle meningeal vessels

Foramen ovale

Foramen spinosum (contents: middle meningeal artery, vein, and meningeal nerve)

Trigeminal ganglion (cranial nerve V)

B

■ **Figure 18–5   Internal view of the middle cranial fossa**

Middle Cranial Fossa and Dural Venous Sinuses—cont'd

■ Identify the following (Figure 18–6):

- **Middle meningeal aa.** – branches of the maxillary aa.; embedded in the outer layer of the dura mater

  - Follow the middle meningeal aa. to the foramen spinosum on each side of the skull's base

- **Greater petrosal n.** (branch of CN VII) – courses from the greater petrosal hiatus on the superior surface of the temporal bone; passes deep to the trigeminal ganglion before it reaches the foramen lacerum; the greater petrosal n. (parasympathetic) joins the deep petrosal n. (sympathetic) to form the nerve of the pterygoid canal

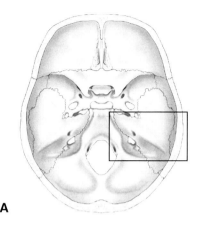

A

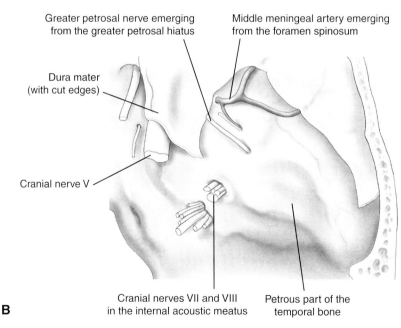

Greater petrosal nerve emerging from the greater petrosal hiatus

Middle meningeal artery emerging from the foramen spinosum

Dura mater (with cut edges)

Cranial nerve V

Cranial nerves VII and VIII in the internal acoustic meatus

Petrous part of the temporal bone

B

■ **Figure 18–6** **Internal view of the petrous part of the temporal bone**

Posterior Cranial Fossa

■ Identify the following structures (Figure 18–7):

- **Temporal bone**
  - **Internal acoustic meatus**
  - **Jugular foramen** – union of notches in the temporal and occipital bones
- **Occipital bone**
  - **Hypoglossal canal** – CN XII exits the cranium through this canal
  - **Foramen magnum** – the spinal cord exits the cranium through this foramen
  - **Clivus** – the inclined surface from the dorsum sellae to the foramen magnum; supports the pons and medulla oblongata
  - **Internal occipital protuberance**
  - **Jugular foramen** – union of notches in the occipital and temporal bones

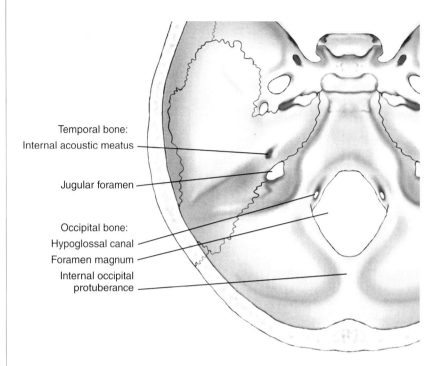

Temporal bone:
Internal acoustic meatus

Jugular foramen

Occipital bone:
Hypoglossal canal
Foramen magnum
Internal occipital
protuberance

■ **Figure 18–7   Internal view of the posterior cranial fossa**

Posterior Cranial Fossa—cont'd

■ Identify the following (Figure 18–8):

- **Foramen magnum** – largest foramen in the skull; transmits the following:

  - **Medulla oblongata** – at the level of the foramen magnum, the medulla oblongata becomes continuous with the spinal cord

  - **Vertebral aa.**

  - **Spinal roots of the spinal accessory nn. (CN XI)**

  - **Venous plexus of the spinal cord and cranial cavity** – these veins have no valves

- **Hypoglossal canal** – superior to the anterolateral margin of the foramen magnum; transmits the hypoglossal n. (CN XII)

- **Jugular foramen** – between the petrous portion of the temporal bone and occipital bone; transmits the following:

  - **Jugular bulb** – origin of the internal jugular v.

  - **Cranial nerves IX, X,** and **XI**

  - **Posterior meningeal a.**

- **Internal acoustic meatus** – superior to the jugular foramen; transmits **CN VII** and **VIII** and the **labyrinthine a.**

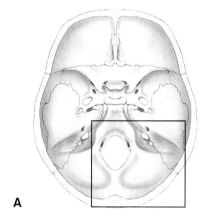

A

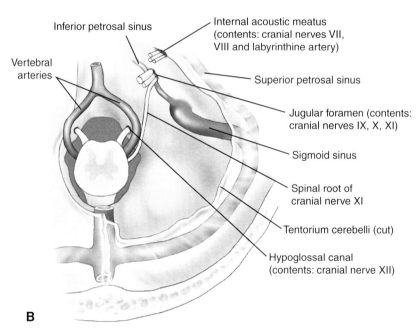

Inferior petrosal sinus

Internal acoustic meatus (contents: cranial nerves VII, VIII and labyrinthine artery)

Vertebral arteries

Superior petrosal sinus

Jugular foramen (contents: cranial nerves IX, X, XI)

Sigmoid sinus

Spinal root of cranial nerve XI

Tentorium cerebelli (cut)

Hypoglossal canal (contents: cranial nerve XII)

B

■ **Figure 18-8   Internal view of the posterior cranial fossa**

Review of the Cranial Nerves

- **Cranial nerves** (12 pairs):

  - Receive their name because they emerge through foramina/fissures in the cranium

  - The cranial nerves are numbered I through XII, from anterior to posterior according to their attachment on the brain

- Identify the following cranial nerves (CNs) (Figure 18–9):

  - **Olfactory n. (CN I)**

  - **Optic n. (CN II)**

  - **Oculomotor n. (CN III)**

  - **Trochlear n. (CN IV)**

  - **Trigeminal n. (CN V)**

  - **Abducens n. (CN VI)**

  - **Facial n. (CN VII)**

  - **Vestibulocochlear n. (CN VIII)**

  - **Glossopharyngeal n. (CN IX)**

  - **Vagus n. (CN X)**

  - **Spinal accessory n. (CN XI)**

  - **Hypoglossal n. (CN XII)**

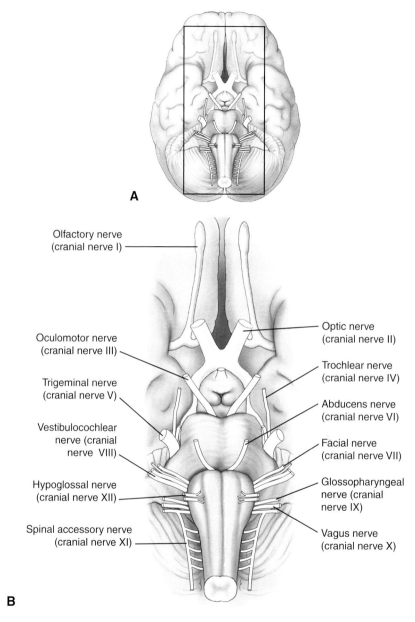

A

Olfactory nerve (cranial nerve I)

Oculomotor nerve (cranial nerve III)

Trigeminal nerve (cranial nerve V)

Vestibulocochlear nerve (cranial nerve VIII)

Hypoglossal nerve (cranial nerve XII)

Spinal accessory nerve (cranial nerve XI)

Optic nerve (cranial nerve II)

Trochlear nerve (cranial nerve IV)

Abducens nerve (cranial nerve VI)

Facial nerve (cranial nerve VII)

Glossopharyngeal nerve (cranial nerve IX)

Vagus nerve (cranial nerve X)

B

- **Figure 18–9   Cranial nerves on the inferior surface of the brain**

Orbit 19

Prior to dissection, you should familiarize yourself with the following structures:

OSTEOLOGY

- Frontal bone
  - Superior orbital margin
  - Frontal sinus
- Lacrimal bone
- Ethmoid bone
  - Ethmoid labyrinth of air cells
  - Anterior and posterior ethmoid foramina
- Sphenoid bone
  - Anterior clinoid processes
  - Optic canal
    - Contents: optic n. and ophthalmic a.
  - Superior orbital fissure
    - Contents: superior ophthalmic v., oculomotor n. (CN III), trochlear n. (CN IV), ophthalmic division of the trigeminal n. (lacrimal, frontal, and nasociliary branches of the CN V-1), and abducens n. (CN VI)
  - Inferior orbital fissure
  - Infraorbital groove
- Zygomatic bone

NERVES

- Optic n. (CN II)

- External nasal n. (CN V-1)
- Frontal n. (CN V-1)
  - Supraorbital n.
  - Supratrochlear n.
- Lacrimal n. (CN V-1)
- Infratrochlear n. (CN V-1)
- Nasociliary n. (CN V-1)
  - Long ciliary n.
  - Anterior and posterior ethmoid nn.
- Trochlear n. (CN IV)
- Abducens n. (CN VI)
- Oculomotor n. (CN III)
  - Superior and inferior divisions
  - Ciliary ganglion
  - Short ciliary nn.

VESSELS

- Ophthalmic a.
  - Branches are named the same as nerves in the orbit
  - Central a. of the retina

GLANDS

- Lacrimal gland
- Tarsal glands

MUSCLES

- Orbicularis oculi m.
- Levator palpebrae superioris m.
- Superior oblique m. (trochlea)
- Inferior oblique m.
- Medial rectus m.
- Lateral rectus m.
- Superior rectus m.
- Inferior rectus m.

EYEBALL (GLOBE)

- Sclera
- Cornea
- Choroid
  - Ciliary body
  - Zonular fibers
- Iris
- Lens
- Chambers
  - Anterior chamber
  - Posterior chamber
  - Vitreous body
- Retina

MISCELLANEOUS

- Periorbita
- Annulus tendineus

Approaches to Dissect the Orbit

■ During this dissection, the right orbit will be dissected via:

1. Superior approach – removal of the orbital roof

2. Anterior (surgical) approach – removal of the eyelids

Superior Approach

■ Follow these instructions to remove the orbital plate over the right eye:

● Remove the dura mater from the orbital plate of the frontal bone

● With a chisel and mallet, carefully break the center of the roof of the orbit (Figure 19–1A)

● Remove bone pieces with forceps (Figure 19–1B)

● Remove the roof of the orbit as far anteriorly as possible, but leave the superior orbital margin intact

● Push a probe inferior to the orbital roof, through the superior orbital fissure; remove the lesser wing of the sphenoid bone that is superior to the probe

● Carefully break away the roof of the optic canal

● Remove the anterior clinoid process (sphenoid bone)

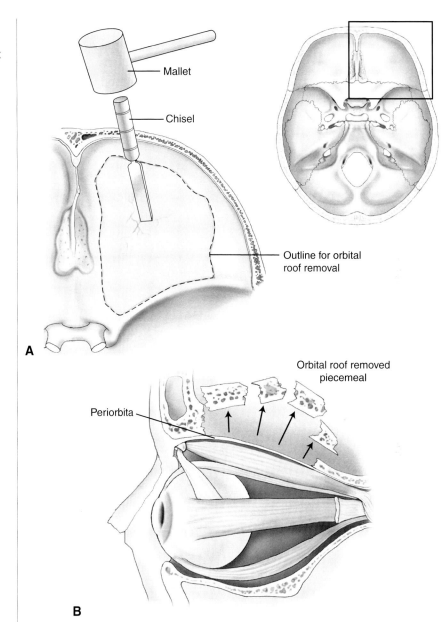

■ **Figure 19–1   Removal of the orbital plate. *A*, Superior view. *B*, Lateral view of a sagittal section**

Superior Approach—cont'd

■ Identify the following:

- **Periorbita** – the periosteal lining of the orbital bones; cut away the exposed periorbita but *do not* damage the frontal n. or other underlying structures (Figure 19–2A)

- **Frontal n. (CN V-1)** – inferior to the periorbita; the nerve branches into the following (Figure 19–2B):

  - **Supraorbital n.** – on the superior border of the levator palpebrae superioris m.

  - **Supratrochlear n.** – a thin nerve that courses superior to the **trochlea** of the superior oblique m.

- **Trochlear n.** (CN IV) – courses along the superior, proximal border of the **superior oblique m.**

- **Lacrimal gland** – located anterolaterally in the superior part of the orbit

- **Lacrimal n. (CN V-1)** – enters the superior orbital fissure lateral to the frontal n.; trace the lacrimal n. anteriorly to the lacrimal region of the eyelid

- **Levator palpebrae superioris m.** – with forceps, pick out lobules of fat to expose this muscle

  - Cut the levator palpebrae superioris m. as far anteriorly as you can; reflect the muscle posteriorly

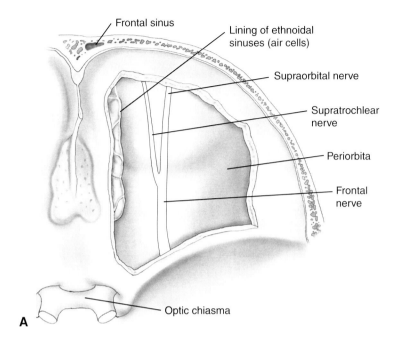

A

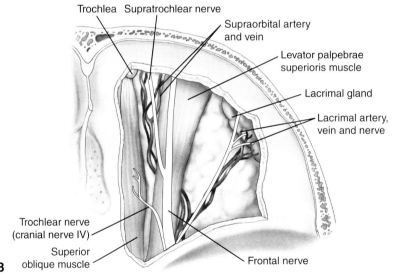

B

■ **Figure 19–2 Superficial view of the orbital contents. *A*, Periorbita. *B*, Periorbita removed**

Superior Approach—cont'd

■ Identify the following muscles (Figure 19–3A):

- **Superior rectus m.** – directly inferior to the levator palpebrae superioris m.

  - Cut the superior rectus m. close to the eyeball to reflect the cut muscle posteriorly

- **Superior oblique m.** – trace anteriorly to the **trochlea,** where the tendon bends at a sharp angle around the trochlea to attach to the lateral, posterior surface of the sclera

- **Medial rectus m.** – inferior to the superior oblique m.; attaches medially on the sclera

- **Lateral rectus m.** – attaches laterally on the sclera

■ Identify the following nerves (Figure 19–3B):

- **Abducens n.** (CN V-1) – enters the deep surface of the lateral rectus m.

- **Superior division of the oculomotor n.** (CN III) – penetrates the deep surface of the levator palpebrae and superior rectus mm.

- **Inferior division of the oculomotor n.** (CN III) – supplies the medial rectus, inferior rectus, and inferior oblique mm.

  - **Ciliary ganglion** – a parasympathetic ganglion (preganglionic from CN V-3) approximately 1–2 mm in diameter; located posteriorly in the orbit, between the optic n. and the lateral rectus m.

  - **Short ciliary n., a., and v.** – postganglionic parasympathetic nerves that arise (synapse) in the ciliary ganglion and course to the posterior region of the eyeball

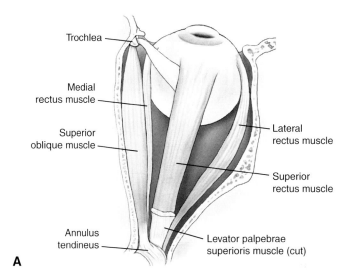

A

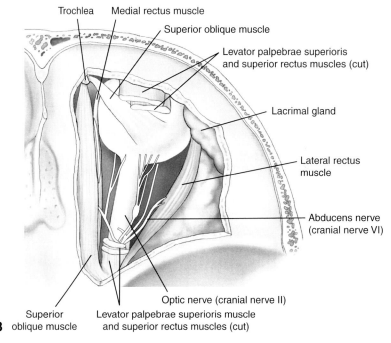

B

■ **Figure 19–3   Intermediate view of the orbital contents.** *A,* **Superior view (superficial).** *B,* **Superior view (deep)**

Superior Approach—cont'd

■ Identify the following (Figures 19–4 and 19–5):

- **Optic n. (CN II)** – deep to the superior rectus m.; the central a. of the retina is located within the optic n.; you may cut the optic n. in cross section to reveal this artery

- **Ophthalmic a.** – enters the orbit via the optic canal, inferolateral to the optic n.; crosses the optic n. (accompanied by the nasociliary n.) between the superior rectus m. and the optic n.; terminal branches are the supratrochlear supraorbital aa. and central artery of the retina

- **Ophthalmic v.** – the superior branch crosses over and the inferior branch crosses under the optic n., and both terminate in the cavernous sinus

- **Nasociliary n.** – course between the optic n. and superior rectus m. in an anterior and medial direction; at the trochlea, the nasociliary n. exits the orbit as the infratrochlear n.

■ Identify the following branches of the nasociliary n. and ophthalmic a. and v.:

- **Anterior ethmoidal n.** – traverse the anterior ethmoidal foramen; nerve exits the nasal cavity as the **external nasal n.**

- **Posterior ethmoidal n., a., and v.** – exit the orbit by the posterior ethmoidal foramen

- **Long ciliary n., a., and v.** – 2 or 3 long ciliary nn. branch from the nasociliary n. as it crosses the optic n.

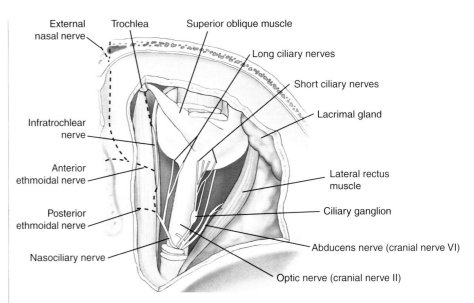

■ **Figure 19–4   Deeper view of the orbital contents**

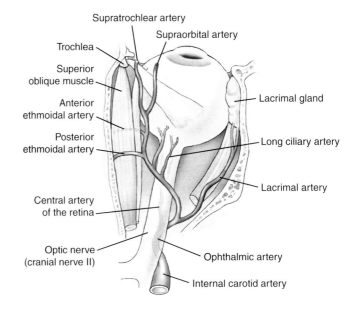

■ **Figure 19–5   Superior view of the ophthalmic artery**

Facial Approach

- To facilitate this dissection, you may remove the upper and lower eyelids and the orbital septum (alternatively, you may simply use a scalpel to cut horizontally through the upper and lower eyelids)

- Identify the following structures in an anterior approach on the right orbit (Figure 19–6):

  - **Lacrimal gland** – superior and lateral to the eyeball
  - **Trochlea** – superior and medial to the eyeball; attached to the internal surface of the bony orbit; the tendon of the superior oblique m. passes through the trochlea
  - **Inferior oblique m.** – inferior and medial to the eyeball
  - Observe (*do not cut*) the insertions of the four recti mm. and two oblique mm. on the **sclera** of the eyeball
    - **Superior rectus m.**
    - **Superior oblique m.**
    - **Medial rectus m.**
    - **Lateral rectus m.**
    - **Inferior rectus m.**
    - **Inferior oblique m.**

Dissection of the Eyeball in Situ (Right Eye Only)

- If you would like to examine the interior of the eyeball, use a scalpel to make a horizontal incision through the eyeball, along its facial (front) surface (Figure 19–7)

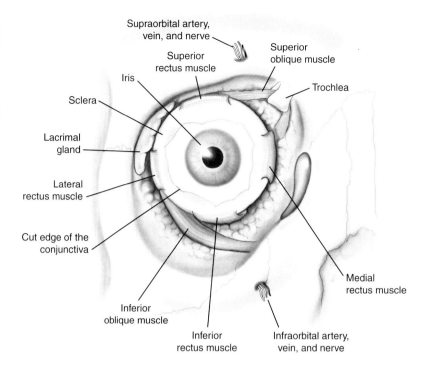

■ **Figure 19–6   Facial/surgical approach to the orbit**

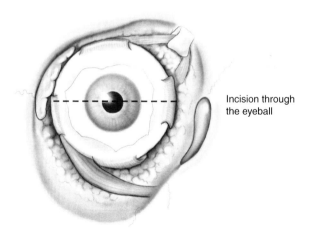

■ **Figure 19–7   Horizontal incision across the right eyeball**

Dissection of the Eyeball in Situ
(Right Eye Only)—cont'd

■ Identify the following (Figure 19–8):

- **Sclera** – the white, external tunic; continuous with the **cornea**

- **Iris** – pigmented drape superficial to the lens

- **Lens** – a transparent, flexible, biconvex, refractive structure located posterior to the iris and anterior to the vitreous humor

- **Choroid** – the dark-colored, middle tunic of the eyeball; blood vessels and ciliary nn. are contained in this layer

  - **Ciliary body** – the anterior termination of the choroid; has protrusions on its internal surface called ciliary processes

  - **Zonular fibers** – attach the lens to the ciliary processes

- **Retina** – internal tunic of the eyeball; in a hydrated eye, this layer is wispy and pinkish in color

  - When you cut open the eye, loss of vitreous pressure may result in detachment of the retina

- **Chambers**

  - **Anterior chamber** – anterior to the lens; located between the cornea and iris; contains aqueous humor; continuous with the posterior chamber at the pupil

  - **Posterior chamber** – anterior to the lens; located between the iris and the lens; contains aqueous humor

- **Vitreous body** – posterior to the lens; contains vitreous humor

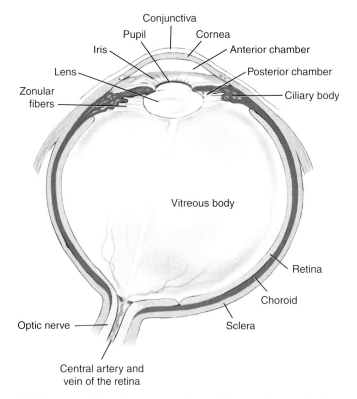

■ **Figure 19–8   Cross-section of the right eyeball**

Superficial Face

Lab 20

Prior to dissection, you should familiarize yourself with the following structures:

MUSCLES

■ Muscles of facial expression
- Frontalis m.
- Orbicularis oculi m.
- Corrugator supercilii m.
- Nasalis m.
- Procerus m.
- Levator labii superioris alaeque nasi m.
- Levator labii superioris m.
- Levator anguli oris m.
- Zygomaticus major and minor mm.
- Orbicularis oris m.
- Platysma m.
- Risorius m.
- Depressor anguli oris m.
- Depressor labii inferioris m.
- Mentalis m.
- Buccinator m.

■ Muscles of mastication
- Temporalis m.
- Masseter m.
- Medial and lateral pterygoid mm.

■ Prevertebral muscles
- Scalene mm. (anterior, middle, and posterior)
- Longus capitis and colli mm.
- Rectus capitis anterior m.
- Rectus capitis lateralis m.

LIGAMENTS

■ Cruciate ligament
- Superior longitudinal ligament
- Inferior longitudinal ligament
- Transverse ligament of the atlas

■ Alar ligaments

NERVES

■ Sympathetic trunk
- Superior, middle, and inferior cervical ganglia

■ Trigeminal n. (CN V)
- V-1 – Ophthalmic division
  - Supraorbital n.
  - Supratrochlear n.
  - Infratrochlear n.
  - Lacrimal n.
  - External nasal n.
- V-2 – Maxillary division
  - Infraorbital n.
  - Zygomaticofacial n.
  - Zygomaticotemporal n.
- V-3 – Mandibular division
  - Auriculotemporal n.
  - Buccal n.
  - Mental n.

■ Facial n. (CN VII)
- Posterior auricular n.
- Temporal branch
- Zygomatic branch
- Buccal branch
- Mandibular branch
- Cervical branch

FASCIA

■ Prevertebral fascia
- Anterior layer (alar part)
  - Retropharyngeal space (danger space)
- Posterior layer

■ Buccopharyngeal fascia

OSTEOLOGY

■ Atlas (C1)
■ Axis (C2)
- Odontoid process (dens)
■ Pharyngeal tubercle

Continued

- Stylomastoid foramen
- Frontal bone
  - Supraorbital margin
  - Supraorbital foramen (notch)
- Maxilla
  - Infraorbital foramen
- Nasal bone
- Lacrimal bone
- Mandible
  - Ramus, neck, and body
  - Mental and mandibular foramina
  - Coronoid and condyloid processes

ARTERIES

- External carotid a.
  - Facial a.
    - Inferior and superior labial aa.
  - Superficial temporal a.
- Transverse facial a.

VEINS

- Internal jugular v.
  - Facial v.
    - Superior and inferior labial vv.
  - Retromandibular v.
    - Superficial temporal v.
    - Transverse facial v.

PAROTID REGION

- Parotid gland, duct, and fascia
- Buccal fat pad

TABLE 20-1   Muscles of Facial Expression

Muscle	Attachments		Action	Innervation
Facial belly of the occipitofrontalis	Epicranial aponeurosis	Skin of the forehead and eyebrows	Elevates the eyebrows and the skin of the forehead	
Procerus	Nasal bone and nasal cartilage	Skin between the eyebrows	Wrinkles the skin over the bridge of the nose	
Corrugator supercilii	Medial region of the supraorbital margin	Skin of the medial half of the eyebrow	Draws the eyebrows downward and medial	
Orbicularis oculi	Frontal and maxillary bones around the orbit	Superficial fascia of the eyelid	Strong closure of the eyelid; sphincter of the eyelids	
Zygomaticus major and minor	Zygomatic arch	Angle of the mouth	Raises lateral corners of the mouth superiorly	
Risorius	Superficial fascia overlying the masseter m.	Angle of the mouth	Draws the corners of the lip laterally	
Levator labii superioris	Frontal process of the maxilla, infraorbital region, and zygomatic bone	Skin of the upper lip	Elevates the lip, dilates the nostril, and raises the angle of the mouth	
Levator labii superioris alaeque nasi	Maxilla inferior to the infraorbital foramen	Alar cartilage of the nose and upper lip	Elevates the nostrils and lip	Facial n. (CN VII)
Depressor labii inferioris	Mandible between the symphysis and the mental foramen	Into the orbicularis oris and skin of the lower lip	Depresses the lower lip	
Depressor anguli oris	Oblique line of the mandible	Angle of the mouth	Turns the corner of the mouth downward	
Orbicularis oris	Surrounds the oral orifice, forming intrinsic muscles of the lips; interlaces with other muscles associated with the lips		Closes and protrudes the lips (kissing muscle)	
Mentalis	Incisive fossa of the mandible	Skin of the chin	Protrudes the inferior lip; wrinkles the chin	
Buccinator	Pterygomandibular raphe, alveolar processes of the jaw	Angle of the mouth	Compresses the cheek (whistling, blowing, and sucking)	
Platysma	Superficial fascia of the pectoral and deltoid regions	Mandible, skin of the neck and cheek, angle of the mouth, and orbicularis oris	Depresses the mandible; tenses the skin of the neck	

TABLE 20-2   Muscles of Mastication

Muscle	Attachments		Action	Innervation
Masseter	Inferior border and deep surface of the zygomatic arch	Lateral surface of the ramus of the mandible	Elevates the mandible	Trigeminal n. (CN V-3)
Temporalis	Temporal fossa and temporal fascia	Coronoid process and anterior border of the ramus of the mandible via a tendon that passes deep to the zygomatic arch	Elevates and retracts the mandible; maintains position of the mandible at rest	
Medial pterygoid	Medial surface of the lateral pterygoid plate of the sphenoid bone	Medial surface of the ramus of the mandible between the mandibular foramen and the angle	Elevates and protracts the mandible; draws the mandible toward the opposite side	
Lateral pterygoid	Lateral surface of the lateral pterygoid plate of the sphenoid bone	Neck of the mandible and capsule of the temporomandibular joint	Protracts and depresses the mandible; draws the mandible toward the opposite side	

TABLE 20–3 Muscles of Prevertebral Region

Muscle	Attachments		Action	Innervation
Scalene muscle group				
• Anterior scalene	Transverse processes of C4–C6 vertebrae	Rib 1	Bilaterally: stabilizes the neck Unilaterally: flexes the neck laterally	Cervical plexus and ventral rami of C4–C6 spinal nerves
• Middle scalene	Posterior tubercles of the transverse processes of C4–C6 vertebrae	Superior surface of rib 1		Ventral rami of cervical spinal nerves
• Posterior scalene		External border of rib 2		Ventral rami of C7–C8 spinal nerves
Longus capitis	Transverse processes of C3–C6 vertebrae	Basilar portion of the occipital bone	Flexes and rotates the head	Ventral rami of C1–C3 spinal nerves
Longus colli	Transverse processes and bodies of C3–T3 vertebrae	Anterior tubercle of C1 (atlas), vertebral bodies of C2–C4, and transverse processes of C5–C6 vertebrae	Flexes the neck and rotates to the opposite side if acting unilaterally	Ventral rami of C2–C6 spinal nerves
Rectus capitis lateralis	Transverse processes of C1 vertebra (atlas)	Jugular process of the occipital bone	Flexes and stabilizes the head	Ventral rami of C1–C2 spinal nerves
Rectus capitis anterior	Lateral mass of C1 vertebra (atlas)	Basilar portion of the occipital bone anterior to the occipital condyles	Flexes and rotates the head	

Special Dissection Instructions – Bisection of the Head and Neck

- To preserve the triangles of the cervical region for study and review, it is recommended that the instructions on this page are followed by every fourth or fifth dissection group; ask for instructions from your laboratory staff

- This dissection is performed to save the cervical triangles and maintain structural continuity among the thorax, neck, and head

- Laboratory staff will help select which cadaver heads will be bisected based on the integrity of the previous dissections

- If you are not instructed to bisect the head and neck, proceed to the next page

- Follow these instructions to bisect the head (Figure 20–1):

  - A member of the laboratory staff will bisect the head, neck, and thorax in a sagittal plane, using a band saw

    - *Note:* If your laboratory is not equipped with a band saw, follow directions from your laboratory staff

  - Proceed to the page entitled **Superficial Face** to continue your dissection

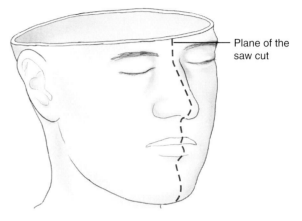

Plane of the saw cut

- **Figure 20–1   Special dissection instructions**

Dissection of the Craniovertebral Joints

■ Place the cadaver prone and peel the dura mater from the basioccipital region and underlying tectorial membrane; reflect the flap of dura to reveal the tectorial membrane

■ Identify the following structures deep to the dura mater:

• **Tectorial membrane** – this strong, longitudinally oriented band within the vertebral canal is the superior continuation of the posterior longitudinal ligament (Figure 20–2*A*)

  ● The dura mater and tectorial membrane may be fused, in which case both may have been peeled away

  ● Cut the tectorial membrane transversely above the anterior border of the foramen magnum; reflect the membrane inferiorly

  ● At this point, the major ligaments of the craniovertebral region are exposed:

    ◆ **Cruciate ligament** – receives its name because the cruciate ligament looks like a cross; composed of the following (Figure 20–2*B*):

      ◆ **Transverse ligament of C1 (atlas)** – attaches to the medial surface of C1 vertebra on both sides

      ◆ The ligament has a superior longitudinal band that extends to the basilar part of the occipital bone and an inferior longitudinal band that extends to the body of C2 vertebra

      ◆ **Alar ligaments** – extend from the sides of the dens to the occipital condyles; immediately superior to the transverse ligament

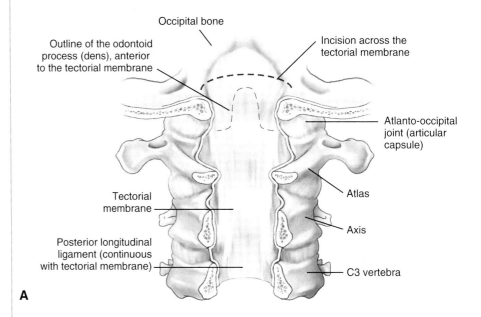

A

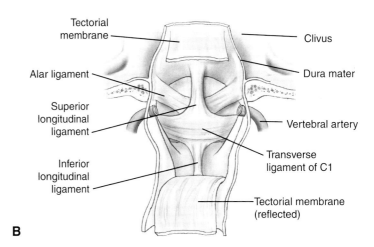

B

■ **Figure 20–2   Craniovertebral joints. *A*, Internal view of the skull base: tectorial membrane incision. *B*, Internal view of the skull base: tectorial membrane reflected**

Dissection of the Retropharyngeal Space

- During this dissection, the head and visceral compartment of the neck will be reflected from the vertebral column; the visceral compartment of the neck will remain attached to the superior mediastinum

- Follow these instructions to dissect the retropharyngeal space (Figure 20–3):

  - Insert a finger or two from both of your hands bilaterally, posterior to the carotid sheaths; push your fingers medially until they meet posterior to the cervical viscera

  - Your fingers are in the **retropharyngeal space**

  - Work your fingers superiorly to reach the base of the skull (superior limit of the retropharyngeal space) and inferiorly to about the level of the T3 vertebra (where the **prevertebral fascia** fuses with the **buccopharyngeal fascia**)

  - Do not damage branches of the subclavian a. at the root of the neck

  - Attempt to leave the cervical sympathetic chains in the prevertebral muscular compartment; look for the cervical sympathetic chain in this compartment (if you do not find the cervical sympathetic chains, examine the posterior surface of the reflected cervical viscera)

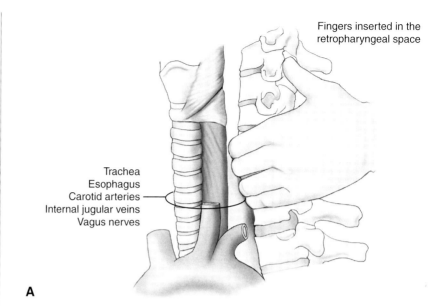

Fingers inserted in the retropharyngeal space

Trachea
Esophagus
Carotid arteries
Internal jugular veins
Vagus nerves

A

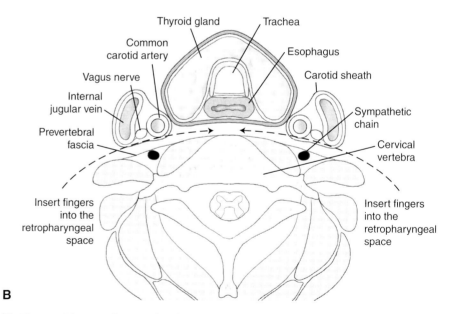

Thyroid gland
Trachea
Common carotid artery
Esophagus
Vagus nerve
Carotid sheath
Internal jugular vein
Sympathetic chain
Prevertebral fascia
Cervical vertebra
Insert fingers into the retropharyngeal space
Insert fingers into the retropharyngeal space

B

- **Figure 20–3** **Fingers in the retropharyngeal space.** *A*, Lateral view. *B*, Cross-section through the thyroid gland

Removal of the Head and Neck Viscera from the Vertebral Column

- *Note:* Do *not* decapitate the cadaver; your goal is to reflect the head and cervical viscera *together* from the vertebral column; in other words, the head will remain attached to the trachea and esophagus (cervical viscera)

- Reflection of the head and neck will be done in two steps:

  - **Transection of the tissue between C1 vertebra and the occipital bone**

    - Circumscribe the foramen magnum with a scalpel between C1 veterbra and the occipital bone, cutting the attachments between the C1 vertebra and the occipital bone (Figure 24–4*A*, *B*)

      - ◆ *Use caution* to avoid cutting CN IX–XII and the internal jugular vv.

    - Pull the cervical viscera anteriorly to cut between the transverse processes of C1 vertebra and the occipital bone, severing the **rectus capitis lateralis mm.** bilaterally (Figure 20–4*A*)

    - Carry the scalpel incision medially, cutting the **rectus capitis anterior** and **longus capitis mm.** (Figure 20–4*B*)

    - Use a chisel to pry the head away from the C1 vertebra; cut remaining attachments with a scalpel (Figure 20–4*C*)

  - **Reflection of the head and cervical viscera from the vertebral column**

    - Place your fingers superiorly into the already defined retropharyngeal space

    - Pull the head and cervical viscera, vessels, and nerves anterior; blunt dissect and cut the remaining attachments (prevertebral and erector spinae mm.) (Figure 20–4*D*)

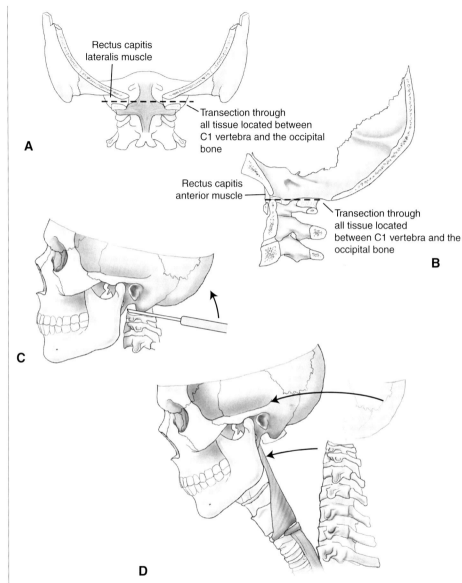

■ **Figure 20–4 Removal of the head from the vertebral column. *A*, Posterior view. *B*, Lateral view. *C*, Reflection of the skull from C1. *D*, Reflection of the head and neck viscera from the vertebral column**

Prevertebral Region

- Examine the prevertebral fascia that covers the prevertebral muscles

- Identify the following (Figure 20–5A):

  - **Prevertebral fascia** – Figure 20–5A shows the fascial planes before dissection; the prevertebral fascia extends between transverse processes of the vertebrae; consists of two layers

    - **Anterior layer (alar part)** – destroyed during previous dissection

    - **Posterior layer**

    - **Retropharyngeal space** – space between the anterior and posterior layers of the prevertebral fascia; also known as the "danger space" because it provides a potential passageway for infection in the cervical region to spread inferiorly into the posterior mediastinum

  - **Prevertebral muscles** (Figure 20–5B):

    - **Rectus capitis lateralis m.** – locate the severed distal attachment

    - **Rectus capitis anterior m.** – locate the severed distal attachment

    - **Longus colli and capitis mm.** – attach along the cervical and superior thoracic vertebrae; the superior attachment of the longus capitis m. is severed

    - **Anterior, middle, and posterior scalene mm.** – attach to the transverse processes of the cervical vertebrae

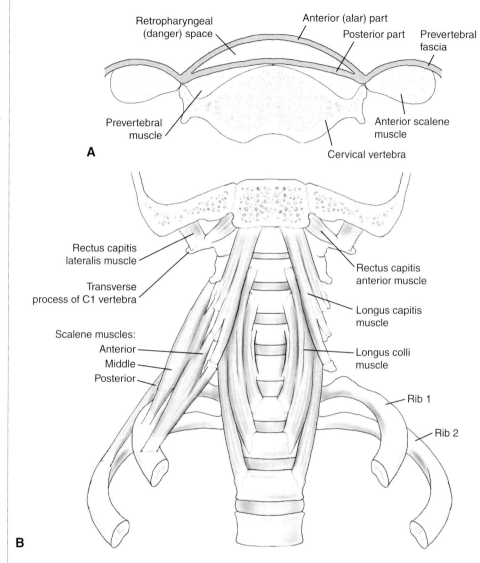

■ **Figure 20–5   Prevertebral structures. *A*, Cross-section. *B*, Coronal section through the temporal bone with the cervical viscera removed**

Superficial Face

- Those who did not bisect the head and neck of their cadaver will bisect the head only to facilitate the following:

  - Dissection of both sides of the superficial face (this laboratory session)

  - Dissection of both sides of the deep face (another laboratory session)

- Bisection allows both sides of the face to be dissected simultaneously

- Follow these instructions to bisect the head (Figure 20–6):

  - Laboratory staff will bisect the head from the base of the skull through the mandible (*not* through the neck)

    - If your laboratory is not equipped with a band saw, follow the instructions of your laboratory staff

  - This approach preserves the pharyngeal constrictors and larynx

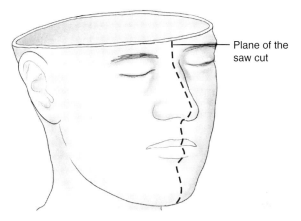

Plane of the saw cut

■ **Figure 20–6   Saw cut through the face**

Skin Removal from the Face

■ Make the skin incisions shown in Figure 20–7 to remove the skin of the face

   ● Leave the mucous membrane on the lips and the skin attached to the eyelids and nostrils

■ Remove *only* the skin because the muscles of the face (muscles of facial expression) attach to the skin, and the nerves and vessels are immediately deep to the skin

■ View the cut edge of the skin along the saw line to gauge the skin's thickness

■ Reflect the skin to the anterior border of the ear and away from the angle of the mandible

■ Skin both sides of the face

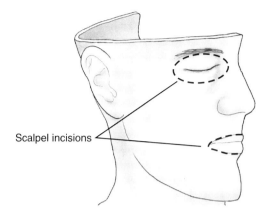

■ **Figure 20–7 Incisions of the facial skin**

Trigeminal Nerve (CN V)

- Identify the three foramina on a dry skull (aligned vertically) through which peripheral branches of CN V emerge:

  - **Supraorbital, infraorbital,** and **mental** foramina

- Identify the cutaneous divisions of **CN V-1** (Figure 20–8):

  - **Supraorbital n.** – originates from the frontal n.; exits the skull through the supraorbital foramen (use a dry skull to orientate); locate by separating the fibers of the corrugator supercilii m. and orbicularis oculi m.

  - **Supratrochlear n.** – also originates from the frontal n.; emerges deep to the corrugator supercilii and frontalis mm. and usually pierces the orbicularis oculi m. to supply the skin of the inferior part of the forehead near the midline

  - **Infratrochlear n.** – originates from the nasociliary n.; exits the orbit inferior to the trochlea and pierces the orbicularis oculi m. to supply the skin of the eyelids and side of the nose

  - **Lacrimal and external nasal nn.** – usually too small to identify

- Vessels that share the same name as the nerve (i.e., supraorbital n., a., and v.) accompany each of these nerves

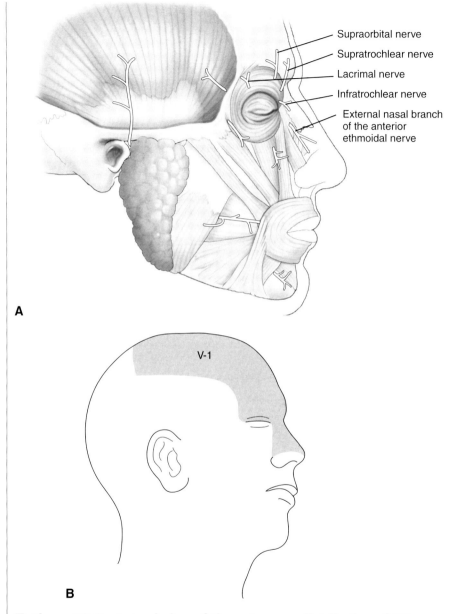

- **Figure 20–8   Lateral view of the cutaneous distribution of CN V-1. *A*, Cutaneous nerves of V-1. *B*, Cutaneous field for CN V-1**

Trigeminal Nerve (CN V)—cont'd

- Identify the cutaneous divisions of **CN V-2** (Figure 20–9):

  - **Infraorbital n.** – locate the nerve by separating the levator labii superioris m. (superficial) from the levator anguli oris m. (deep)

  - **Zygomaticofacial n.** – originates from the zygomatico-orbital n.; exits the skull through the zygomaticofacial foramen; pierces the orbicularis oculi m. to supply the skin over the cheek

  - **Zygomaticotemporal n.** – also originates from the zygomatico-orbital n.; exits the skull through the zygomaticotemporal foramen; pierces the temporal fascia less than an inch above the zygomatic arch to supply the skin of the temple

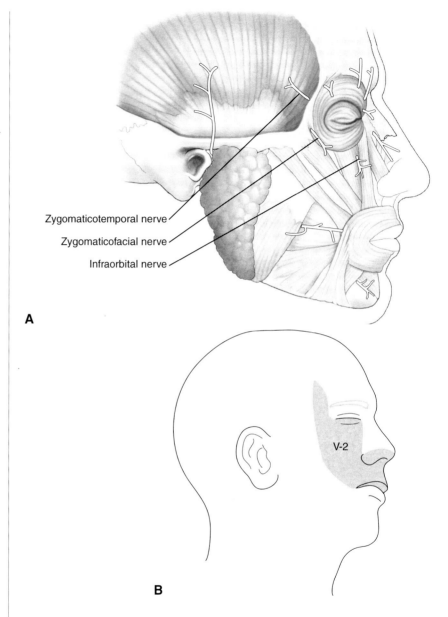

Zygomaticotemporal nerve

Zygomaticofacial nerve

Infraorbital nerve

A

V-2

B

- **Figure 20–9   Lateral view of the cutaneous distribution of CN V-2. *A,* Cutaneous nerves of V-2. *B,* Cutaneous field for CN V-2**

Trigeminal Nerve (CN V)—cont'd

■ Identify the cutaneous divisions of CN V-3 (Figure 20–10):

- **Auriculotemporal n.** – emerges posterior and inferior to the temporomandibular joint and ascends posterior to the superficial temporal vessels

- **Buccal n.** – courses between the two bellies of the lateral pterygoid m. and descends deep to the temporalis tendon; passes deep to the masseter m. to supply the skin superficial to, and the mucous membrane deep to, the buccinator m.

- **Mental n.** – originates from the inferior alveolar n.; exits the mandible via the mental foramen; locate by separating the fibers of the mentalis m. from the depressor anguli oris m.

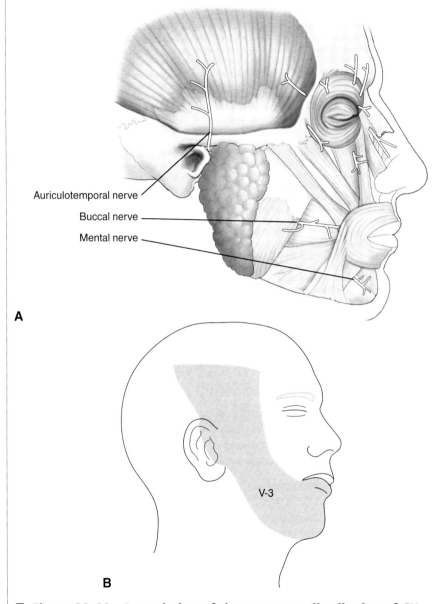

Auriculotemporal nerve

Buccal nerve

Mental nerve

A

B

■ **Figure 20–10   Lateral view of the cutaneous distribution of CN V-3. *A*, Cutaneous nerves of V-3. *B*, Cutaneous field for CN V-3**

Parotid Region

■ Identify the following (Figure 20–11):

- **Parotid gland** – located inferior to the zygomatic arch and anterior to the ear

- **Masseter m.** – muscle of mastication; deep to the parotid gland

- **Parotid duct** – emerges 3 cm inferior and parallel to the zygomatic arch; courses anteriorly from the parotid gland toward the cheek (mouth); enters the oral cavity near the second maxillary molar

- **Buccal fat pad** – follow the parotid duct to the anterior border of the masseter m., where the duct dives into the buccal fat pad and buccinator m. to reach the mouth

- **Buccinator m.** – remove the buccal fat pad to expose the buccinator m., which is pierced by the parotid duct

- **Transverse facial a.** – located parallel and superior to the parotid duct

- **Facial a.** and **v.** – both cross the mandible (at the anterior border of the masseter m.) and course medially and superiorly toward the angle of the eye

- **Retromandibular v.** – deep to the parotid gland and inferior to the ear; reveal that this vein unites with the **facial v.,** forming the **common facial v.** (a tributary to the internal jugular v.); not shown

- **Superficial temporal a.** and **v.** – branch of the external carotid a. and a tributary to the retromandibular v., respectively; located anterior to the ear; they course superficially over the temporalis m.

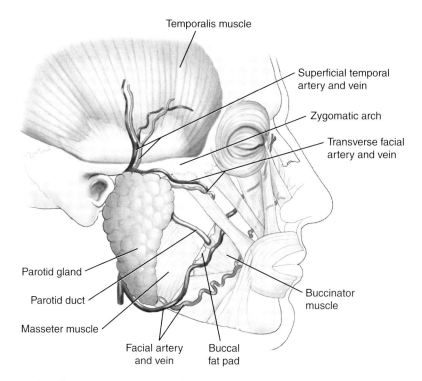

Temporalis muscle

Superficial temporal artery and vein

Zygomatic arch

Transverse facial artery and vein

Buccinator muscle

Parotid gland

Parotid duct

Masseter muscle

Facial artery and vein

Buccal fat pad

■ **Figure 20–11   Parotid region and facial vasculature**

Facial Nerve in the Parotid Region

■ To identify the branches of the facial n., follow these instructions (Figure 20–12):

- Look for white, flattened branches of the facial n. that emerge through the substance of (or deep to) the parotid gland; follow the branches through the parotid gland toward the stylomastoid foramen

- Remove the parotid gland, piece by piece, to reveal the continuity of CN VII to the stylomastoid foramen

- The branches of the **facial n.** are named by the region that they innervate (may be several branches to each region):

  - **Temporal branches**

  - **Zygomatic branches**

  - **Buccal branches**

  - **Mandibular branches**

  - **Cervical branches**

  - **Posterior auricular n.** – attempt to identify this nerve where it courses along the posterior border of the ear

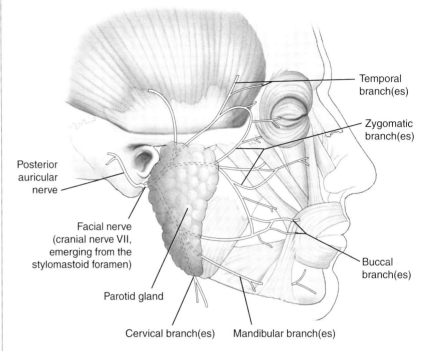

Posterior auricular nerve

Facial nerve (cranial nerve VII, emerging from the stylomastoid foramen)

Parotid gland

Temporal branch(es)

Zygomatic branch(es)

Buccal branch(es)

Cervical branch(es)   **Mandibular branch(es)**

■ **Figure 20-12   Branches of the facial nerve**

Muscles of Facial Expression

■ Identify the muscles of facial expression (Figure 20–13):

- **Facial belly of the occipitofrontalis m. (1)**
- **Auricularis mm.** – anterior, superior, and posterior divisions
- **Corrugator supercilii m. (2)**
- **Orbicularis oculi m. (3)**
- **Nasalis m. (4)**
- **Procerus m. (5)**
- **Risorius m. (6)**
- **Zygomaticus major m. (7)**
- **Zygomaticus minor m. (8)**
- **Levator labii superioris alaeque nasi m. (9)**
- **Orbicularis oris m. (10)**
- **Levator labii superioris m. (11)**
- **Levator anguli oris m. (12)**
- **Depressor anguli oris m. (13)**
- **Depressor labii inferioris m. (14)**
- **Mentalis m. (15)** .
- **Buccinator m. (16)** – deep to the buccal fat pad
- **Platysma m.** – not shown; extends from the inferior border of the mandible inferiorly over the clavicles

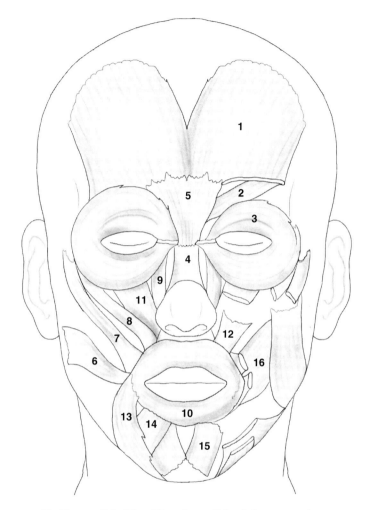

■ **Figure 20–13   Muscles of facial expression**

Muscles of Mastication

■ Identify the muscles of mastication (Figure 20–14):

- **Masseter m.** – attached to the zygomatic arch and external surface of the ramus of the mandible

- **Temporalis m.** – attached to the temporal fossa; courses deep to the zygomatic arch; inserts on the coronoid process of the mandible

- **Medial pterygoid m.** – will be dissected in another laboratory session

- **Lateral pterygoid m.** – will be dissected in another laboratory session

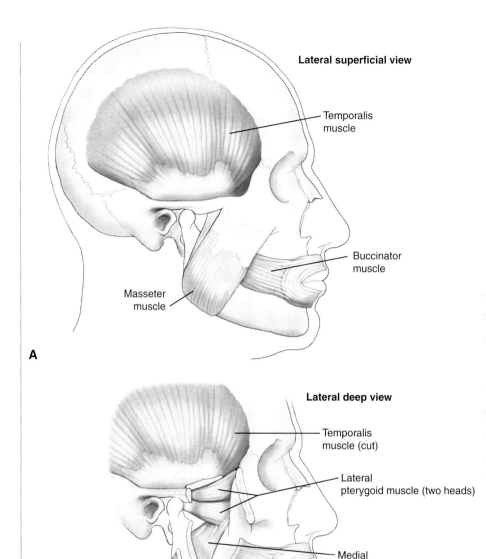

Lateral superficial view

Temporalis muscle

Buccinator muscle

Masseter muscle

A

Lateral deep view

Temporalis muscle (cut)

Lateral pterygoid muscle (two heads)

Medial pterygoid muscle

Masseter muscle (cut)

B

■ **Figure 20–14   Muscles of mastication.** *A*, Lateral superficial view. *B*, Lateral deep view

Deep Face and Pharynx

Lab 21

Prior to dissection, you should familiarize yourself with the following structures:

OSTEOLOGY

- Hyoid bone
  - Greater horns
  - Lesser horns
- Zygomatic bone and arch
- Temporal bone
  - Mastoid process
  - Styloid process
- Mandible
  - Ramus, body, and neck
  - Lingula
  - Coronoid and condylar processes
  - Mandibular notch
  - Mandibular foramen
  - Mylohyoid line
- Sphenoid bone
  - Lateral and medial pterygoid plates
    - Pterygoid hamulus
    - Pterygoid canal
- Maxilla
  - Palatine process
- Palatine bone
  - Perpendicular plate
  - Horizontal plate
- Occipital bone
  - Pharyngeal tubercle

INFRATEMPORAL FOSSA CONTENTS

- Muscles of mastication
  - Temporalis m.
  - Masseter m.
  - Medial pterygoid m.
  - Lateral pterygoid m.
- Pterygoid plexus of vv.
- Maxillary artery and branches
  - Maxillary a.
    - Middle meningeal a.
    - Deep temporal a.
    - Inferior alveolar a.
    - Posterior superior alveolar a.
    - Infraorbital a.
    - Muscular branches
    - Buccal a.
- Nerves
  - Trigeminal n.
    - CN V-2
      - Posterior superior alveolar n.
    - CN V-3
      - Mandibular n.
      - Mylohyoid n.
      - Inferior alveolar n.
      - Lingual n.
      - Buccal n.
      - Auriculotemporal n.
  - Facial n. (CN VII)
    - Chorda tympani n.
  - Glossopharyngeal n. (CN IX)
    - Otic ganglion
  - Vagus n. (CN X)
    - Superior laryngeal n.
      - Internal laryngeal n.
      - External laryngeal n.
    - Recurrent laryngeal n.
  - Cervical ganglia (superior, middle, and inferior)

MISCELLANEOUS

- Pterygomaxillary fissure
  - Pterygoid fossa
- Thyroid cartilage
  - Oblique line
- Cricoid cartilage
- Pterygomandibular raphe
- Temporomandibular joint (TMJ)
  - Joint capsule

- Sphenomandibular ligament
- Petrotympanic fissure
- Stylohyoid ligament

PHARYNX

- Pharyngeal mm.
  - Superior pharyngeal constrictor m.
  - Middle pharyngeal constrictor m.
  - Inferior pharyngeal constrictor m.
  - Stylopharyngeus m.
  - Salpingopharyngeus m.
  - Palatopharyngeus m.

NASOPHARYNX

- Auditory (Eustachian) tube
- Torus tubarius
- Salpingopharyngeal fold
- (Naso)pharyngeal tonsils (adenoids)

TABLE 21-1 Muscles of Mastication

Muscle	Attachments		Action	Innervation
Masseter	Inferior border and deep surface of the zygomatic arch	Lateral surface of the ramus of the mandible	Elevates the mandible	Trigeminal n. (CN V-3)
Temporalis	Temporal fossa and temporal fascia	Coronoid process and anterior border of the ramus of the mandible via a tendon that passes deep to the zygomatic arch	Elevates and retracts the mandible	
Medial pterygoid	Medial surface of the lateral pterygoid plate of the sphenoid bone	Medial surface of the ramus of the mandible between the mandibular foramen and the angle	Elevates and protracts the mandible; draws the mandible toward the opposite side	
Lateral pterygoid	Lateral surface of the lateral pterygoid plate of the sphenoid bone	Neck of the mandible and capsule of the temporomandibular joint	Protracts and depresses the mandible; draws the mandible toward the opposite side	

TABLE 21–2   Muscles of the Pharynx

Muscle	Attachments		Action	Innervation
Pharyngeal constrictor mm.				
• Superior constrictor	Pterygoid hamulus, pterygomandibular raphe, and lateral surface of the tongue	Median raphe of the pharynx and pharyngeal tubercle on the basilar portion of the occipital bone	Constrict the wall of the pharynx during swallowing	Cranial root of the spinal accessory n. (CN XI) via the pharyngeal branch of the vagus n. and pharyngeal plexus
• Middle constrictor	Stylohyoid ligament and greater and lesser horns of the hyoid bone	Median raphe of the pharynx		Cranial root of the spinal accessory n. (CN XI) via the pharyngeal branch of the vagus n. and pharyngeal plexus plus branches of the external and recurrent laryngeal nn.
• Inferior constrictor	Oblique line of the thyroid cartilage and side of the cricoid cartilage			
Palatopharyngeus	Hard palate and palatine aponeurosis	Posterior lamina of the thyroid cartilage and lateral side of the pharynx and esophagus	Elevate (shorten and widen) the pharynx and elevate the larynx during swallowing and speaking	Cranial root of the spinal accessory n. (CN XI) via the pharyngeal branch of the vagus n. and pharyngeal plexus
Salpingopharyngeus	Cartilaginous part of the auditory tube	Blends with the palatopharyngeus m.		
Stylopharyngeus	Styloid process of the temporal bone	Posterior and superior borders of the thyroid cartilage with the palatopharyngeus m.		Glossopharyngeal n. (CN IX)

Infratemporal Fossa – Overview

■ Identify the following on a dry skull (Figure 21–1A):

● **Infratemporal fossa** – an irregularly shaped space deep and inferior to the zygomatic arch and continuous with the temporal fossa; the fossa is located between these structures:

  ● Lateral – deep surface of the **ramus of the mandible**

  ● Medial – **lateral pterygoid plate of the sphenoid bone**

  ● Anterior – posterior surface of the **maxilla**

  ● Posterior – **temporomandibular joint, carotid sheath,** and **styloid process**

■ During this dissection, you will identify the following contents of the infratemporal fossa:

● **Muscles of mastication**

  ● Temporalis, medial, and lateral pterygoid mm. (and the masseter m., which is superficial to the infratemporal fossa)

● **Branches of the maxillary a.**

  ● Middle meningeal, masseteric, inferior alveolar, deep temporal, posterior superior alveolar, sphenopalatine, and infraorbital aa.

● **Nerves** (Figure 21–1B)

  ● Mandibular, inferior alveolar, lingual, buccal, deep temporal, masseteric, posterior superior alveolar, and the chorda tympani nn., with its related submandibular ganglion

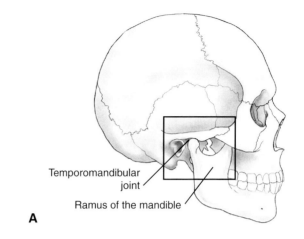

Temporomandibular joint

Ramus of the mandible

A

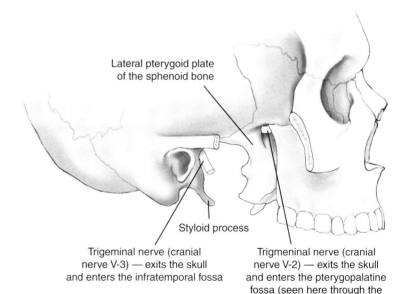

Lateral pterygoid plate of the sphenoid bone

Styloid process

Trigeminal nerve (cranial nerve V-3) — exits the skull and enters the infratemporal fossa

Trigmeninal nerve (cranial nerve V-2) — exits the skull and enters the pterygopalatine fossa (seen here through the pterygomaxillary fissure)

B

■ **Figure 21-1   Infratemporal fossa. *A,* Lateral view. *B,* Lateral view close-up**

Dissection of the Parotid Gland and Masseter Muscle

- Perform this dissection on the half of the head displaying the fewest muscles of facial expression and branches of the facial n.

- To access the infratemporal fossa, you must first remove the following:

  - Parotid gland
  - Masseter m.
  - Zygomatic arch
  - Ramus of the mandible

- To begin this dissection, follow these instructions:

  - **Remove the parotid gland** – cut the parotid duct where it exits the parotid gland and remove the gland (Figure 21–2A)

  - **Cut the masseter m.** – cut along the inferior border of the zygomatic arch and reflect the muscle inferiorly; preserve the facial n.

- Identify the following (Figure 21–2B):

  - **Masseteric n., a., and v.** – after you reflect the masseter m. from the ramus of the mandible, look for the masseteric neurovascular bundle in the mandibular notch

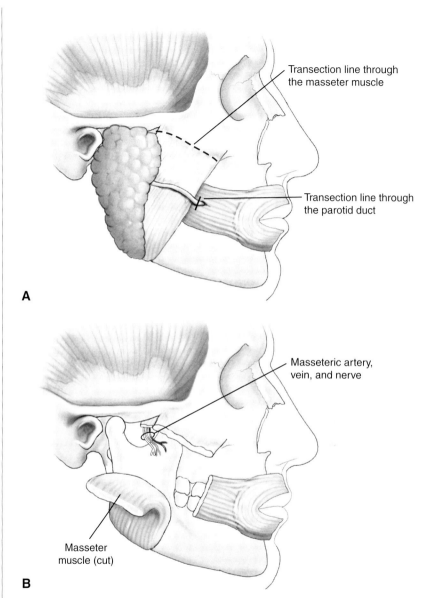

A

B

■ **Figure 21–2   Dissection of the parotid gland and masseter m. *A*, Incisions. *B*, Reflection of the masseter muscle**

Zygomatic Arch and Ramus of the Mandible

- *Before* you make any saw cuts, you may want to push a probe deep to the bone you will cut to avoid cutting underlying structures; follow these instructions:

  - **Zygomatic arch** – saw through the zygomatic arch, using Figure 21-3*A* as a guide; remove the cut section of the arch; if necessary, use a scalpel to cut the attachments of the temporalis m. from the zygomatic arch

  - **Ramus of the mandible** – make the following three cuts; use Figure 21-3*B* as a guide

    - **Neck of the mandibular condyle**

    - **Coronoid process**

    - **Ramus of the mandible** – stay superior to the **lingula** of the mandible (a bony spike at the mandibular foramen); do not cut the inferior alveolar n., a., and v.

  - **Temporalis m.** – can remain attached to the coronoid process of the mandible; reflect the temporalis m. superiorly; use a pin to hold the reflected muscle away from the dissection field

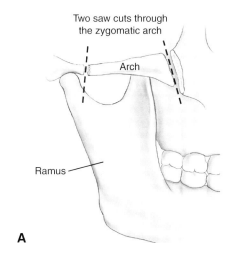

A

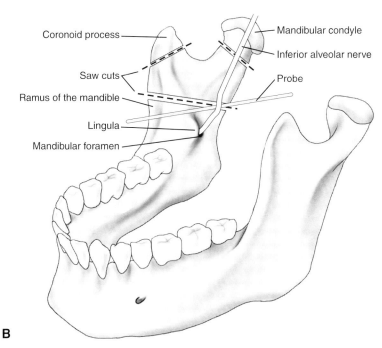

B

■ **Figure 21-3   Instructions to reveal the infratemporal fossa. *A*, Removal of the zygomatic arch. *B*, Removal of the mandibular ramus**

Infratemporal Fossa – Superficial Contents

- Remove fat from the infratemporal fossa; this fat is continuous with the buccal fat pad; as you pick away the fat, note the **pterygoid plexus of vv.**

- Identify the following (Figure 21–4):

  - **Lateral pterygoid m.** – oriented horizontally from the neck of the mandible to the lateral pterygoid plate; has two heads

  - **Medial pterygoid m.** – muscle fibers course parallel to the masseter m. on the internal surface of the ramus of the mandible; deep to the lateral pterygoid m. and oriented obliquely upward from the angle of the mandible to the lateral pterygoid plate

  - **Inferior alveolar n. (CN V-3), a.,** and **v.** – trace distally, where they enter the mandibular foramen

    - Trace the inferior alveolar n. proximally to the inferior border of the lateral pterygoid m.

  - **Sphenomandibular ligament** – courses between the spine of the sphenoid bone to the lingula of the mandible

  - **Lingual n.** (CN V-3) – courses anterior to the inferior alveolar n., then enters the oral cavity between the medial pterygoid m. and the ramus of the mandible

  - **Maxillary a.** – arises from the external carotid a., posterior to the neck of the mandible; crosses superficial or deep to the lateral pterygoid m. en route to the pterygopalatine fossa

    - **Deep temporal nn.** and **aa.** – branches of the maxillary a. that course deep to the temporalis m.

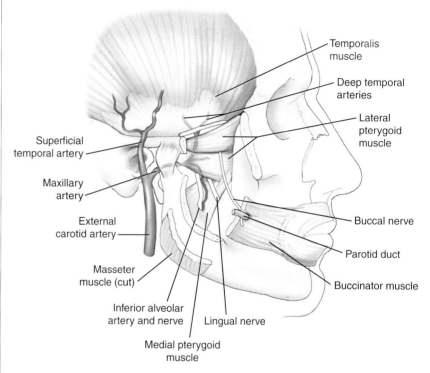

■ **Figure 21–4   Lateral view of the infratemporal fossa for superficial dissection. (Only part of the mandibular ramus is shown removed.)**

Removal of the Lateral Pterygoid Muscle

- To obtain a complete view of the infratemporal region, the lateral pterygoid m. must be removed

- Follow these directions (Figure 21–5):

  - Insert a probe between the lateral and medial pterygoid mm.

    - *Note:* The inferior alveolar and lingual nn. mark the plane of separation between the lateral and medial pterygoid mm.

  - Remove the lateral pterygoid m. completely in a piece-meal fashion to preserve the nerves, arteries, and veins that course around or through the lateral pterygoid m.

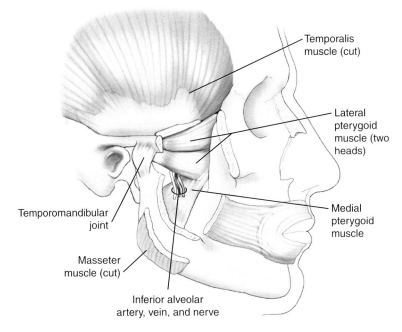

- **Figure 21-5 Lateral view of the removal of the lateral pterygoid muscle. (Only part of the mandibular ramus is shown removed.)**

Infratemporal Fossa – Deep Contents

- Identify the following nerves (Figure 21–6A) and arteries (Figure 21–6B):

  - **Mylohyoid n.** – branch of the inferior alveolar n. to the mylohyoid m.

  - **Chorda tympani n.** – a branch of CN VII; emerges from the **petrotympanic fissure** (medial to the temporomandibular joint); joins the lingual n. from posterior

  - **Buccal n. (CN V-3)** and **buccal a.** – course anteriorly to pierce the buccinator m. (the buccal n. and a. may have been destroyed while removing the buccal fat pad)

  - **Auriculotemporal n. (CN V-3)** – usually splits and rejoins (forms a loop) around the middle meningeal a.

  - **Middle meningeal a.** – branch of the maxillary a.; courses superiorly through the foramen spinosum

  - **Posterior superior alveolar n. (CN V-2)** and **a.** – enters the posterior surface of the maxilla

  - **Sphenopalatine** and **infraorbital aa.** – terminal branches of the maxillary a. (dissected in a future laboratory session)

- Branches of CN V-3 and the maxillary a. that may be destroyed or are too small to identify include **anterior tympanic, deep auricular, muscular branches to the pterygoids,** and **accessory meningeal a.**

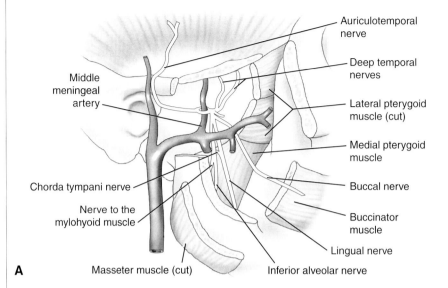

A

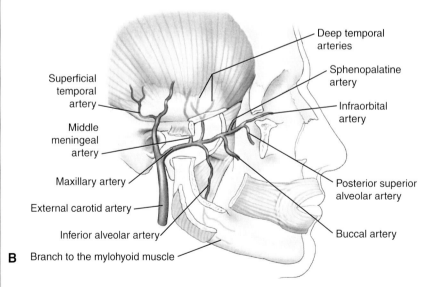

B

- **Figure 21–6   Lateral view of the infratemporal fossa deep dissection. _A_, Nerves. _B_, Arteries**

Temporomandibular Joint

■ Identify the following:

- **Joint capsule** – the fibrous capsule surrounding the temporomandibular joint; attaches to the articular area of the temporal bone and around the neck of the mandible (Figure 21–7A)

  ● Make a horizontal incision through the joint capsule to reveal the internal structures

- **Articular disc** – a fibrocartilaginous disc that divides the joint cavity into superior and inferior compartments; fused with the articular capsule surrounding the joint (Figure 21–7B)

■ Review the following ligaments (see Figure 21–7A):

- **Sphenomandibular ligament** – attaches to the spine of the sphenoid bone and lingula of the mandible

- **Stylomandibular ligament** – attaches to the styloid process of the temporal bone and the internal surface of the mandible

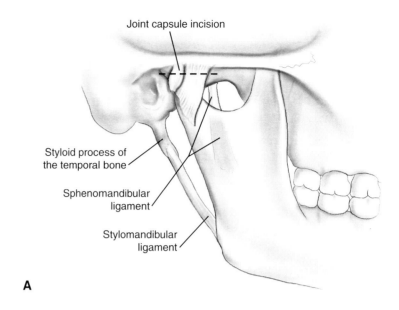

A

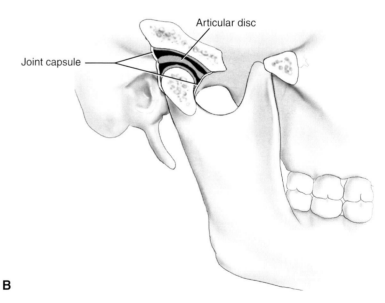

B

■ **Figure 21-7   Lateral views of the temporomandibular joint. _A_, Superficial view. _B_, Deep view. (The mandible is shown intact.)**

Pharynx (Posterior Approach)

■ The three pharyngeal constrictor mm. meet in a median raphe on the posterior surface of the pharynx and receive support from an internal layer of pharyngobasilar fascia

■ Identify the following (Figure 21–8):

- **Superior pharyngeal constrictor m.** – arises from the medial pterygoid plates and attaches to the posterior median raphe; inferiorly, the fibers disappear deep to the middle pharyngeal constrictor m.

- **Middle pharyngeal constrictor m.** – easiest to identify because its fibers arise from the greater horns of the hyoid bone and attach to the posterior median raphe; inferiorly, the fibers disappear deep to the inferior pharyngeal constrictor m.

- **Inferior pharyngeal constrictor m.** – arises from the thyroid and cricoid cartilages, attaches to the posterior median raphe and esophagus; inferiorly, the fibers disappear deep to the musculature of the esophagus

- **Salpingopharyngeus m.** – internal to the constrictors; studied later in this laboratory session

- **Palatopharyngeus m.** – internal to the constrictors; studied later in this laboratory session

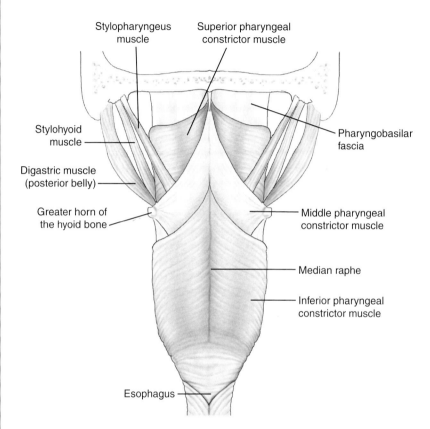

■ **Figure 21–8** **Posterior view of the pharyngeal constrictor muscles**

Pharynx – Posterior and Lateral Aspects

■ Identify the following structures (Figure 21–9A, B):

- **Stylopharyngeus m.** and **glossopharyngeal n.** (CN IX) – pass between the internal and external carotid aa. and enter the pharynx by passing between the superior and middle pharyngeal constrictor mm.

- **Vagus n.** – trace the vagus n. inferiorly, between the internal jugular v. and internal carotid a.

  - **Inferior ganglion of the vagus n.** – a 2-cm long swelling on the vagus n., inferior to the jugular foramen

  - **Internal laryngeal n.** and the **superior laryngeal vessels** – pass between the middle and inferior pharyngeal constrictor mm. to perforate the thyrohyoid membrane to reach the larynx

  - **Recurrent laryngeal n.** – enters the pharyngeal wall at the inferior border of the inferior pharyngeal constrictor m. and ascends to the larynx

- **Cervical sympathetic trunks** – posterior to the prevertebral fascia or on the dissected posterior surface of the pharynx

- **Superior, middle,** and **inferior cervical ganglia**

  - When the inferior cervical ganglion is fused with the superior thoracic ganglion, the fused structure is called the stellate ganglion

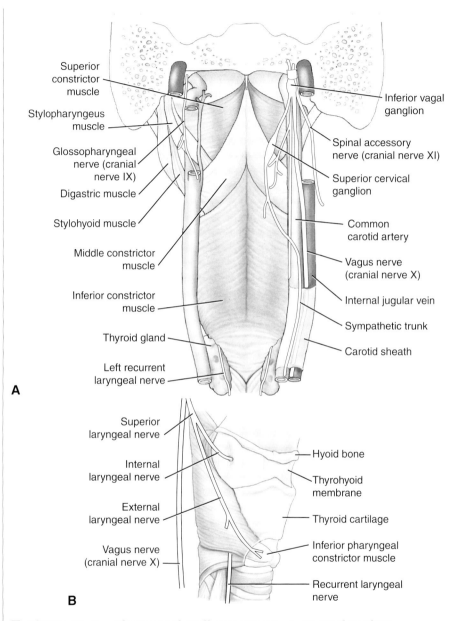

Figure 21–9   Pharyngeal wall structures. *A,* Posterior view. *B,* Lateral view

Nasopharynx

- To study the nasopharynx, choose the side of the bisected head without the nasal septum

- The borders of the nasopharynx (Figure 21–10A):

  - Anterior – choana

  - Roof – mucous membrane attached to the basilar portions of the sphenoid and occipital bones

  - Lateral and posterior walls – superior pharyngeal constrictors

  - Inferior – soft palate

- Identify the following (Figure 21–10B):

  - **Torus tubarius**

  - **Opening of the auditory tube**

  - **Salpingopharyngeal fold**

  - **Nasopharyngeal tonsils (adenoids)**

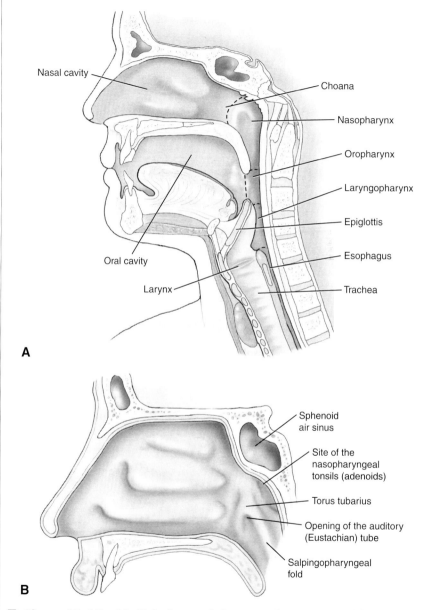

A

B

- **Figure 21–10   Medial views of the nasopharynx. *A*, Midsagittal view. *B*, Sagittal view of the lateral nasal wall**

Oropharynx

■ Identify the following (Figure 21–11):

● **Soft palate** – located along the posterior border of the hard palate; mobile muscular partition projecting between the nasopharynx and oropharynx

  ● **Uvula** – a conical midline projection of the soft palate

  ● **Palatoglossal arch** – the anterior fold coursing between the soft palate and the tongue; covers the palatoglossal m.

  ● **Palatopharyngeal arch** – the posterior fold coursing between the soft palate and the pharynx; covers the palatopharyngeus m.

● **Palatine tonsil** – a lymphoid aggregate located between the palatoglossal and palatopharyngeal arches

● **Palatopharyngeus m.** – arises from the hard palate and attaches to the thyroid cartilage; blends with the other vertical muscles of the pharynx; the muscle and its overlying mucosa form a ridge called the **palatopharyngeal fold**

● **Salpingopharyngeus m.** – arises from the cartilage of the auditory tube and blends with the palatopharyngeus m.; the muscle and its overlying mucosa form a ridge called the **salpingopharyngeal fold**

● **Stylopharyngeus m.** – arises from the styloid process of the temporal bone, passes between the superior and middle pharyngeal constrictor mm. with CN IX, and blends with the palatopharyngeus m.; attaches to the thyroid cartilage

● **Glossopharyngeal n. (CN IX)** – find the nerve by stripping the mucous membrane from the posterior wall of the pharynx immediately posterior to the base of the tongue

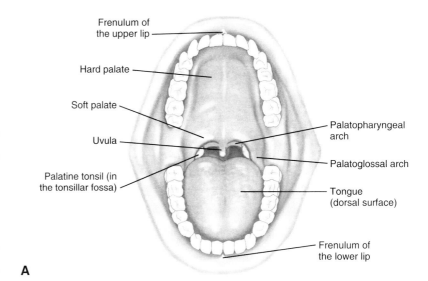

Frenulum of the upper lip

Hard palate

Soft palate

Uvula

Palatine tonsil (in the tonsillar fossa)

Palatopharyngeal arch

Palatoglossal arch

Tongue (dorsal surface)

Frenulum of the lower lip

A

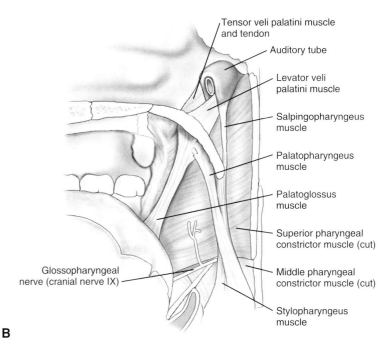

Tensor veli palatini muscle and tendon

Auditory tube

Levator veli palatini muscle

Salpingopharyngeus muscle

Palatopharyngeus muscle

Palatoglossus muscle

Superior pharyngeal constrictor muscle (cut)

Glossopharyngeal nerve (cranial nerve IX)

Middle pharyngeal constrictor muscle (cut)

Stylopharyngeus muscle

B

■ **Figure 21-11   Pharyngeal arches. *A*, Anterior view. *B*, Medial view**

Nasal Cavity, Palate, and Oral Cavity

Prior to dissection, you should familiarize yourself with the following structures:

OSTEOLOGY

- Ethmoid bone
  - Ethmoid labyrinth of air cells
  - Perpendicular plate
  - Superior nasal concha (superior meatus)
  - Middle nasal concha (middle meatus)
  - Ethmoidal bulla
  - Sphenoethmoidal recess
- Inferior nasal concha (inferior meatus)
- Palatine bone
  - Greater and lesser palatine foramina
  - Palatine canal
- Maxilla
  - Palatine process of the maxilla
  - Incisive canal
  - Maxillary sinus
  - Infraorbital foramen and canal
  - Posterior superior alveolar foramen
- Sphenoid bone
  - Sphenoid sinus
  - Pterygomaxillary fissure
  - Sphenopalatine foramen
  - Pterygoid canal
  - Pterygoid hamulus

- Zygomatic bone
  - Zygomaticofacial foramen
  - Zygomaticotemporal foramen
- Mandible
  - Mylohyoid line
  - Sublingual fossa
- Temporal bone
  - Greater and lesser petrosal foramina/grooves

STRUCTURES ASSOCIATED WITH THE NASAL CAVITY

- Nasal septum
  - Perpendicular plate of the ethmoid bone
  - Vomer bone
  - Septal cartilage
- Vestibule
- Atrium
- Nasolacrimal duct opening
- Hiatus semilunaris

NERVES

- Trigeminal n. (CN V)
  - V-2 – Maxillary division

- Zygomatic n.
    - Zygomaticotemporal n.
    - Zygomaticofacial n.
  - Pterygopalatine ganglion
  - Greater and lesser palatine nn.
  - Nasopalatine n.
  - Superior, middle, and posterior lateral alveolar nn.
  - Infraorbital n.
  - V-3 – Mandibular division
    - Lingual n.
    - Nerve to the mylohyoid m.
- Facial n. (CN VII)
  - Nerve of the pterygoid canal (Vidian n.)
- Glossopharyngeal n. (CN IX)
- Hypoglossal n. (CN XII)

ARTERIES

- External carotid a.
  - Maxillary a.
    - Lingual a.
    - Sphenopalatine a.
      - Posterior lateral nasal aa. (superior, middle, inferior)

Continued

PALATE

- Palatal muscles
  - Levator veli palatini m.
  - Tensor veli palatini m.
  - Uvula
- Palatine aponeurosis
- Soft palate
- Palatoglossal and palatopharyngeal arches
- Palatine tonsil

TONGUE

- Tongue muscles
  - Genioglossus m.
  - Styloglossus m.
  - Hyoglossus m.
- Vallate (circumvallate) papillae (taste buds)
- Filiform/fungiform papillae (taste buds)
- Sulcus terminalis
- Median and lateral glossoepiglottic folds
- Vallecula
- Foramen cecum
- Lingual tonsils

MISCELLANEOUS

- Pterygomandibular raphe
- Auditory tube
- Pharyngobasilar fascia
- Plica sublingualis
- Sublingual gland
- Submandibular salivary gland and duct
- Submandibular ganglion
- Geniohyoid m.
- Mylohyoid m.
- Digastric m.
- Superior, middle, and inferior pharyngeal constrictor mm.
- Stylopharyngeal m.

TABLE 22–1 Muscles of the Tongue

Muscle	Attachments		Action	Innervation
Genioglossus	Superior aspect of the genial tubercle of the mandible	Dorsal surface of the tongue and body of the hyoid bone	Depresses the tongue; its posterior aspect protracts the tongue	Hypoglossal n. (CN XII)
Hyoglossus	Body and greater horn of the hyoid bone	Lateral and inferior surface of the tongue	Depresses and retracts the tongue	
Styloglossus	Styloid process of the temporal bone and stylohyoid ligament		Retracts the tongue; draws the tongue up to create a trough for swallowing	
Palatoglossus	Palatine aponeurosis of the soft palate	Lateral side of the tongue	Elevates the posterior part of the tongue	Cranial root of the spinal accessory n. (CN XI) via the pharyngeal branch of the vagus n. (CN X) and pharyngeal plexus

TABLE 22–2 Muscles of the Soft Palate

Muscle	Attachments		Action	Innervation
Tensor veli palatini	Scaphoid fossa of the medial pterygoid plate, spine of the sphenoid bone and cartilage of the auditory tube	Palatine aponeurosis	Tenses the soft palate and opens the mouth of the auditory tube during swallowing and yawning	CN V-3
Levator veli palatini	Cartilage of the auditory tube and petrous part of the temporal bone		Elevates the soft palate during swallowing and yawning	Cranial part of the spinal accessory n. (CN XI) through the pharyngeal branch of the vagus n. (CN X) via the pharyngeal plexus
Palatoglossus	Palatine aponeurosis and the soft palate	Side of the tongue	Elevates the posterior aspect of the tongue and draws the soft palate onto the tongue	
Palatopharyngeus	Hard palate and palatine aponeurosis	Lateral wall of the pharynx	Tenses the soft palate and pulls walls of the pharynx superiorly, anteriorly, and medially during swallowing	

Paranasal Sinuses - Overview

■ The paranasal sinuses are air-filled extensions of the nasal cavity located in the following bones (Figure 22–1A):

- Frontal
- Ethmoid
- Sphenoid
- Maxilla

■ Paranasal sinuses are easily identified on radiographs (Figure 22–1B):

- **Frontal sinuses**
- **Ethmoidal labyrinth of air cells**
- **Sphenoid sinus**
- **Maxillary sinuses**

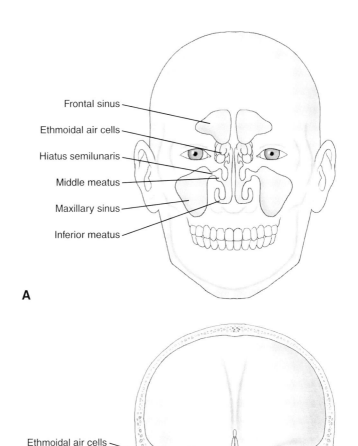

A

B

■ **Figure 22–1** **Paranasal sinuses.** *A,* **Anterior view.** *B,* **Coronal sectional view**

Nasal Cavity

- **Boundaries** – the two sides of the nasal cavity are separated by the nasal septum

- Examine the half of the bisected head containing the nasal septum to identify the following (Figure 22–2):

  - **Choana** – posterior opening into the nasopharynx; each nasal cavity begins at each nostril and ends posteriorly at the choana
  - **Nasal septum** – composed of three structures:
    - **Perpendicular plate of the ethmoid bone**
    - **Vomer bone**
    - **Septal cartilage**

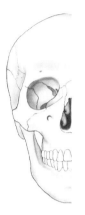

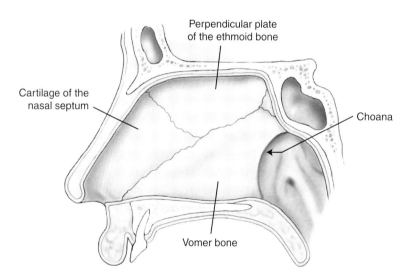

Perpendicular plate of the ethmoid bone

Cartilage of the nasal septum

Choana

Vomer bone

■ **Figure 22–2   Medial view of the nasal septum**

Nasal Septum

■ Identify the following (Figure 22–3):

- **Nasopalatine n.** – accompanied by vessels that have a different name **(sphenopalatine a. and v.)**

  - Look for septal branches in the mucous membrane covering the nasal septum

- **Incisive canal** – anterior portion of the hard palate, just posterior to the incisors; follow the nasopalatine n. and sphenopalatine a. through the incisive canal, where they reach the hard palate

- **Palatine branches of the nasopalatine n. and sphenopalatine a.** – peel the mucoperiosteum from the anterior portion of the hard palate to locate these structures

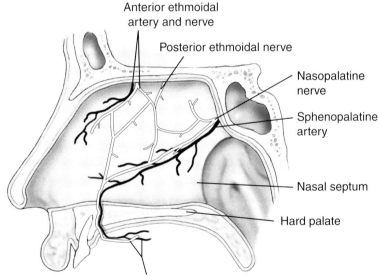

Anterior ethmoidal artery and nerve

Posterior ethmoidal nerve

Nasopalatine nerve

Sphenopalatine artery

Nasal septum

Hard palate

Palatine branches of the nasopalatine nerve and sphenopalatine artery

■ **Figure 22–3   Medial view of the nerves and vessels of the nasal septum**

Lateral Nasal Wall

- Identify the following on the lateral wall of the nasal cavity (Figure 22–4):

  - **Inferior concha** – a separate bone (*not* part of the ethmoid bone); ends 1 cm anterior to the auditory tube

    - **Inferior meatus** – the space inferior to the inferior concha

  - **Middle concha** – part of the ethmoid bone

    - **Middle meatus** – the space inferior to the middle concha

  - **Superior concha** – also part of the ethmoid bone; anterior to the sphenoid sinus

    - **Superior meatus** – the space inferior to the superior concha

  - **Sphenoethmoidal recess** – space posterosuperior to the superior concha

  - **Vestibule** – located superior to the nostril and anterior to the inferior meatus; note the presence of hairs (vibrissae)

  - **Atrium** – located superior to the vestibule and anterior to the middle meatus

  - **Opening of the nasolacrimal duct** (tear duct) – use scissors to cut away the inferior concha to identify the duct in the inferior meatus (Figure 22–5)

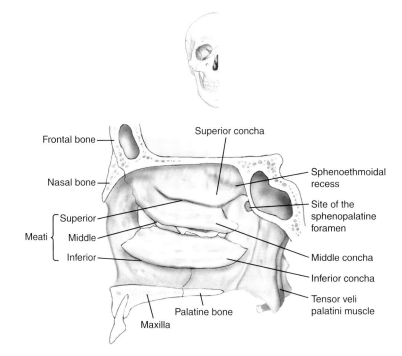

■ **Figure 22–4   View of the lateral nasal wall**

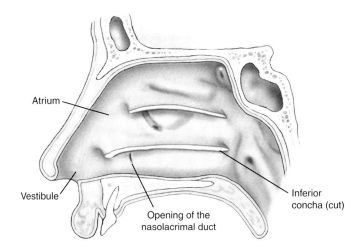

■ **Figure 22–5   Opening of the nasolacrimal duct in the lateral nasal wall**

Lateral Nasal Wall—cont'd

■ Use scissors to cut away the middle concha

■ Identify the following (Figure 22–6):

- **Hiatus semilunaris** – a curved slit inferior to the middle concha (middle meatus) that drains the maxillary air sinus

  - Note an opening (infundibulum) at the anterior margin of the hiatus semilunaris; the frontal air sinus drains through the infundibulum (middle meatus)

  - Uncinate process – forms the medial edge of the hiatus semilunaris and helps form the medial wall of the maxillary sinus

- **Ethmoidal bulla** – an elevation over the hiatus semilunaris

- **Anterior** and **middle ethmoidal air cells** – openings are located on the superior aspect of the ethmoid bulla

- **Posterior ethmoidal air cells** – communicate with the superior meatus

- **Sphenoid air sinus** – look for its opening into the **sphenoethmoidal recess** (space superior to the superior concha)

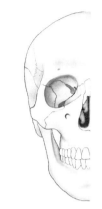

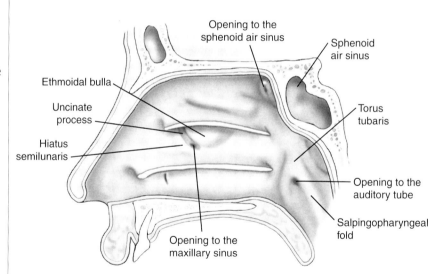

■ **Figure 22–6   View of the lateral nasal wall**

Vasculature of the Lateral Nasal Cavity

■ Strip the mucoperiosteum from the lateral wall of the nasal cavity to identify the following arteries (Figure 22–7):

- **Sphenopalatine a.** – originates from the maxillary a.; enters the nasal cavity via the sphenopalatine foramen (accompanied by the posterior superior nasal nn.); bifurcates into lateral and septal (medial) posterior nasal aa.

  - **Posterior lateral nasal aa.** – originate from the sphenopalatine a.; divide into superior, middle, and inferior branches, giving each concha an arterial branch

  - Uncinate process – forms the medial edge of the hiatus semilunaris and helps form the medial wall of the maxillary sinus

- **Anterior ethmoidal a.** – originates from the ophthalmic a. and enters the nasal cavity via the anterior ethmoidal foramen; anastomoses with the posterior lateral nasal branches of the sphenopalatine a.

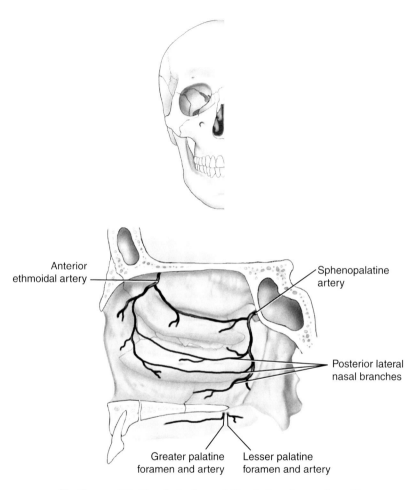

Anterior ethmoidal artery

Sphenopalatine artery

Posterior lateral nasal branches

Greater palatine foramen and artery

Lesser palatine foramen and artery

■ **Figure 22–7   Arteries of the lateral nasal wall**

Contents of the Palatine Canal

- Strip the remaining mucoperiosteum from the posterolateral wall of the nasal cavity

- Identify the following (Figure 22–8):

  - **Palatine canal** – a vertical channel in the posterior wall of the nasal cavity that courses the height of the lateral nasal wall; its wall is opalescent

    - Use a probe or scissors to break through the opalescent wall; the underlying channel is called the palatine canal, which contains the following, all of which are in a sleeve of periosteum (use a probe or tip of scissors to open the periosteal sleeve):

      - ◆ **Pterygopalatine ganglion** (parasympathetic) – located at the superior end of the palatine canal

      - ◆ **Greater** and **lesser palatine nn.** – course through the palatine canal inferiorly from the pterygopalatine ganglion; follow both palatine nn. through the palatine foramina to enter the mucosal lining on the inferior surface of the hard palate of the maxilla and palatine bone; note the accompanying vessels, which are named the same as the nerves

- Dissect the mucoperiosteum from the oral part of the hard palate to identify the following:

  - **Greater palatine n., a., and v.** – course in an anterior direction; branches meet with the palatine branches of the nasopalatine n. and sphenopalatine a. and v.

  - **Lesser palatine n., a., and v.** – course in a posterior direction over the soft palate

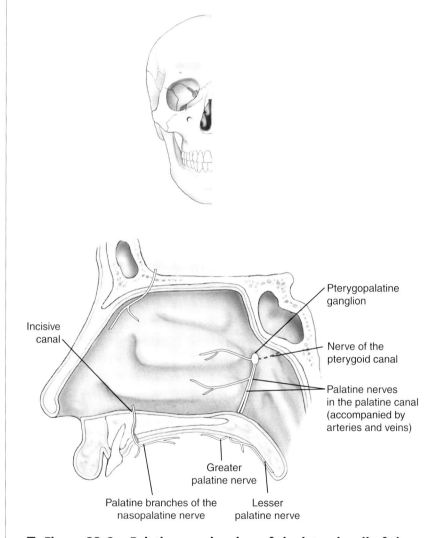

■ **Figure 22–8   Palatine canal – view of the lateral wall of the nasal cavity**

Greater Petrosal Nerve and Nerve to the Pterygoid Canal

■ Identify the following (Figure 22–9A):

● **Greater petrosal n.** – emerges from the greater petrosal hiatus; the greater petrosal n. courses anteriorly in a groove toward the foramen lacerum, where the nerve passes through the foramen before it reaches the pterygoid canal

● **N. of the pterygoid canal (Vidian n.)** – formed by union of the greater petrosal n. (parasympathetic) and the deep petrosal n. (sympathetic nerves that follow the internal carotid a.); courses through the pterygoid canal to enter the pterygopalatine ganglion

  ● Look for the n. of the pterygoid canal in the floor of the sphenoid sinus by breaking away the bone covering the nerve (Figure 22–9B)

  ● A helpful approach is to follow the n. of the pterygoid canal posteriorly from the pterygopalatine ganglion by pulling the pterygopalatine ganglion anteriorly to locate the nerve

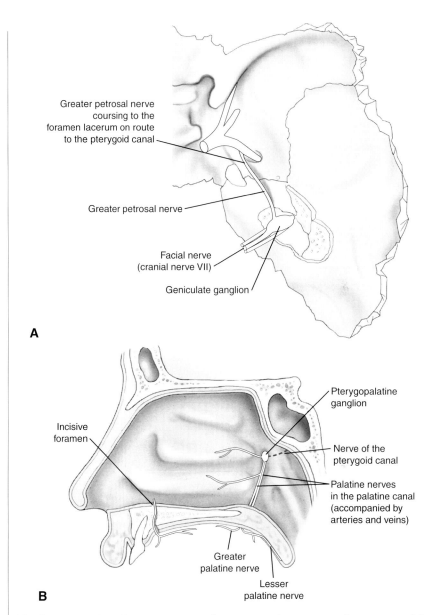

A

B

■ **Figure 22–9   Greater petrosal nerve and nerve to the pterygoid canal. *A*, Superior view of the temporal bone with the tegmen tympani portion of the temporal bone removed. *B*, View of the lateral nasal wall**

Dissection of the Infraorbital Canal

■ Identify the following:

- **Maxillary sinuses** – paired air sinuses on the lateral sides of the nasal cavity; to identify, remove mucoperiosteum and push a probe through the thin bone on the lateral wall of the nasal cavity (do this on the side with the removed conchae)

- **Infraorbital canal** – a channel that courses from posterior to anterior along the roof of the opened maxillary sinus (Figure 22–10A)

- **Infraorbital n., a., and v.** – course through the infraorbital canal; break open the roof of the opened maxillary sinus with a probe to expose the infraorbital canal; follow the infraorbital n. posteriorly to the foramen rotundum

    - **Anterior and middle superior alveolar n., a., and v.** – branches of the infraorbital n., a., and v. (Figure 22–10B); they supply the corresponding maxillary teeth; examine the floor of the maxillary air sinus to locate the roots of teeth that may project into the sinus

- **Posterior superior alveolar n., a., and v.** – course along the lateral and posterior surface of the maxillary sinus to enter the posterior superior alveolar foramen in the infratemporal fossa

■ The pharyngeal and zygomatico-orbital branches of CN V-2 will not be explored

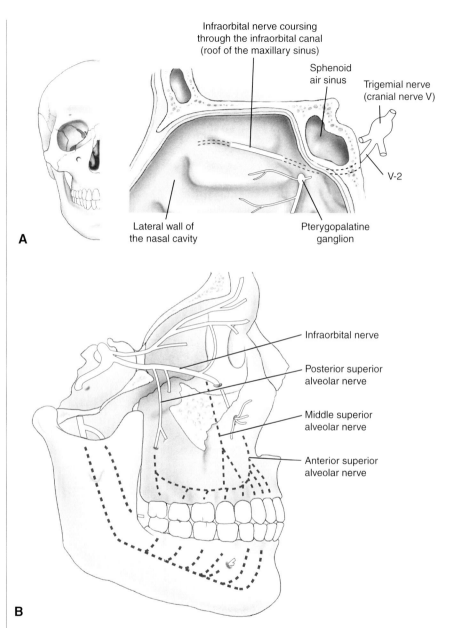

A, Infraorbital nerve coursing through the infraorbital canal (roof of the maxillary sinus)

Sphenoid air sinus

Trigemial nerve (cranial nerve V)

V-2

Lateral wall of the nasal cavity

Pterygopalatine ganglion

B, Infraorbital nerve

Posterior superior alveolar nerve

Middle superior alveolar nerve

Anterior superior alveolar nerve

■ **Figure 22–10   CN V-2. *A*, View of the lateral nasal wall. *B*, View of the maxilla from the infratemporal fossa**

Palatal Structures

- Identify the following (Figure 22–11A):

  - **Palatine tonsils** (if present; usually atrophied in the aged) – located in the triangular (tonsillar) fossa, between the palatoglossal and palatopharyngeal arches

  - **Palatoglossal arch** – extends from the tip of the uvula to the base of the tongue

  - **Palatopharyngeal arch** – posterior to the palatoglossal arch; extends from the tip of the uvula posterolaterally to the pharynx

- Identify the following muscles by removing the mucous membrane covering them (Figure 22–11B):

  - **Levator veli palatini m.** – located between the floor of the auditory tube and the superior aspect of the soft palate

  - **Tensor veli palatini m.** – descends vertically; anterior and lateral to the levator veli palatini m.; the tensor veli palatini m. changes direction (90°) at the **pterygoid hamulus,** becoming the palatine aponeurosis in the soft palate

    - **Palatine aponeurosis** – stabilizing foundation of the soft palate for palatine muscles to act upon

  - **Salpingopharyngeus m.** – courses from the auditory tube to the pharynx

  - **Palatopharyngeus m.** – courses from the soft palate to the pharynx

  - **Palatoglossus m.** – courses from the soft palate to the tongue

  - **Superior pharyngeal constrictor m.** – arises from the pterygoid hamulus, pterygomandibular raphe, and mylohyoid line of the mandible (posterior to the third molar); also attaches to the pharyngeal tubercle

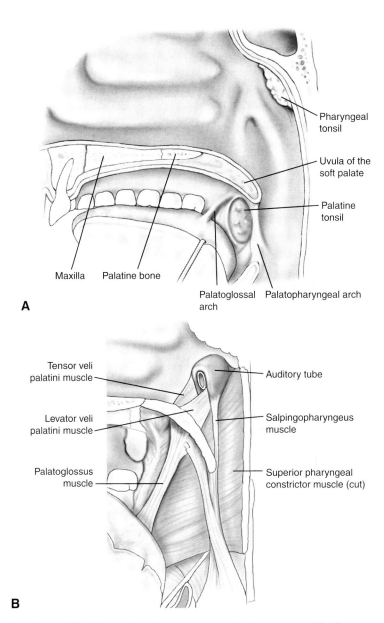

A

B

- **Figure 22–11   Palatal structures. A, Pharynx with the mucous membrane intact. B, Pharynx with the mucous membrane removed**

Oral Cavity – Tongue

■ Identify the following muscles (Figure 22–12):

• **Genioglossus m.** – superior to the geniohyoid m.;
  largest extrinsic muscle of the tongue

• **Geniohyoid m.** – superior to the mylohyoid m.

• **Mylohyoid m.** – forms the floor of the oral cavity

• **Intrinsic muscles of the tongue** – four groups: superior and inferior longitudinal, transverse, and vertical;
  the three directions in which these muscles course
  allow them to alter the shape and contour of the
  tongue

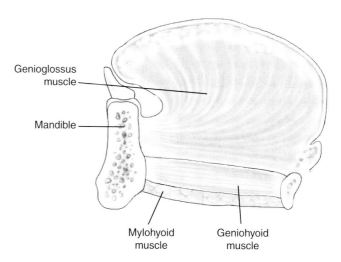

Genioglossus
muscle

Mandible

Mylohyoid
muscle

Geniohyoid
muscle

■ **Figure 22–12   Medial view of the extrinsic
tongue muscles**

Oral Cavity – Tongue—cont'd

■ To dissect deeper structures within the oral cavity, follow these instructions (Figure 22–13A):

- Incise the mucous membrane between the tongue and mandible

  • Start at the lingual frenulum and carry the incision posteriorly, but not beyond the mandibular molar teeth

  • Carefully remove the mucous membrane in the depression between the mandibular molar teeth and the tongue

- **Sublingual gland** – located superior to the anterior part of the mylohyoid m.; the gland is separated from the genioglossus m. by the lingual n. and submandibular duct (Figure 22–13B)

- **Submandibular duct** – located in the deep part of the gland; courses diagonally across the medial aspect of the sublingual gland to the side of the lingual frenulum; the duct opens posterior to the mandibular front teeth

- **Lingual n.** – located posterior to the third molar of the mandible; courses between the ramus of the mandible and medial pterygoid m.; trace it anteriorly to observe it spiraled around the submandibular duct

  • **Submandibular ganglion** – hangs from the lingual n. in the vicinity of the third molar tooth of the mandible

  • **Chorda tympani n.** – joins the lingual n. in the infratemporal fossa

- **Hypoglossal n.** – courses anteriorly between the submandibular gland and the hyoglossus m., inferior to the lingual n.; follow the hypoglossal n. to the musculature of the tongue

- **Lingual a.** – courses medial to the hyoglossus m.; divides into the **deep lingual a.** and **sublingual a.**

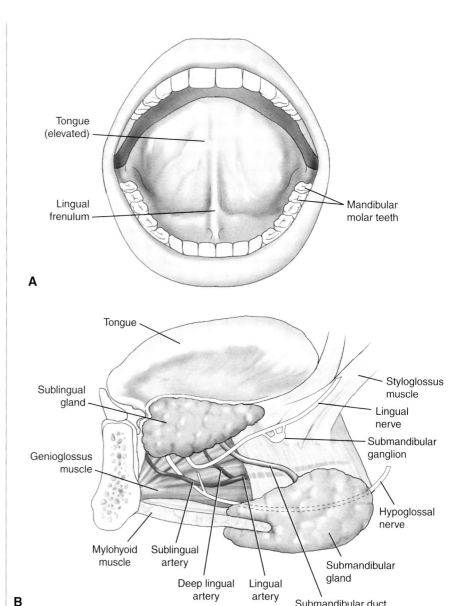

■ **Figure 22–13   Oral cavity. *A,* Incisions. *B,* Nerves and vessels of the tongue; salivary glands**

Muscles of the Tongue

■ Identify the following extrinsic muscles of the tongue (Figure 22–14):

- **Genioglossus m.** – arises from the internal surface of the mental symphysis; forms the bulk of the tongue

- **Hyoglossus m.** – attached to the hyoid bone and tongue; look for the **styloglossus m.,** which interdigitates with the posterior fibers of the hyoglossus m.

- **Styloglossus m.** – follow the muscle posterior to the styloid process of the temporal bone

- **Palatoglossus m.** – previously discussed

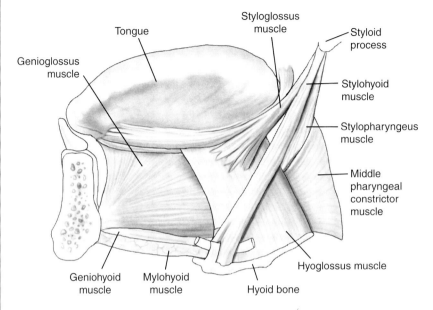

■ **Figure 22–14   Extrinsic muscles of the tongue**

Tongue

■ Identify the following (Figure 22–15):

- **Oral part of the tongue** – anterior two-thirds; taste sensation is carried by the chorda tympani n. (CN VII)

  - **Vallate papillae** – the largest of the taste buds; located anterior to the sulcus terminalis

  - **Sulcus terminalis** – separates the oral and pharyngeal parts of the tongue

  - **Filiform/fungiform papillae** – small elevations that give the tongue its rough surface

- **Pharyngeal part of the tongue** – posterior one-third; taste sensation carried by the glossopharyngeal n. (CN IX)

  - **Medial** and **lateral glossoepiglottic folds**

  - **Vallecula** – depressions between the glossoepiglottic folds

  - **Foramen cecum** – remnant of the opening of the embryonic thyroglossal duct; the thyroid gland in the embryo was attached to the tongue by this duct, which normally disappears, leaving only this small pit in the tongue

  - **Lingual tonsils** – cause the surface elevations of the posterior one-third of the tongue

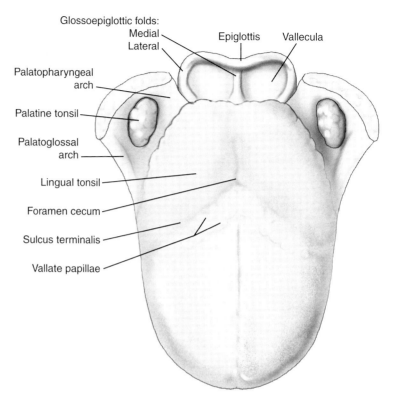

■ **Figure 22–15 View of the superior surface of the tongue**

Review of the Osteology of the Pterygopalatine Fossa

■ With a dry skull and broom straws, probe the various foramina radiating from the pterygopalatine fossa (Figure 22–16A and B):

- **Infraorbital fissure**
- **Infraorbital foramen**
- **Zygomaticofacial foramen**
- **Zygomaticotemporal foramen**
- **Posterior superior alveolar foramina**
- **Pterygomaxillary fissure**
  - Deep to this fissure is the pterygopalatine fossa

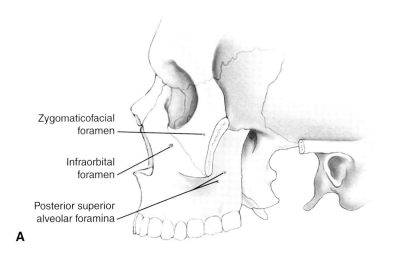

A

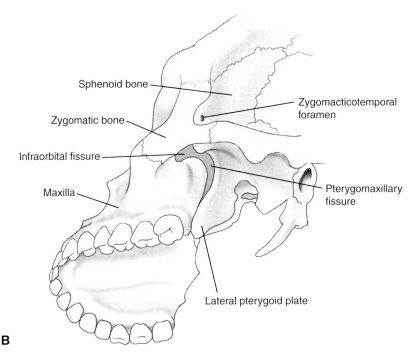

B

■ **Figure 22–16   Foramina of the skull for branches of CN V-2. A, Lateral view. B, Skull tilted to show an inferior view**

Pterygopalatine Fossa

■ Identify the following on a dry skull (Figure 22–17):

- **Pterygopalatine fossa** – a small pyramidal space inferior to the orbital apex; the fossa has the following borders:
    - Posterior – pterygoid plates of the sphenoid bone
    - Anterior – maxilla
    - Medial – perpendicular plate of the palatine bone
    - Lateral – open to the infratemporal fossa (pterygomaxillary fissure)

- The pterygopalatine fossa communicates with the following (refer to Table 22–3a):
    - Lateral communication – infratemporal fossa via the pterygomaxillary fissure
    - Medial communication – nasal cavity via the sphenopalatine foramen
    - Anterior communication – orbit via the medial end of the inferior orbital fissure
    - Posterior communication – the foramen rotundum is located in the posterior wall, the latter is traversed by the maxillary n. (CN V-2) (refer to Table 22–3b)

- Contents of the pterygopalatine fossa:
    - Maxillary n. (CN V-2)
    - Pterygopalatine ganglion
    - Terminal branches of the maxillary a.

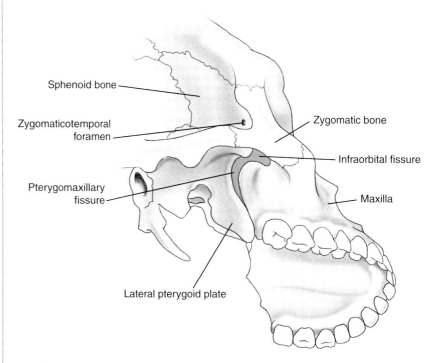

■ **Figure 22–17   Pterygopalatine fossa. Right side of the skull tilted to show an inferior view**

Pterygopalatine Fossa and CN V-2
• During this dissection, you will find the following branches of CN V-2 (and the associated vessels) in the nasal and oral cavities:

a. Schematic of the pterygopalatine fossa

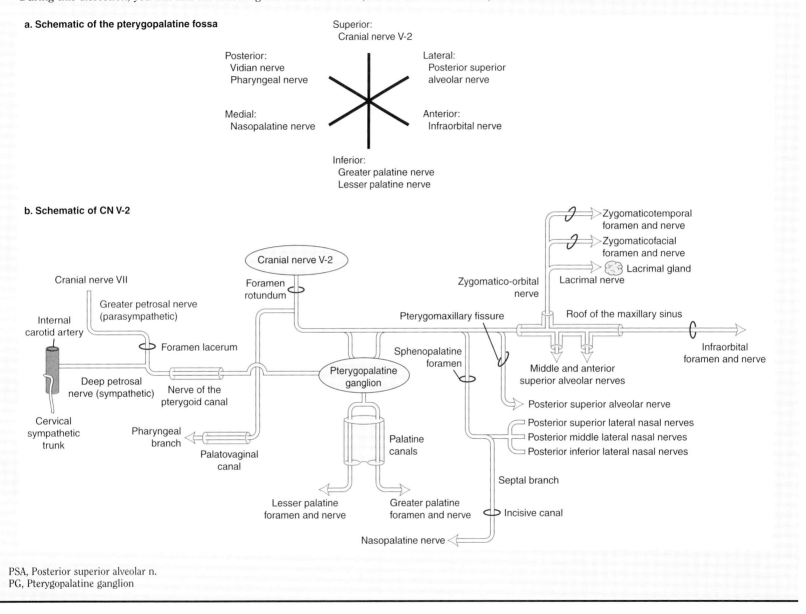

Superior:
  Cranial nerve V-2

Posterior:
  Vidian nerve
  Pharyngeal nerve

Lateral:
  Posterior superior
  alveolar nerve

Medial:
  Nasopalatine nerve

Anterior:
  Infraorbital nerve

Inferior:
  Greater palatine nerve
  Lesser palatine nerve

b. Schematic of CN V-2

Cranial nerve V-2

Cranial nerve VII

Greater petrosal nerve
(parasympathetic)

Internal
carotid artery

Foramen
rotundum

Zygomaticotemporal
foramen and nerve

Zygomaticofacial
foramen and nerve

Lacrimal gland

Zygomatico-orbital   Lacrimal nerve
nerve

Pterygomaxillary fissure   Roof of the maxillary sinus

Foramen lacerum

Deep petrosal
nerve (sympathetic)

Cervical
sympathetic
trunk

Nerve of the
pterygoid canal

Pharyngeal
branch

Palatovaginal
canal

Pterygopalatine
ganglion

Sphenopalatine
foramen

Palatine
canals

Lesser palatine
foramen and nerve

Greater palatine
foramen and nerve

Nasopalatine nerve

Septal branch

Incisive canal

Middle and anterior
superior alveolar nerves

Posterior superior alveolar nerve

Posterior superior lateral nasal nerves
Posterior middle lateral nasal nerves
Posterior inferior lateral nasal nerves

Infraorbital
foramen and nerve

PSA, Posterior superior alveolar n.
PG, Pterygopalatine ganglion

Larynx and Middle Ear

Lab 23

Prior to dissection, you should familiarize yourself with the following structures:

LARYNGEAL CARTILAGES AND JOINTS

- Thyroid cartilage
  - Laryngeal prominence
  - Right and left laminae
  - Superior and inferior thyroid notches
  - Oblique line
  - Superior and inferior horns
  - Thyrohyoid membrane
- Cricoid cartilage
- Cricothyroid joint
  - Median cricothyroid ligament
- Arytenoid cartilages
- Cricoarytenoid joint
  - Cricoarytenoid ligament
- Corniculate cartilages
- Cuneiform cartilages
- Epiglottic cartilage

LARYNGEAL MUSCLES

- Cricothyroid m.
- Posterior cricoarytenoid m.
- Lateral cricoarytenoid m.
- Vocalis m.
- Thyroarytenoid m.
- Transverse and oblique arytenoid mm.
- Aryepiglotticus m.
- Thyroepiglottic m.
- Thyrohyoid m.

LARYNGEAL NERVES

- Superior laryngeal n.
  - Internal laryngeal n.
  - External laryngeal n.
- Recurrent laryngeal n.

LARYNGEAL CAVITY

- Laryngeal inlet
  - Aryepiglottic fold
    - Corniculate tubercle
    - Cuneiform tubercle
- Laryngeal vestibule
  - Vestibular folds (false vocal cords)
- Laryngeal ventricle
  - Laryngeal saccule
- Glottis
  - Vocal folds (true vocal cords)
    - Rima glottidis
- Fibroelastic membrane of the larynx
  - Vocal ligament

MISCELLANEOUS

- Laryngopharynx
  - Piriform fossa/recess
    - Fold over the superior laryngeal n.

EAR

- External ear
  - Auricle; pinna
    - Lobule of the auricle
    - Auricular cartilage
    - Helix
    - Antihelix
      - Triangular fossa
    - Scaphoid fossa
    - Concha of the auricle
    - Tragus
    - Antitragus
    - External auditory canal
    - Tympanic membrane

Continued

- Middle ear (tympanic cavity)
  - Tegmen tympani
    - Epitympanic recess
  - Aditus to the mastoid antrum
  - Mastoid antrum
    - Mastoid air cells
  - Ear ossicles
    - Malleus (hammer)
    - Incus (anvil)
    - Stapes (stirrup)
  - Muscles of the auditory tube
    - Stapedius m.
    - Tensor tympani m.
  - Chorda tympani n. (CN VII)
  - Mastoid air cells
  - Auditory (pharyngotympanic) tube
    - Bony and cartilaginous parts
- Inner ear
  - Vestibule
  - Semicircular canals
  - Cochlea
  - Internal acoustic meatus
  - Vestibulocochlear n. (CN VIII)

TABLE 23-1   Muscles of the Larynx

Muscle	Attachments		Action	Innervation
Cricothyroid	Anterolateral part of the cricoid cartilage	Inferior region of the thyroid cartilage	Stretches and tenses the vocal fold	External laryngeal n. (CN X)
Posterior cricoarytenoid	Posterior surface of the cricoid cartilage		Abducts the vocal fold	Recurrent laryngeal n. (CN X)
Lateral cricoarytenoid	Arch of cricoid cartilage	Muscular process of the arytenoid cartilages	Adducts the vocal fold (interligamentous region)	
Thyroarytenoid	Posterior surface of the thyroid cartilage		Relaxes the vocal fold	
Transverse and oblique arytenoids	The muscles course between the two arytenoid cartilages		Close the intercartilaginous region of the rima glottidis	
Vocalis	Vocal processes of the arytenoid cartilages	Vocal ligaments	Relaxes the posterior part of the vocal ligament while maintaining (or increasing) tension of the anterior part	

External Surface of the Larynx – Review

■ Identify the following (Figure 23–1):

- **Thyrohyoid mm.** – lateral to the larynx; paired muscles that attach the inferior border of the hyoid bone to the lateral surface of the thyroid cartilage

- **Cricothyroid mm.** – anterior and lateral to the larynx; paired muscles that attach the cricoid cartilage to the inferior border and inferior horns of the thyroid cartilage

- **Thyrohyoid membrane** – connective tissue sheet that spans between the superior border of the thyroid cartilage and the inferior border of the hyoid bone

- **Cricothyroid membrane** – fan-shaped connective tissue sheet that spans between the superior border of the cricoid cartilage and inferior border of the thyroid cartilage

- **Vagus n.** (CN X) – descends between the carotid a. and internal jugular v. and divides into the following branches to the larynx:

  - **Internal laryngeal n.** – descends and pierces the thyrohyoid membrane

  - **External laryngeal n.** – descends and pierces the cricothyroid m.

- **Recurrent laryngeal n.** – ascends to reach the posteromedial aspect of the thyroid gland (inferior pole); ascends in the tracheoesophageal groove to supply the intrinsic laryngeal mm.

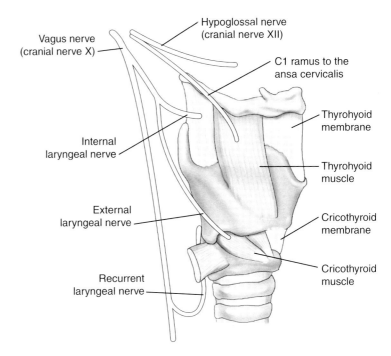

■ **Figure 23–1   Lateral view of the laryngeal muscles and membranes**

External Surface of the Larynx – Skeleton

- The larynx has a cartilaginous skeleton

- Identify the following structures (Figure 23–2):

  - **Thyroid cartilage** – largest of the cartilages

    - **Laryngeal prominence** – at the anterior median plane; union of the two laminae of the thyroid cartilage ("Adam's apple")

    - **Superior and inferior horns** – superior and inferior projections from the posterior surfaces of the thyroid cartilage

  - **Cricoid cartilage** – inferior to the thyroid cartilage; shaped like a "signet ring," with the band facing anteriorly; the only complete ring of the respiratory tract

  - **Arytenoid cartilages** – paired, pyramidal-shaped cartilages located on the superior border of the body of the cricoid cartilage (posterior surface of the larynx)

  - **Corniculate cartilages** – small cartilages attached to the apex of the arytenoid cartilages

  - **Epiglottic cartilage** – a leaf-shaped cartilage that extends superiorly from the thyroid cartilage, posterior to the hyoid bone and the base of the tongue; the base of the epiglottic cartilage is attached to the inside of the thyroid cartilage at the midline by the thyroepiglottic ligament

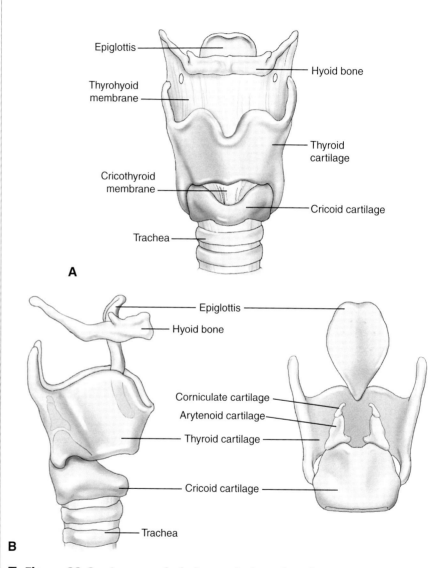

- **Figure 23–2   Laryngeal skeleton. *A*, Anterior view. *B*, Lateral and posterior views**

Internal Features of the Larynx

■ Inspect the interior of the larynx from its superior aspect (as in looking down a tube) to identify the following (Figure 23–3):

- **Glottis** – refers to the vocal folds, the rima glottis, and the space superior to the vocal folds

  - **Rima glottis** – the aperture between the vocal folds

  - **Vocal folds** – true vocal cords; these folds produce sound; the apex of each wedge-shaped vocal fold projects posteromedially; its base lies against the lamina of the thyroid cartilage

  - **Vocal ligament** – thickened edges of the cricothyroid ligament complex; forms the fibrous core of the true vocal fold

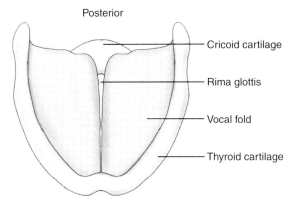

■ **Figure 23–3   Superior view down into the larynx**

Laryngopharynx

- The laryngopharynx is posterior to the larynx, extending from the superior border of the epiglottis to the inferior border of the cricoid cartilage

- View the posterior aspect of the larynx to identify the following (Figure 23–4):

  - **Tongue**

  - **Epiglottis** – a leaf-shaped plate of elastic cartilage covered with mucous membrane

  - **Greater horn of the hyoid bone** – projects postero-superiorly; the tip may be seen through the mucous membrane on the posterior aspect of the pharynx

  - **Thyroid cartilage** – superior horn; projects posterosuperiorly and may be seen through the mucous membrane on the posterior aspect of the pharynx

  - **Cricoid cartilage** – prominence may be seen through the mucous membrane

  - **Piriform fossa** – located in the posterior part of the laryngopharynx, lateral to the mucous membrane that covers the thyroid and cricoid cartilages

  - **Recurrent** and **superior laryngeal nn.** – strip the mucosa from the laryngopharyngeal aspect of the larynx; look for the recurrent and superior laryngeal nn. in the inferior and superior portions of the piriform recess, respectively

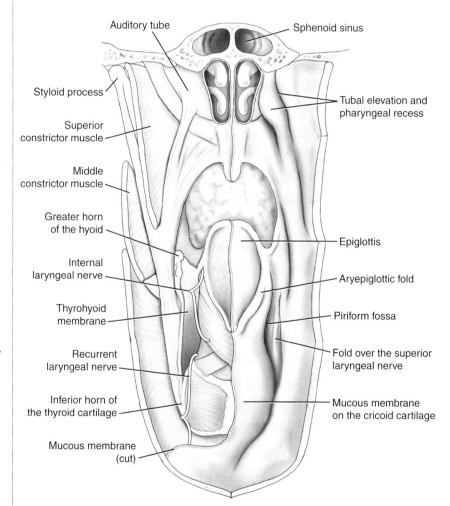

Auditory tube

Sphenoid sinus

Styloid process

Tubal elevation and pharyngeal recess

Superior constrictor muscle

Middle constrictor muscle

Greater horn of the hyoid

Epiglottis

Internal laryngeal nerve

Aryepiglottic fold

Thyrohyoid membrane

Piriform fossa

Recurrent laryngeal nerve

Fold over the superior laryngeal nerve

Inferior horn of the thyroid cartilage

Mucous membrane on the cricoid cartilage

Mucous membrane (cut)

■ Figure 23–4   **Posterior view of the pharynx and larynx**

Larynx – Intrinsic Muscles: Posterior View

- Cut and peel the mucous membrane from the posterior surface of the larynx

- There is no fascial separation between the muscles on the internal wall of the larynx; the muscles are named by their attachments

- Identify the following (Figure 23–5):

  - **Posterior cricoarytenoid mm.** – posterior to the larynx; paired muscles that attach the posterior lamina of the cricoid cartilage to the muscular processes of the arytenoid cartilages

  - **Transverse arytenoid mm.** – posterior to the larynx; paired muscles that attach horizontally to both arytenoid cartilages

  - **Oblique arytenoid mm.** – posterior to the larynx; paired muscles that course diagonally and attach to both arytenoid cartilages; the muscle fibers continue superiorly to the epiglottis **(aryepiglotticus mm.)**

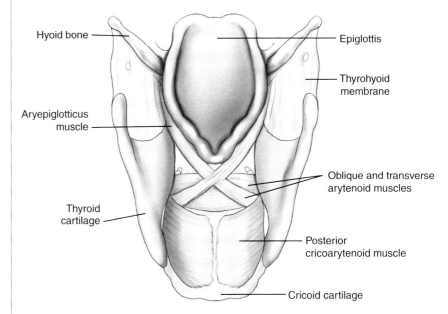

■ **Figure 23–5   Posterior view of the laryngeal muscles**

Internal Features of the Larynx

■ Make a longitudinal incision through the posterior aspect of the cricoid cartilage (Figure 23–6A) to identify the following (Figure 23–6B):

- **Vestibular folds** – false vocal cords; superior to the vocal folds (true vocal cords)

- **Vestibule** – the space superior to the vestibular folds

- **Ventricle** – the space inferior to the vestibular folds; may extend laterally and superiorly, forming a recess called the **saccule**

- **Vocal folds** – true vocal cords; inferior to the ventricle; more closely opposed than the false vocal folds

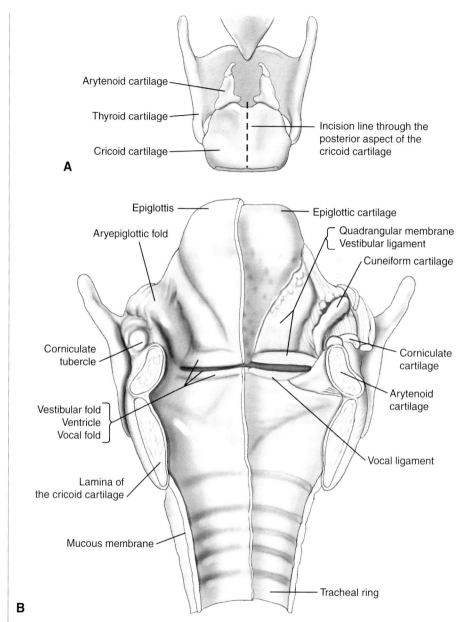

■ **Figure 23–6   Posterior view of the internal features of the larynx. A, Cricoid cartilage incision. B, Internal view of the larynx**

Larynx Intrinsic Muscles

■ Identify the following:

- **Cricothyroid joint** (Figure 23–7A) – lateral sides of the larynx; synovial articulation of the cricoid and thyroid cartilages

- Cut through the cricothyroid joint on one side of the larynx and separate the two cartilages by reflecting the thyroid cartilage anteriorly to identify the following two muscles:

  - **Lateral cricoarytenoid mm.** – lateral to the larynx; paired muscles that course from the superior border of the cricothyroid ligament to the muscular processes of the arytenoid cartilages (Figure 23–7B)

  - **Thyroarytenoid mm.** – lateral to the larynx; paired muscles that are superior to the lateral cricoarytenoid mm.; course from the thyroid cartilage anteriorly to the arytenoid cartilages posteriorly; their superior and most medial fibers are the **vocalis m.**

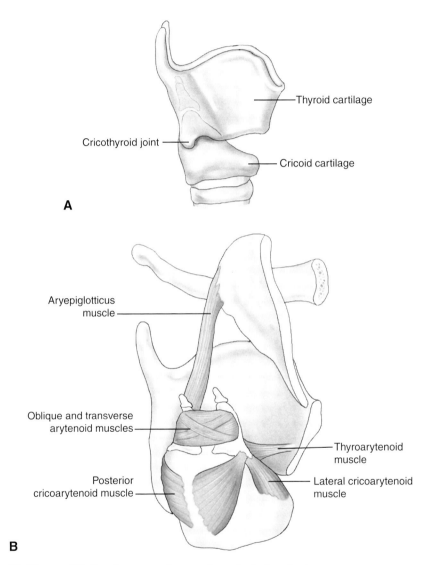

A

B

■ **Figure 23–7 Laryngeal muscles.** *A,* **Lateral view.**
B, **Posterolateral view; right side of thyroid cartilage is removed**

Ear Dissection

- Check with your instructor prior to performing this dissection

- You will identify the following during this dissection (Figure 23–8):

  - **Pinna**
  - **External ear**
    - **External auditory canal**
  - **Middle ear**
    - **Tympanic membrane**
    - **Ear ossicles**
      - ◆ **Malleus**
      - ◆ **Incus**
      - ◆ **Stapes**
    - **Stapedius m.**
    - **Tensor tympani m.**
    - **Chorda tympani n. (CN VII)**
    - **Mastoid air cells**
    - **Auditory tube**
  - **Inner ear**
    - **Cochlea**
    - **Semicircular canals**
    - **Vestibule**
    - **Vestibulocochlear n. (CN VIII)**

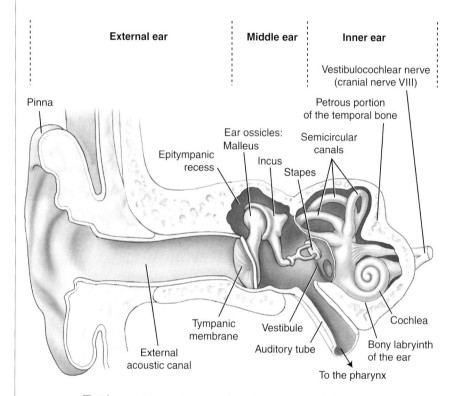

■ **Figure 23–8   Coronal section through the ear**

External Ear

- The external (outer) ear extends from the auricle (pinna) medially, via the external auditory canal, to the tympanic membrane

- Identify the following (Figure 23–9):

  - **Auricle** – the shell-like part of the external ear; consists of elastic cartilage, skin, hairs, sweat glands, and sebaceous glands; acts as a collection funnel for sound waves

  - **Helix**

  - **Scaphoid fossa**

  - **Antihelix**

  - **Tragus**

  - **Antitragus**

  - **Lobule (ear lobe)** – consists of fibrous tissue, fat, and blood vessels

  - **Concha** – central depression of the ear leading to the external acoustic meatus

  - **Opening of the external acoustic canal** – the canal extends from the concha of the auricle to the tympanic membrane; about 2.5 cm long in the adult

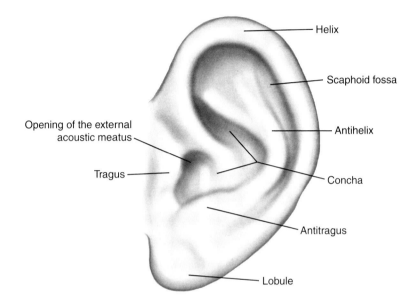

Helix

Scaphoid fossa

Opening of the external acoustic meatus

Antihelix

Tragus

Concha

Antitragus

Lobule

■ **Figure 23-9   External ear**

Middle Ear Dissection – Removal of the Tegmen Tympani

■ The middle ear is an air-filled chamber sandwiched between the tympanic membrane (lateral) and the oval and round windows (medial) of the vestibulocochlear apparatus

  ● Dissection of the middle ear is performed to demonstrate the basic spatial relationships of the external ear, middle ear, internal ear, auditory tube, and mastoid air cells

■ To observe the contents of the middle ear you will remove the **tegmen tympani** (bony roof of the middle ear [Figure 23–10]):

  ● Use a mallet and chisel to break the **tegmen tympani** from the petrous portion of the temporal bone about:

    ● 1.5 cm from the squamous part of the temporal bone

    ● 2.5 cm from the posterior ridge of the petrous part of the temporal bone

  ● Remove the bony fragments from the tegmen tympani to reveal the air-filled chamber of the middle ear

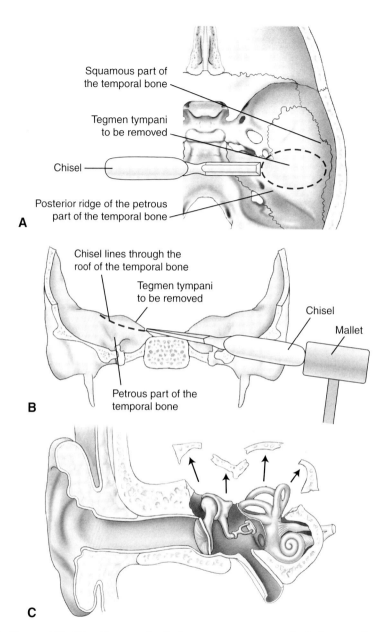

■ **Figure 23–10** **Removal of the tegmen tympani.** *A,* **Internal view of the skull base.** *B,* **Coronal section in an anterior view.** **C, Tegmen tympani of the temporal bone removed**

Middle Ear Dissection - Ossicles

■ Identify the following (Figure 23–11A, B):

- **Epitympanic recess** – the region between the ear ossi-cles and the tegmen tympani

- **Ossicles** – these three tiny bones form a chain across the middle ear (tympanic cavity), from the tympanic membrane to the oval widow; look for smooth, condy-lar surfaces in the exposed compartment; these struc-tures are the malleus and incus bones

  - **Malleus** – attached to the tympanic membrane; gently insert a probe into the external auditory canal to gently push against the tympanic membrane; because the malleus is attached to the tympanic membrane, and all of the ear ossicles are attached together, you will see the ossicles move when you gently push on and release the tympanic membrane

  - **Incus** – located between and attached to both the malleus and the stapes

  - **Stapes** – attached to the oval window; you will prob-ably not see this ossicle

- **Tensor tympani m.** – look for the tensor tympani m.'s tendon passing anterior to the head of the malleus

- **Mastoid antrum** – the large cavity that continues into the mastoid air cells inferiorly; note the superior com-munication between the mastoid air cells and the middle ear

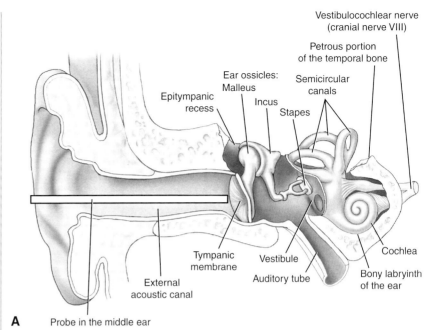

A

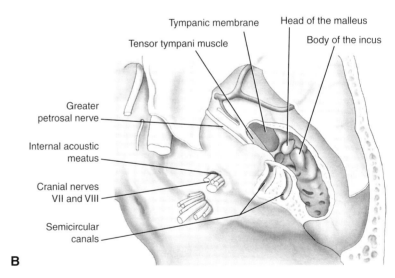

B

■ **Figure 23–11   Middle ear dissection. *A*, Anterior view of a coronal section through the ear. *B*, Internal view of the skull base**

Middle Ear Dissection – Geniculate Ganglion

- To identify the geniculate ganglion, follow these instructions (Figure 23–12A):

  - Push a probe into the internal acoustic meatus to use as a guide
  - Use a mallet and chisel to break the bony roof above the probe
  - Remove the bone fragments to identify the facial n.

- Identify the following (Figure 23–12B):

  - **Geniculate ganglion of the facial n.** – the facial n. enters the internal acoustic meatus and travels laterally through the petrous portion of the temporal bone into the cavity of the middle ear

    - Within the middle ear, the facial n. turns sharply posterior, forming a knee-shaped bend (genu)
    - Here lies the geniculate (sensory) ganglion of CN VII

  - **Greater petrosal n.** – if the nerve is still intact, follow it posteriorly to the geniculate ganglion

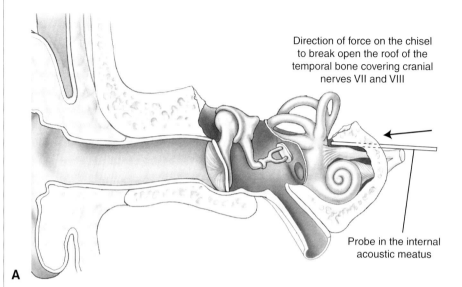

Direction of force on the chisel to break open the roof of the temporal bone covering cranial nerves VII and VIII

Probe in the internal acoustic meatus

A

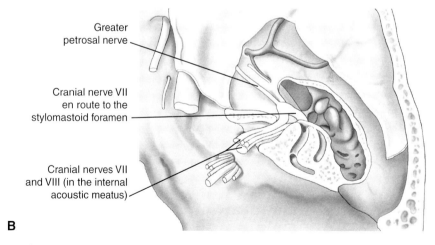

Greater petrosal nerve

Cranial nerve VII en route to the stylomastoid foramen

Cranial nerves VII and VIII (in the internal acoustic meatus)

B

■ **Figure 23–12   Dissection of CN VII and the geniculate ganglion.**
A, **Anterior view of a coronal section through the middle ear.**
B, **Internal view of the skull base, showing the course of CN VII**

Middle Ear Dissection – Auditory Tube

- To demonstrate the communication between the naso-pharynx, middle ear, and mastoid air cells, follow these instructions:

  - Starting from the nasopharynx, push a probe into the auditory tube (Figure 23–13A) as far as you can to reveal the course of the auditory tube (Figure 23–13B)

  - Using a mallet and chisel, extend the removal of the tegmen tympani anteriorly, until you see the tip of the probe

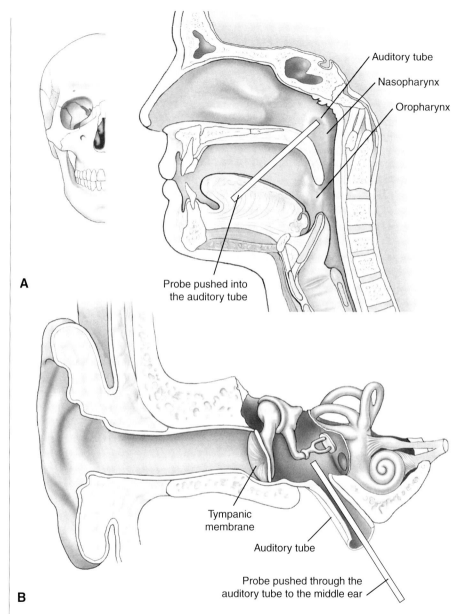

Auditory tube

Nasopharynx

Oropharynx

A

Probe pushed into the auditory tube

Tympanic membrane

Auditory tube

B

Probe pushed through the auditory tube to the middle ear

- **Figure 23-13   Dissection of the auditory tube. _A_, Medial view of a sagittal section through the head. _B_, Anterior view of a coronal section through the ear**

Unit **3** Neck and Head Overview

At the end of Unit 3, you should be able to identify the following structures on cadavers, skeletons, and/or radiographs:

Osteology

- ■ Parietal bone
  - ● Groove for the superior sagittal sinus
  - ● Grooves for the middle meningeal a.
- ■ Frontal bone
  - ● Glabella
  - ● Supraorbital margin
    - ▪ Supraorbital foramen
  - ● Frontal sinus
  - ● Foramen cecum
- ■ Occipital bone
  - ● Foramen magnum
  - ● Basioccipital region
    - ▪ Clivus
    - ▪ Pharyngeal tubercle
    - ▪ Grooves for the inferior petrosal sinuses
  - ● Occipital condyle
  - ● Hypoglossal canal
  - ● Condylar canal
  - ● Jugular notch (for the jugular foramen)
  - ● External occipital protuberance
  - ● Internal occipital crest
  - ● Internal occipital protuberance
  - ● Groove for the superior sagittal sinus
  - ● Grooves for the transverse sinuses
  - ● Groove for the occipital sinus

- ■ Temporal bone
  - ● Petrous part
    - ▪ Mastoid process
    - ▪ Mastoid notch
    - ▪ Grooves for the sigmoid sinuses
    - ▪ Mastoid foramen
    - ▪ Facial canal
    - ▪ Carotid canal
    - ▪ Canal for the auditory tube
    - ▪ Tegmen tympani
    - ▪ Hiatus and groove for the greater petrosal n.
    - ▪ Internal acoustic meatus
    - ▪ Grooves for the inferior petrosal sinuses
    - ▪ Jugular notch (for the jugular foramen)
    - ▪ Styloid process
    - ▪ Stylomastoid foramen
- ■ External acoustic meatus
- ■ Groove for the middle meningeal a.
- ■ Zygomatic process
- ■ Mandibular fossa
- ■ Petrotympanic fissure
- ■ Foramen lacerum
- ■ Sphenoid bone
  - ● Sella turcica

  - ▪ Hypophysial fossa
  - ▪ Dorsum sellae
  - ▪ Posterior clinoid processes
  - ● Carotid sulcus
  - ● Sphenoid sinus
  - ● Lesser wing
    - ▪ Optic canal
    - ▪ Anterior clinoid process
    - ▪ Supraorbital fissure
  - ● Greater wing
    - ▪ Foramen rotundum
    - ▪ Foramen ovale
    - ▪ Foramen spinosum
  - ● Pterygoid process
    - ▪ Lateral plate
    - ▪ Medial plate
  - ● Pterygoid notch and fossa
  - ● Pterygoid hamulus
  - ● Pterygoid canal
  - ● Sphenopalatine foramen
  - ● Inferior orbital fissure
  - ● Foramen lacerum
  - ● Sphenoethmoidal recess
- ■ Ethmoid bone
  - ● Cribriform plate
    - ▪ Cribriform foramina
    - ▪ Crista galli
    - ▪ Perpendicular plate

- Ethmoidal air cells
- Anterior and posterior ethmoid foramina
  - Superior nasal concha
    - Superior meatus
  - Middle nasal concha
    - Middle meatus
  - Ethmoidal bulla
  - Uncinate process
  - Hiatus semilunaris
  - Sphenoethmoidal recess
  - Atrium
- Inferior nasal concha
  - Inferior meatus
- Lacrimal bone
- Nasal bone
- Vomer
- Zygomatic bone
  - Temporal process
  - Zygomaticofacial foramen
  - Zygomaticotemporal foramen
- Palatine bone
  - Perpendicular plate
    - Sphenopalatine notch
    - Greater and lesser palatine canals
  - Horizontal plate
    - Greater palatine foramen
    - Lesser palatine foramen
- Maxilla
  - Infraorbital foramen and canal
  - Alveolar foramen
  - Maxillary sinus
  - Incisive canal and foramen
  - Pterygomaxillary fissure
  - Anterior nasal spine
  - Palatine process
  - Greater palatine groove
  - Greater palatine foramen

- Mandible
  - Ramus, angle, neck, and body
  - Mental protuberance
  - Mental foramen
  - Oblique line
  - Digastric fossa
  - Superior and inferior mental spines
  - Mylohyoid line
  - Sublingual fossa
  - Mandibular foramen
    - Lingula
    - Mandibular canal
  - Coronoid and condylar processes
  - Mandibular notch
- Hyoid bone
  - Lesser horn
  - Greater horn
- Cranial fossae
  - Anterior
  - Middle
  - Posterior
- Sutures and landmarks
  - Sagittal
  - Coronal
  - Pterion
  - Lambdoidal
  - Squamous
  - Bregma/lambda
- Auditory ossicles
  - Malleus, incus, stapes
- Vertebral column
  - C1 (atlas)
  - C2 (axis)
    - Odontoid process (dens)
  - Transverse foramina
- Sternum
  - Manubrium
    - Jugular notch

- Clavicle
- Scapula
  - Acromion
  - Superior angle of the scapula

Joints
- Atlanto-occipital joint
- Atlantoaxial joint
- Temporomandibular joint (TMJ)
- Cricothyroid joint

Ligaments
- Cruciate ligament
  - Superior and inferior longitudinal bands
  - Transverse ligament of atlas (C1)
- Alar ligaments
- Sphenomandibular ligament
- Stylomandibular ligament

Muscles
- Trapezius m.
- Splenius capitis m.
- Levator scapulae m.
- Muscles of the neck
  - Platysma m.
  - Longus colli m.
  - Longus capitis m.
  - Scalene mm.
    - Anterior, middle, and posterior
  - Sternocleidomastoid m.
  - Suboccipital muscles
    - Rectus capitis lateralis m.
  - Suprahyoid muscles
    - Digastric m.
      - Anterior belly
      - Posterior belly
    - Stylohyoid m.
    - Mylohyoid m.

- Geniohyoid m.
- Infrahyoid muscles
  - Sternohyoid m.
  - Omohyoid m.
    - ◆ Superior belly
    - ◆ Inferior belly
  - Sternothyroid m.
  - Thyrohyoid m.
- Prevertebral mm.
  - Longus colli m.
  - Rectus capitis anterior m.
  - Longus capitis m.
  - Longus coli m.

- **Muscles of mastication**
  - Masseter m.
  - Temporalis m.
  - Lateral pterygoid m.
  - Medial pterygoid m.

- **Facial muscles**
  - Occipitofrontalis
  - Procerus m.
  - Nasalis m.
  - Orbicularis oculi m.
    - Palpebral part
    - Orbital part
  - Corrugator supercilii m.
  - Auricularis mm.
  - Orbicularis oris m.
  - Depressor anguli oris m.
  - Risorius m.
  - Zygomaticus major m.
  - Zygomaticus minor m.
  - Levator labii superioris m.
  - Levator labii superioris alaeque nasi m.
  - Depressor labii inferioris m.
  - Levator anguli oris m.
  - Buccinator m.
  - Mentalis m.

- **Extraocular muscles and associated structures**
  - Levator palpebrae superioris m.
  - Superior and inferior oblique mm.
  - Medial, lateral, superior, and inferior rectus mm.
  - Common tendinous ring
  - Trochlea

- **Muscles of the soft palate**
  - Levator veli palatini m.
  - Tensor veli palatini m.
  - Musculus uvulae m.
  - Palatoglossus m.
  - Palatopharyngeus m.
  - Structures associated with the palate
    - Uvula
    - Palatoglossal arch
    - Palatopharyngeal arch
    - Palatine aponeurosis

- **Pharyngeal muscles and related structures**
  - Superior, middle, and inferior pharyngeal constrictor mm.
  - Stylopharyngeus m.
  - Salpingopharyngeus m.
  - Palatopharyngeus m.
  - Buccopharyngeal fascia
  - Pharyngeal raphe
  - Pterygomandibular raphe

- **Muscles of the tongue**
  - Genioglossal m.
  - Hyoglossus m.
  - Styloglossus m.
  - Superior and inferior longitudinal and oblique mm.
  - Palatoglossus m.

- **Intrinsic muscles of the larynx**
  - Cricothyroid m.
  - Posterior cricoarytenoid m.
  - Lateral cricoarytenoid m.
  - Vocalis m.
  - Thyroarytenoid m.
  - Oblique arytenoid m.
  - Transverse arytenoid m.

Nerves

- **12 pairs of cranial nerves (CN)**
  - Olfactory n. (CN I)
  - Optic n. (CN II)
  - Oculomotor n. (CN III)
    - Branch to the ciliary ganglion (parasympathetic root of the ciliary ganglion)
    - Long (sensory) and short (parasympathetic) ciliary nn.
  - Trochlear n. (CN IV)
  - Trigeminal n. (CN V)
    - Motor root
    - Sensory root
      - ◆ Trigeminal ganglion
    - V-1 – Ophthalmic division of the trigeminal n.
      - ◆ Lacrimal n.
      - ◆ Frontal n.
        - ◇ Supraorbital n.
        - ◇ Supratrochlear n.
      - ◆ Nasociliary n.
      - ◆ External nasal n.
      - ◆ Anterior and posterior ethmoidal nn.
      - ◆ Infratrochlear n.
    - V-2 – Maxillary division of the trigeminal n.
      - ◆ Pterygopalatine ganglion
      - ◆ Nasopalatine n.
      - ◆ Greater and lesser palatine nn.
      - ◆ Superior, middle, and posterior superior alveolar n.
      - ◆ Zygomatic n.

- ◆ Zygomaticotemporal n.
- ◆ Zygomaticofacial n.
- ◆ Infraorbital n.
- ■ V-3 – Mandibular division of the trigeminal n.
  - ● Meningeal n.
  - ● Nerves to the masseter and medial and lateral pterygoid mm.
  - ● Deep temporal nn.
  - ● Buccal n.
  - ● Auriculotemporal n.
    - ▪ Superficial temporal nn.
  - ● Lingual n.
  - ● Inferior alveolar n.
    - ▪ Nerve to the mylohyoid m.
    - ▪ Inferior dental plexus
- ■ Abducens n. (CN VI)
- ■ Facial n. (CN VII)
  - ● Posterior auricular n.
  - ● Temporal branch(es)
  - ● Zygomatic branch(es)
  - ● Buccal branch(es)
  - ● Mandibular branch(es)
  - ● Cervical branches
  - ● Intermediate n.
    - ▪ Geniculate ganglion
    - ▪ Greater petrosal n. (parasympathetic root to the pterygopalatine ganglion)
      - ◆ Joins with the deep petrosal (sympathetic) nerve to become the nerve of the pterygoid canal (Vidian n.)
    - ▪ Chorda tympani n. (carries the parasympathetic root to the submandibular ganglion)
- ■ Vestibulocochlear n. (CN XIII)
- ■ Glossopharyngeal n. (CN IX)
  - ● Carotid body
  - ● Lesser petrosal n. (parasympathetic root

- to the otic ganglion)
- ■ Vagus n. (CN X)
  - ● Superior ganglion of the vagus n.
  - ● Inferior ganglion of the vagus n.
  - ● Pharyngeal n.
  - ● Superior laryngeal n.
    - ▪ Internal laryngeal n.
    - ▪ External laryngeal n.
  - ● Right and left recurrent laryngeal nn.
- ■ Spinal accessory n. (CN XI)
- ■ Hypoglossal n. (CN XII)
- ■ Cervical plexus
  - ● Ansa cervicalis
    - ▪ Superior and inferior roots
  - ● Lesser occipital n.
  - ● Great auricular n.
  - ● Transverse cervical n.
  - ● Supraclavicular nn.
  - ● Phrenic n.
- ■ Brachial plexus
  - ● Suprascapular n.
- ■ Autonomics
  - ● Sympathetic
    - ▪ Superior cervical sympathetic ganglion
      - ◆ Deep petrosal n. (sympathetic root)
    - ▪ Middle cervical sympathetic ganglion
    - ▪ Inferior cervical sympathetic ganglion
  - ● Parasympathetic
    - ▪ Oculomotor n. (CN III)
    - ▪ Facial n. (CN VII)
    - ▪ Glossopharyngeal n. (CN IX)
    - ▪ Vagus n. (CN X)

Arteries

- ■ Common carotid a.
  - ● Carotid body

- ■ Common carotid a.
  - ● External carotid a.
    - ▪ Superior thyroid a.
      - ◆ Superior laryngeal a.
    - ▪ Ascending pharyngeal a.
    - ▪ Lingual a.
      - ◆ Sublingual a.
      - ◆ Deep lingual a.
    - ▪ Facial a.
      - ◆ Submental a.
      - ◆ Superior labial a.
      - ◆ Inferior labial a.
      - ◆ Angular a.
    - ▪ Occipital a.
    - ▪ Posterior auricular a.
    - ▪ Superficial temporal a.
      - ◆ Transverse facial a.
    - ▪ Maxillary a.
      - ◆ Inferior alveolar a.
      - ◆ Middle meningeal a.
      - ◆ Masseteric a.
      - ◆ Deep temporal aa.
      - ◆ Buccal a.
        - ◆ Posterior superior alveolar a.
      - ◆ Infraorbital a.
        - ◆ Anterior and middle superior alveolar aa.
      - ◆ Descending palatine a.
        - ◆ Greater and lesser palatine aa.
      - ◆ Sphenopalatine a.
        - ◆ Posterior lateral nasal aa.
        - ◆ Posterior septal aa.
- ■ Common carotid a.
  - ● Internal carotid a.
    - ▪ Carotid sinus
    - ▪ Ophthalmic a.
      - ◆ Central retinal a.
      - ◆ Lacrimal a.
      - ◆ Ciliary aa. (long and short)
      - ◆ Supraorbital a.

- ◆ Anterior and posterior ethmoidal aa.
- ◆ Supratrochlear a.
- Subclavian a.
  - ▪ Thyrocervical trunk
    - ◆ Transverse cervical a.
    - ◆ Suprascapular a.
    - ◆ Inferior thyroid a.
  - ▪ Vertebral a.
  - ▪ Costocervical trunk
    - ◆ Supreme intercostal a.
    - ◆ Deep cervical a.
- Cerebral arterial circle (of Willis)
  - ▪ Internal carotid a.
    - ◆ Anterior cerebral a.
      - ◆ Anterior communicating a.
    - ◆ Middle cerebral a.
    - ◆ Posterior communicating a.
  - ▪ Basilar a.
    - ◆ Posterior cerebral a.
  - ▪ Vertebral a.
    - ◆ Posterior meningeal a.
    - ◆ Posterior inferior cerebellar a.
      - ◆ Posterior spinal a.
    - ◆ Anterior spinal a.
  - ▪ Basilar a.
    - ◆ Anterior inferior cerebellar a.
      - ◆ Labyrinthine a.
    - ◆ Pontine aa.
    - ◆ Superior cerebellar a.
    - ◆ Posterior cerebral a.
- Subclavian a.
  - ● Vertebral a.
  - ● Thyrocervical trunk
    - ▪ Inferior thyroid a.
      - ◆ Inferior laryngeal a.
    - ▪ Ascending cervical a.
    - ▪ Suprascapular a.
    - ▪ Transverse cervical a.
      - ◆ Superficial cervical a.
      - ◆ Dorsal scapular a.

- ● Costocervical trunk
  - ▪ Deep cervical a.

Lymphatics

- ■ Submental lymph nodes
- ■ Submandibular lymph nodes
- ■ Superficial cervical lymph nodes
- ■ Deep cervical lymph nodes
- ■ Thoracic lymphatic duct
- ■ Palatine tonsil
- ■ Lingual tonsils
- ■ Pharyngeal tonsils (adenoids)

Veins

- ■ Superior vena cava
  - ● Brachiocephalic v.
    - ▪ Inferior thyroid v.
    - ▪ Inferior laryngeal v.
    - ▪ Vertebral v.
    - ▪ Deep cervical v.
  - ● Internal jugular v.
    - ▪ Superior bulb of the jugular v.
    - ▪ Superior thyroid v.
    - ▪ Middle thyroid v.
    - ▪ Superior laryngeal v.
    - ▪ Facial v.
      - ◆ Supratrochlear, supraorbital, superior labial, inferior labial, and deep facial vv.
    - ▪ Retromandibular v.
      - ◆ Superficial temporal, middle, temporal, transverse facial, and maxillary vv.
      - ◆ Pterygoid plexus of veins
  - ● External jugular v.
    - ▪ Posterior auricular v.
    - ▪ Anterior jugular v.
      - ◆ Jugular venous arch

- ▪ Suprascapular v.
- ▪ Transverse cervical v.
- ■ Cerebral veins
  - ● Superficial cerebral vv.
  - ● Deep cerebral vv.
  - ● Veins of the brainstem
  - ● Cerebellar vv.
- ■ Orbital veins
  - ● Superior ophthalmic v.
  - ● Ciliary vv.
  - ● Central vein of the retina
  - ● Inferior ophthalmic v.

Dural Venous Sinuses

- ■ Superior sagittal sinus
- ■ Inferior sagittal sinus
- ■ Straight sinus
- ■ Transverse sinuses
- ■ Occipital sinus
- ■ Confluence of sinuses
- ■ Sigmoid sinuses
- ■ Inferior petrosal sinuses
- ■ Superior petrosal sinuses
- ■ Basilar plexus
- ■ Cavernous sinuses
- ■ Sphenoparietal sinus

Pharynx

- ■ Nasopharynx
  - ● Choanae
  - ● Pharyngeal tonsil
  - ● Openings of the auditory tubes
  - ● Torus tubarius
    - ▪ Salpingopharyngeal fold
    - ▪ Salpingopalatine fold
- ■ Oropharynx

- Epiglottic vallecula
■ Laryngopharynx
  - Piriform recess/fossa
    - Fold over the superior laryngeal n.
  - Pharyngobasilar fascia
■ Pharyngeal muscles (see above)

Glands

■ Thyroid gland
  - Right lobe, left lobe, and isthmus
■ Parathyroid glands
■ Submandibular glands
■ Parotid glands (parotid ducts)
■ Sublingual glands

Tongue

■ Frenulum
■ Filiform papillae
■ Fungiform papillae
■ Vallate papillae
■ Sulcus terminalis
■ Foramen cecum
■ Lingual tonsil
■ Median and lateral glossoepiglottic folds
■ Vallecula
■ Extrinsic muscles of the tongue (see above)
■ Plica sublingualis

Larynx

■ Laryngeal inlet
  - Aryepiglottic fold
  - Vestibular folds (false vocal cords)
  - Glottis
    - Vocal folds (true vocal cords)
      ◆ Rima glottis
  - Quadrangular membrane

- Vestibular ligament
- Vocal ligament
■ Laryngeal skeleton
  - Thyroid cartilage
    - Laryngeal prominence
    - Superior and inferior horns
  - Cricoid cartilage
  - Arytenoid cartilages
  - Corniculate cartilages
  - Cuneiform cartilages
■ Thyrohyoid membrane
■ Cricothyroid ligament
■ Intrinsic muscles of the larynx (see above)

Eye and Related Structures

■ Sclera and cornea
■ Choroid
  - Ciliary body
    - Ciliary mm. (meridional fibers)
■ Iris and pupil
■ Retina
■ Central artery and vein of the retina
■ Lens
■ Chambers of the eyeball
  - Aqueous humor
    - Anterior and posterior chambers
  - Vitreous body
    - Vitreous chamber
■ Periorbita
■ Conjunctiva
■ Lacrimal glands (nasolacrimal ducts)
■ Tarsal glands
■ Extraocular muscles (see above)

Brain

■ Cerebrum

- Frontal, temporal, and occipital poles
- Frontal, temporal, parietal, and occipital lobes
- Lateral sulcus and central sulcus
- Cerebral gyri and sulci
■ Corpus callosum
■ Diencephalon
■ Midbrain
■ Pons
■ Medulla oblongata
■ Cerebellum and vermis
■ Ventricular system
  - Lateral ventricles
    - Interventricular foramen
  - Third ventricle
    - Cerebral aqueduct
  - Fourth ventricle
  - Central canal

Ear

■ Outer ear
  - Auricle; pinna
  - Lobule
  - Helix
  - Antihelix
    - Triangular fossa
  - Scapha and concha of the auricle
  - Antitragus and tragus
  - External auditory canal
  - Tympanic membrane
■ Middle ear
  - Tegmen tympani
    - Epitympanic recess
  - Mastoid antrum
    - Mastoid air cells
  - Auditory ossicles
    - Malleus, incus, and stapes
  - Muscles of the auditory ossicles
    - Stapedius m.

- Tensor tympani m.
- Auditory tube
  - Tympanic opening
  - Bony and cartilaginous parts
- Inner ear
  - Vestibule
  - Semicircular canals
  - Cochlea
  - Internal acoustic meatus

Meninges
- Dura mater
  - Falx cerebri and cerebelli
  - Tentorium cerebelli
- Arachnoid mater
  - Subarachnoid space
  - Arachnoid granulations (villi)

- Pia mater

Fascia/Membranes
- Cervical fascia
  - Investing fascia; superficial fascia
  - Pretracheal fascia
  - Prevertebral fascia
  - Carotid sheath
- Alar fascia
- Palatine aponeurosis
- Pterygomandibular raphe
- Retropharyngeal space
- Thyrohyoid membrane
- Cricothyroid membrane
- Tectorial membrane

- Buccopharyngeal fascia
- Parotid fascia
- Temporal fascia

Miscellaneous
- Buccal fat pad
- Triangles of the neck
  - Anterior triangle of the neck
    - Muscular triangle
    - Submandibular triangle
    - Submental triangle
    - Carotid triangle
  - Posterior triangle of the neck
    - Supraclavicular triangle
    - Suboccipital triangle
    - Interscalene triangle

Bone	Foramen	Corresponding Structure(s)
Sphenoid	Optic canal	Optic n.
	Superior orbital fissure	Ophthalmic, oculomotor, abducens, and trochlear nn.
	Foramen rotundum	Maxillary n.
	Foramen ovale	Mandibular n.
	Foramen spinosum	Middle meningeal vessels
	Foramen venosum	Emissary v.
	Pterygoid canal	Nerve of the pterygoid canal (Vidian n.)
Sphenoid/temporal	Foramen lacerum	Emissary vessels
Temporal	Internal acoustic meatus	Vestibulocochlear and facial nn.
	External acoustic meatus	Sound waves traveling to the tympanic membrane
	Stylomastoid foramen	Facial n.
	Carotid canal	Internal carotid a. and sympathetic nerve plexus
	Hiatus for the greater petrosal n.	Greater petrosal n.
	Hiatus for the lesser petrosal n.	Lesser petrosal n.
	Mastoid foramen	Emissary v.
	Petrotympanic fissure	Chorda tympani n.
Temporal/occipital	Jugular foramen	Internal jugular v., glossopharyngeal, vagus, and spinal accessory nn.
Occipital	Foramen magnum	Spinal cord and vertebral aa.
	Hypoglossal canal	Hypoglossal n.
	Condylar canal	Emissary v.
Mandible	Mandibular canal	Inferior alveolar n.
	Mental foramen	Mental n.
Ethmoid	Foramina cribrosa	Olfactory nn.
Ethmoid/frontal	Anterior ethmoidal foramen	Anterior ethmoid n.
	Posterior ethmoidal foramen	Posterior ethmoid n.
Frontal	Supraorbital foramen	Supraorbital vessels and n.
	Frontal foramen	Supratrochlear vessels and n.
Maxilla	Incisive foramen	Greater palatine vessels and nasopalatine nn.
	Infraorbital foramen	Infraorbital vessels and n.
Maxilla/palatine	Greater palatine foramen	Greater palatine vessels and nn.
Palatine	Lesser palatine foramen	Lesser palatine vessels and nn.

Foramina of the Skull—cont'd

Bone	Foramen	Corresponding Structure(s)
Palatine, vomer, and sphenoid	Palatovaginal canal	Pharyngeal branch from the pterygopalatine ganglion
Zygomatic	Zygomaticofacial foramen Zygomaticotemporal foramen	Zygomaticofacial vessels and n. Zygomaticotemporal vessels and n.
Parietal	Parietal foramen	Emissary v.

Limbs

Upper Limb – Osteology and Superficial Structures

Prior to dissection, you should familiarize yourself with the following structures:

OSTEOLOGY

- Scapula
  - Subscapular fossa
  - Spine
  - Supraspinous fossa
  - Infraspinous fossa
  - Acromion
  - Medial and lateral borders
  - Superior angle
    - Suprascapular notch
  - Inferior angle
  - Glenoid cavity
  - Supraglenoid and infraglenoid tubercles
  - Coracoid process
- Clavicle
  - Acromial and sternal ends
  - Conoid tubercle
- Humerus
  - Head
  - Anatomic and surgical necks
  - Greater and lesser tubercles
  - Intertubercular (bicipital) groove
  - Radial groove
  - Deltoid tuberosity
  - Lateral and medial supracondylar ridges
  - Lateral and medial epicondyles
  - Trochlea

- Capitulum
- Coronoid and olecranon fossae
- Radius
  - Head
  - Neck
  - Radial tuberosity
  - Interosseous border
  - Radial styloid process
  - Ulnar notch
- Ulna
  - Olecranon
  - Coronoid process
    - Radial notch
  - Trochlear notch
  - Interosseous border
  - Supinator crest
  - Ulnar head
    - Ulnar styloid process
- Hand
  - Carpals (8)
    - Proximal row
      - Pisiform
      - Triquetrum
      - Lunate
      - Scaphoid
    - Distal row
      - Trapezium

      - Trapezoid
      - Capitate
      - Hamate
        - Hook of the hamate
  - Metacarpals (5)
  - Phalanges (14)
    - Proximal, middle, and distal phalanges

SUPERFICIAL STRUCTURES OF THE UPPER LIMB

- Superficial veins
  - Cephalic v.
    - Thoracoacromial v.
  - Basilic v.
  - Median cubital v.
  - Dorsal venous arch
  - Superficial venous palmar arch
- Lymphatic system
  - Axillary lymph nodes
- Superficial nerves
  - Intercostobrachial n.
  - Medial cutaneous n. of the arm
  - Lateral cutaneous n. of the arm
  - Posterior cutaneous n. of the arm
  - Medial cutaneous n. of the forearm
  - Lateral cutaneous n. of the forearm
  - Posterior cutaneous n. of the forearm

TABLE 24–1   Osteology of the Upper Limb

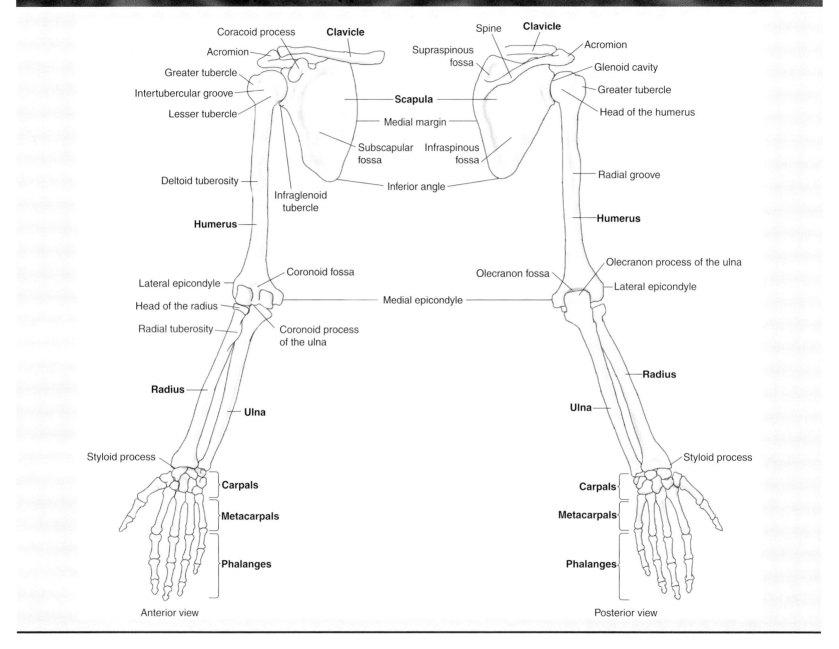

Coracoid process
Clavicle
Acromion
Greater tubercle
Intertubercular groove
Lesser tubercle

Spine
Clavicle
Supraspinous fossa
Acromion
Glenoid cavity
Greater tubercle
Head of the humerus

Scapula
Medial margin
Subscapular fossa
Infraspinous fossa
Inferior angle

Deltoid tuberosity
Infraglenoid tubercle

Radial groove

Humerus

Humerus

Coronoid fossa
Lateral epicondyle
Head of the radius
Radial tuberosity
Coronoid process of the ulna

Olecranon fossa
Medial epicondyle

Olecranon process of the ulna
Lateral epicondyle

Radius

Radius
Ulna

Ulna

Styloid process

Styloid process

Carpals

Carpals

Metacarpals

Metacarpals

Phalanges

Phalanges

Anterior view

Posterior view

Upper Limb – Skin Removal

■ Make the following circular incisions (Figure 24–1):

- Proximal on the arm (brachium); *A*

- Proximal on the forearm (antebrachium); *B*

- Wrist (*C*)

- About 5 cm distal to the wrist on the palmar surface of the hand and along the metacarpophalangeal joints on the posterior surface of the hand (*D*)

■ Join the four circular incisions with a longitudinal incision on the anterior aspect of the upper limb (*A–D*)

■ Remove the skin of the upper limb *but* leave the superficial fascia intact

- Do *not* cut the superficial nerves and veins in the superficial fascia

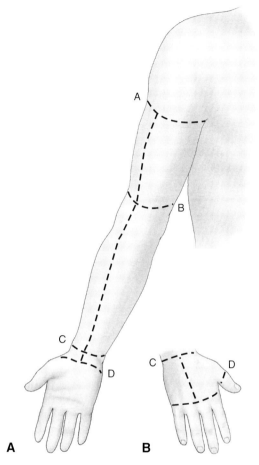

■ **Figure 24–1   Right upper limb skin incisions. *A*, Anterior view. *B*, Posterior view of the hand**

Upper Limb – Cutaneous Veins

■ Identify the following veins in the superficial fascia; you may want to prop up the shoulder to improve the view of structures on both the anterior and posterior surfaces:

● **Cephalic v.** – twists proximally, from the **dorsal venous arch** (network of veins) on the posterior aspect (dorsum) of the hand (Figure 24–2*B*) to the posterior, lateral side of the forearm and then to the anterior surface of the arm (Figure 24–2*A*)

  ● Anterior to the elbow, the cephalic v. communicates with the **basilic v.** via the **median cubital v.**

  ● Ascends on the lateral surface of the biceps brachii m. to pierce the deep fascia of the deltopectoral triangle, emptying into the **axillary v.**

● **Basilic v.** – twists proximally, from the dorsal venous arch of the hand to the posterior, medial side of the forearm (see Figure 24–2*B*), and then to the anterior side of the elbow, where it anastomoses with the cephalic v. via the median cubital v.

  ● Courses superiorly on the medial surface of the arm, then pierces the deep fascia at midarm to become the **axillary v.** (see Figure 24–2*A*)

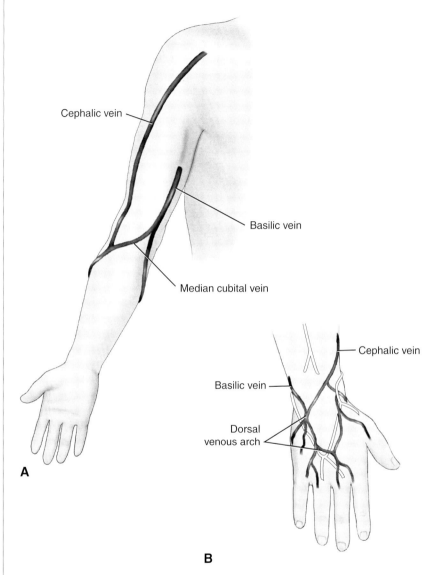

A

B

■ **Figure 24–2   Cutaneous veins of the right upper limb. *A*, Anterior view. *B*, Posterior view**

Upper Limb – Cutaneous Nerves

■ Identify the following nerves in the superficial fascia (Figure 24–3):

- **Intercostobrachial n.** – arises from the second intercostal n. (lateral cutaneous branch) and courses through the second intercostal space, piercing the serratus anterior m.; crosses the axilla; communicates with the medial cutaneous n. of the arm

- **Medial cutaneous n. of the arm** – smallest branch of the brachial plexus (may have been removed with the skin); pierces the fascia in the axilla and courses along the medial side of the axillary v.; may communicate with the intercostobrachial n.

- **Medial cutaneous n. of the forearm** – courses between the axillary a. and v.; descends medial to the brachial a., pierces the deep fascia with the basilic v. and continues distally on the ulnar aspect of the elbow

- **Lateral cutaneous n. of the forearm** – cutaneous branch of the musculocutaneous n.; pierces the deep fascia lateral to the biceps brachii tendon at the elbow

  - Passes deep to the cephalic v., descends along the radial border of the forearm to the wrist

- **Palmar cutaneous branch of the ulnar n.** – pierces the deep fascia of the forearm, proximal to the wrist

- **Palmar cutaneous branch of the median n.** – pierces the deep fascia near the flexor retinaculum and supplies the skin over the thenar compartment of the hand

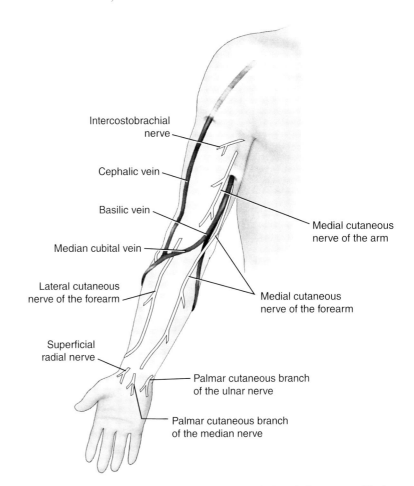

■ **Figure 24-3   Cutaneous nerves of the right upper limb**

Upper Limb – Cutaneous Nerves—cont'd

■ Identify the following superficial cutaneous nerves along the posterior aspect of the upper limb (Figure 24–4):

- **Superior lateral cutaneous n. of the arm** – the terminal branch of the axillary n.; emerges from the posterior border of the deltoid m. to supply the skin over the lower border of the deltoid m. and the skin around the deltoid tuberosity

- **Posterior cutaneous n. of the arm** – a branch of the radial n.; pierces the deep fascia on the posterior aspect of the arm, inferior to the posterior border of the deltoid m.

- **Inferior lateral cutaneous n. of the arm** – a branch of the radial n.; perforates the lateral head of the triceps brachii m., distal to the deltoid tuberosity, and supplies the inferior, lateral skin of the arm

- **Posterior cutaneous n. of the forearm** – a branch of the radial n.; perforates the lateral, distal surface of the lateral head of the triceps brachii m.; descends along the lateral aspect of the forearm and then along the posterior surface of the hand

- **Superficial branch of the radial n.** – descends in the forearm, deep to the brachioradialis m.; emerges at the distal part of the forearm and crosses over the anatomical snuff box (a landmark at the wrist) to provide cutaneous innervation to the posterolateral surface of the hand

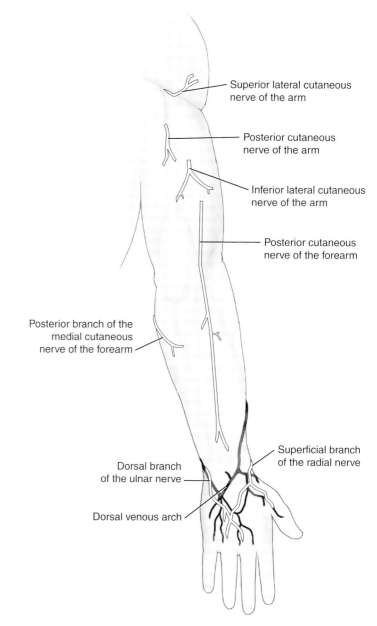

Superior lateral cutaneous nerve of the arm

Posterior cutaneous nerve of the arm

Inferior lateral cutaneous nerve of the arm

Posterior cutaneous nerve of the forearm

Posterior branch of the medial cutaneous nerve of the forearm

Superficial branch of the radial nerve

Dorsal branch of the ulnar nerve

Dorsal venous arch

■ **Figure 24–4   Posterior view of the cutaneous nerves of the right upper limb**

Upper Limb – Musculature

Prior to dissection, you should familiarize yourself with the following structures:

MUSCLES

- **Shoulder**
  - Deltoid m.
  - Trapezius m.
  - Levator scapulae m.
  - Serratus anterior m.
  - Rhomboid major m.
  - Rhomboid minor m.
  - Subclavius m.
  - Rotator cuff muscles
    - Supraspinatus m.
    - Infraspinatus m.
    - Teres minor m.
    - Subscapularis m.
  - Teres major m.
  - Pectoralis major m.
  - Pectoralis minor m.
  - Latissimus dorsi m.
- **Arm**
  - Coracobrachialis m.
  - Biceps brachii m.
    - Long head
    - Short head
  - Brachialis m.
  - Triceps brachii m.
    - Long head
    - Lateral head
    - Medial head
  - Anterior compartment of the forearm
    - Pronator teres m.
    - Flexor carpi radialis m.
    - Flexor carpi ulnaris m.
    - Palmaris longus m.
    - Flexor digitorum superficialis m.
    - Flexor digitorum profundus m.
    - Flexor pollicis longus m.
    - Pronator quadratus m.
- **Posterior compartment of the forearm**
  - Brachioradialis m.
  - Extensor carpi radialis longus m.
  - Extensor carpi radialis brevis m.
  - Extensor digitorum m.
  - Extensor digiti minimi m.
  - Extensor carpi ulnaris m.
  - Extensor indicis m.
  - Extensor pollicis longus m.
  - Extensor pollicis brevis m.
  - Abductor pollicis longus m.
  - Anconeus m.
  - Supinator m.

FASCIA

- Supraspinous fascia
- Infraspinous fascia
- Axillary fascia
  - Suspensory ligament of the axilla
- Brachial fascia
- Medial intermuscular septum of the arm
- Lateral intermuscular septum of the arm
- Interosseous membrane
- Antebrachial fascia
- Extensor retinaculum

TABLE 25-1   Muscles of the Upper Limb

Muscle	Proximal Attachment	Distal Attachment	Action	Innervation
Subclavius	Rib 1 and costal cartilage	Inferior medial surface of the clavicle	Depresses the clavicle	Nerve to the subclavius (C5)
Trapezius	Occipital bone, nuchal ligament, C7–T12 vertebrae	Spine, acromion, and lateral end of the clavicle	Elevates, retracts, rotates, and depresses the scapula	Spinal accessory n. and C3 and C4 spinal nerves
Latissimus dorsi	Spinous process of vertebrae T7–T12, sacrum, thoracolumbar fascia	Intertubercular groove of the humerus	Adducts, extends, and medially rotates the humerus	Thoracodorsal n. (C6–C8)
Levator scapulae	Transverse processes of C1–C4 vertebrae	Superior angle of the scapula	Elevates the scapula	Dorsal scapular n. (C5) and C3 and C4 spinal nerves
Rhomboid minor	Spinous processes of C7–T1 vertebrae	Medial margin of the scapula	Retracts and elevates the scapula	Dorsal scapular n. (C4 and C5)
Rhomboid major	Spinous processes of T2–T5 vertebrae			
Serratus anterior	External and lateral surfaces of ribs 1–8	Medial margin of the scapula	Protracts and rotates the scapula	Long thoracic n. (C5–C7)
Pectoralis major	Clavicle, sternum and costal cartilage	Intertubercular groove of the humerus	Adducts, medially rotates, extends, and flexes the humerus	Medial and lateral pectoral nn. (C5–T1)
Pectoralis minor	Ribs 3–5	Coracoid process of the scapula	Protracts and depresses the scapula	Medial pectoral n. (C8–T1)
Deltoid	Spine, acromion, and lateral end of the clavicle	Deltoid tuberosity of the humerus	Flexes, extends, and abducts the humerus	Axillary n. (C5–C6)
Supraspinatus	Supraspinous fossa		Stabilizes the shoulder joint and abducts the humerus	Suprascapular n. (C4–C6)
Infraspinatus	Infraspinous fossa	Greater tubercle of the humerus	Stabilizes the shoulder joint and laterally rotates the humerus	Suprascapular n. (C5–C6)
Teres minor	Lateral margin of the scapula		Stabilizes the shoulder joint and laterally rotates the humerus	Axillary n. (C5–C6)
Teres major	Inferior angle of the scapula	Intertubercular groove of the humerus	Adducts, extends, and medially rotates the humerus	Lower subscapular n. (C6–C7)

TABLE 25-1   Muscles of the Upper Limb—cont'd

Muscle	Proximal Attachment	Distal Attachment	Action	Innervation
Subscapularis	Subscapular fossa	Lesser tubercle of the humerus	Stabilizes the shoulder joint and medially rotates the humerus	Upper and lower subscapular nn.
Biceps brachii	Long head: supraglenoid tubercle Short head: coracoid process	Radial tuberosity	Flexes and supinates the elbow	Musculocutaneous n. (C5–C6)
Brachialis	Distal, ventral surface of the humerus	Coronoid process of the ulna	Flexes of the forearm	
Coracobrachialis	Coracoid process of the scapula	Medial surface of the humerus	Flexes and adducts the humerus	Musculocutaneous n. (C5–C7)
Triceps brachii	Long head: infraglenoid tubercle Lateral head: posterior surface of the humerus Medial head: posterior surface of the humerus	Olecranon process of the ulna	Extends the forearm	Radial n. (C6–C8)
Anconeus	Lateral epicondyle of the humerus	Olecranon process of the ulna	Extends the forearm	Radial n. (C7–T1)
Pronator teres	Medial epicondyle of the humerus	Middle of the lateral surface of the radius	Pronates and flexes the forearm	Median n. (C6–C7)
Flexor carpi radialis		Base of the second metacarpal bone	Flexes and abducts the hand	
Palmaris longus		Flexor retinaculum and the palmar aponeurosis	Flexes the hand and tightens the palmar aponeurosis	Median n. (C7–C8)
Flexor carpi ulnaris	Medial epicondyle and the olecranon process	Pisiform, hamate, and fifth metacarpal bones	Flexes and adducts the hand	Ulnar n. (C7–C8)
Flexor digitorum superficialis	Medial epicondyle, coronoid process of the ulna, and anterior border of the radius	Bodies of the middle phalanx of digits 2–5	Flexes the hand, metacarpophalangeal and proximal interphalangeal joints	Median n. (C7–T1)
Flexor digitorum profundus	Proximal $^3/_4$ of the medial surfaces of the ulna and interosseous membrane	Distal phalanx of digits 2–5	Flexes the hand and distal interphalangeal joints	Medial part: ulnar n. (C8–T1) Lateral part: median n. (C8–T1)

Continued

TABLE 25–1   Muscles of the Upper Limb—cont'd

Muscle	Proximal Attachment	Distal Attachment	Action	Innervation
Flexor pollicis longus	Anterior surface of the radius	Distal phalanx of digit 1	Flexes the thumb	Anterior interosseous n. from the median n. (C8–T1)
Pronator quadratus	Distal anterior surface of the ulna	Distal anterior surface of the radius	Pronates the forearm	
Brachioradialis	Lateral supracondylar ridge of the humerus	Styloid process of the radius	Flexes the forearm	Radial n. (C5–C7)
Extensor carpi radialis longus		Base of the second metacarpal bone	Extends and abducts the hand	Radial n. (C6–C7)
Extensor carpi radialis brevis	Lateral epicondyle of the humerus	Base of the third metacarpal bone	Extends the hand	Posterior interosseous n. (C7–C8), the continuation of the deep branch of the radial n.
Extensor digitorum		Extensor expansion of digits 2–5	Extends the hand and fingers	
Extensor digiti minimi		Extensor expansion of digit 5	Extends digit 5	
Extensor carpi ulnaris		Base of the fifth metacarpal bone	Extends and adducts the hand	
Supinator	Lateral epicondyle and supinator crest of the ulna	Distal to the radial tuberosity	Supinates the forearm	Deep branch of the radial n. (C5–C6)
Abductor pollicis longus	Posterior surface of the ulna, radius, and interosseous membrane	Base of the first metacarpal bone	Abducts and extends the thumb	Posterior interosseous n. (C7–C8), the continuation of the deep branch of the radial n.
Extensor pollicis brevis	Posterior surface of the radius and interosseous membrane	Base of the proximal phalanx of digit 1	Extends the thumb at the carpometacarpal joint	
Extensor pollicis longus	Posterior surface of the ulna and interosseous membrane	Distal phalanx of digit 1	Extends the thumb	
Extensor indicis		Extensor expansion of digit 2	Extends digit 2	

Deltoid Muscle

■ Turn the cadaver prone

■ Identify the following (Figure 25–1):

- **Deltoid m.** – observe the striations of this multipen-nate muscle; courses from the spine of the scapula, acromion, and lateral third of the clavicle to the deltoid tuberosity of the humerus

■ To better study the muscles that act on the shoulder joint, reflect the deltoid m. from the scapula and clavicle by making the following incisions (Figure 25–1A, B):

- Along the spine of the scapula (A)
- Along the acromion (B)
- Along the clavicle (C)

■ Reflect the cut deltoid m.

- Be careful during this step to preserve the **axillary n.** and **posterior circumflex humeral a.** on the deep, posterior surface of the deltoid m.

■ Identify the following (Figure 25–1C):

- **Subacromial bursa** – located between the deltoid m., acromion, and the supraspinatus m. and tendon

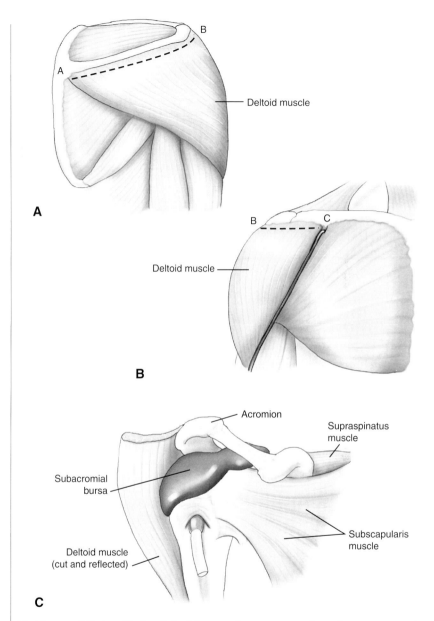

■ **Figure 25–1   Right deltoid muscle. *A*, Posterior view. *B*, Anterior view. *C*, Anterior view of the shoulder with the deltoid muscle cut and reflected**

Scapular Region

- Review the cut attachments of the **trapezius** and **deltoid mm.** to the clavicle, acromion, and spine of the scapula

- Review the following muscles on the medial border of the scapula (Figure 25–2):

  - **Levator scapulae m.** – attached near the superior angle of the scapula

  - **Rhomboid minor m.** – attached at the level of the spine of the scapula

  - **Rhomboid major m.** – attached to the medial border of the scapula, inferior to the spine; inferior to the rhomboid minor m.

- Identify the following; "rotator cuff" muscles are designated with *:

  - **Supraspinatus m.*** – named according to its location in the supraspinous fossa

  - **Infraspinatus m.*** – named according to its location in the infraspinous fossa

  - **Teres minor m.*** – located inferior to the infraspinatus m.

  - **Teres major m.** – located inferior to the teres minor m.

  - **Latissimus dorsi m.** – located inferior to the teres major m.; note its superior attachment wraps around the teres major m. to attach near the intertubercular groove of the humerus

  - **Subscapularis m.*** – studied further in another laboratory session

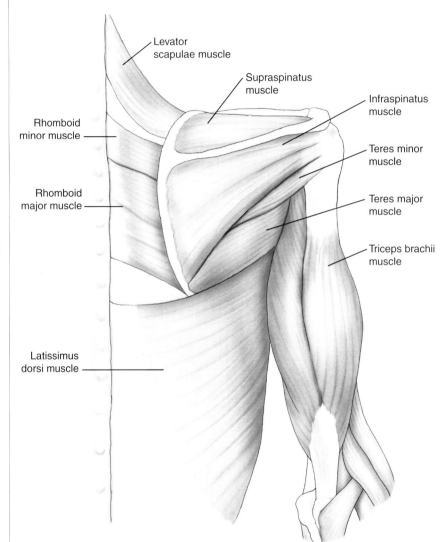

■ **Figure 25–2   Posterior view of the right shoulder muscles**

Posterior Muscles of the Arm

- Identify the following (Figure 25–3):

  - **Triceps brachii m.** – positioned on the posterior aspect of the humerus; has three heads:

    - **Long head** – separates the teres major and minor mm.

      - ◆ Teres major m. is anterior to the long head of the triceps brachii m.

      - ◆ Teres minor m. is posterior to the long head of the triceps m.

    - **Lateral head** – the most lateral head of the three proximal attachments

    - **Medial head** – deep to the long and lateral heads

  - The distal attachment of the triceps brachii m. is to the olecranon process of the ulna

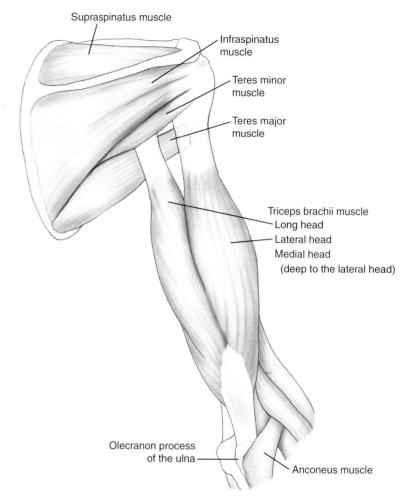

Supraspinatus muscle

Infraspinatus muscle

Teres minor muscle

Teres major muscle

Triceps brachii muscle
Long head
Lateral head
Medial head
(deep to the lateral head)

Olecranon process of the ulna

Anconeus muscle

■ **Figure 25-3   Posterior view of the right arm muscles**

Anterior Muscles of the Shoulder and Arm

■ Turn the cadaver supine

■ Identify the following muscles (Figure 25–4):

- **Pectoralis major m.** – the most superficial muscle on the anterior thoracic wall

- **Pectoralis minor m.** – located deep to the pectoralis major m.

- **Subclavius m.** – a small muscle attached to the inferior border of the clavicle and the first rib

- **Deltoid m.** – triangular-shaped muscle attached to the spine of the scapula, the acromion, and lateral one third of the clavicle; the deltoid m. is superficial to the brachial and rotator cuff muscles

- **Serratus anterior m.** – attached laterally along the first 8 ribs; courses posteriorly to attach to the medial margin of the scapula

- **Teres major m.** – distal attachment is on the intertubercular groove

- **Latissimus dorsi m.** – distal attachment is on the intertubercular groove

- **Subscapularis m.** – a deep muscle attached to the subscapular fossa; the brachial plexus and axillary a. course anterior to this muscle and as such you may not be able to see much of it until a future laboratory session

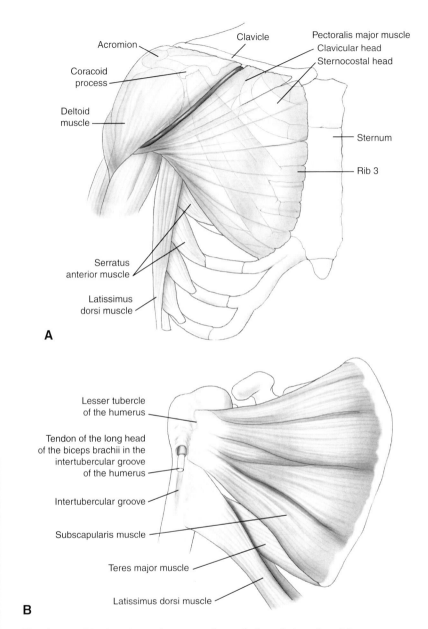

A

B

■ **Figure 25–4   Anterior muscles of the right shoulder. A, Superficial view. B, Deep view**

Anterior Muscles of the Arm

- Identify the following:

  - **Biceps brachii m.** (Figure 25–5*A*):

    - **Long head** – the tendon of the long head courses through the intertubercular groove of the humerus and arches over the head of the humerus to reach the scapula

      - ◆ **Transverse humeral ligament** – the tendon of the long head passes through an opening in the capsular ligament called the transverse humeral ligament

    - **Short head** – attaches to the coracoid process of the scapula

  - **Coracobrachialis m.** – medial to the short head of the biceps m.; its belly is pierced by the musculocutaneous n.

- Reflect the biceps brachii m. to reveal the brachialis m. (Figure 25–5*B*):

  - **Brachialis m.** – located deep to the biceps brachii m.; attached to the coronoid process of the ulna

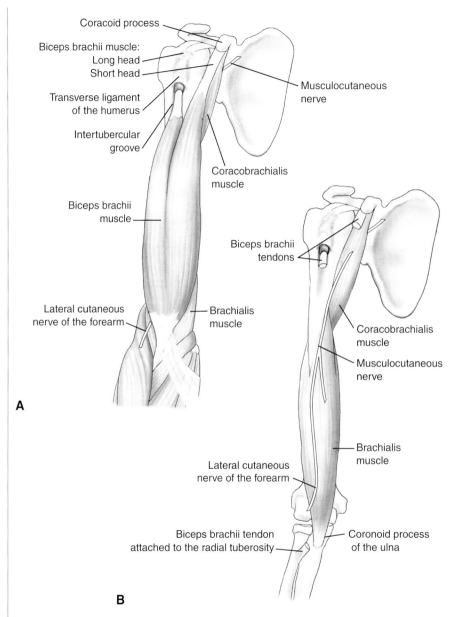

A

B

- **Figure 25-5   Anterior muscles of the right arm. *A*, Superficial view. *B*, Deep view; the biceps brachii m. is removed**

Anterior Muscles of the Arm—cont'd

- Identify the following (Figure 25–5C):

  - **Bicipital aponeurosis** – near the insertion on the radial tuberosity, the biceps m. gives off a fascial sheet (bicipital aponeurosis) that passes at an inferomedial angle over the brachial artery to fuse with the deep fascia of the forearm

  - **Cubital fossa** – a triangular region between the epicondyles of the humerus, pronator teres m., and brachioradialis m. (Figure 25–5C)

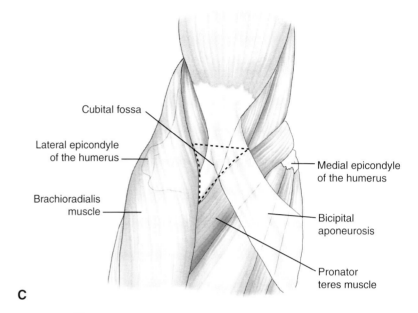

Cubital fossa

Lateral epicondyle of the humerus

Brachioradialis muscle

Medial epicondyle of the humerus

Bicipital aponeurosis

Pronator teres muscle

C

■ **Figure 25–5 continued.** *C,* **Bicipital fossa**

Anterior Muscles of the Forearm – Superficial Layer

■ Turn the forearm supine

■ Many of the muscles in the anterior compartment of the forearm attach to the medial epicondyle of the humerus

■ Identify the following (Figure 25–6):

- **Pronator teres m.** – deep to the bicipital aponeurosis; identify the superior border of the pronator teres m. by locating the tendon of the biceps brachii m.

- **Flexor carpi radialis m.** – inferior to the pronator teres m.; attaches distally to the base of the second and third metacarpals on their palmar surface

- **Palmaris longus m.** – if present, it is medial to the flexor carpi radialis m.; attached distally in the transverse fibers of the palmar aponeurosis (palmar carpal ligament) and the palmar aponeurosis

- **Flexor carpi ulnaris m.** – the most medial muscle on the anterior surface of the forearm; attached distally to the pisiform bone and base of the fifth metacarpal on its palmar surface

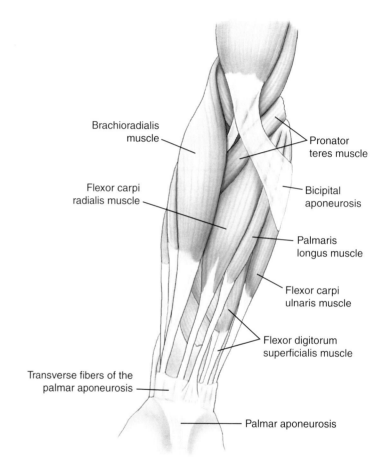

Brachioradialis muscle

Flexor carpi radialis muscle

Pronator teres muscle

Bicipital aponeurosis

Palmaris longus muscle

Flexor carpi ulnaris muscle

Flexor digitorum superficialis muscle

Transverse fibers of the palmar aponeurosis

Palmar aponeurosis

■ **Figure 25–6 Anterior view of the superficial muscles of the right forearm**

Anterior Muscles of the Forearm – Intermediate Layer

■ To reveal the flexor digitorum superficialis m., you may follow either of these instructions:

1. Flex the hand at the wrist and flex the forearm at the elbow; the intervening muscles will bow and therefore may be pushed away from the deeper muscles

2. Transect the flexor carpi radialis, palmaris longus, and flexor carpi ulnaris mm. on one forearm

■ Be careful to leave vessels and nerves intact

■ Identify the muscle in the intermediate layer of the anterior region of the forearm (Figure 25–7):

• **Flexor digitorum superficialis m.** – deep to the flexor carpi radialis, palmaris longus, and flexor carpi ulnaris mm.

  ● Inserts distally on the middle phalanx of digits 2–5 on their palmar surface; you will not study the distal attachments until the hand dissection

• **Flexor pollicis longus m.** – attached to the distal phalanx of digit 1 (thumb); located lateral and deep to the flexor digitorum superficialis m. on the radial side of the forearm

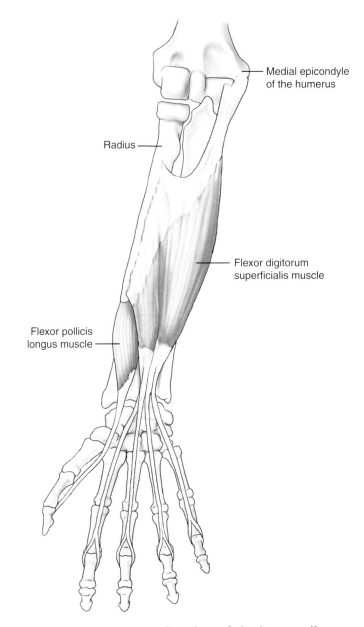

Medial epicondyle of the humerus

Radius

Flexor digitorum superficialis muscle

Flexor pollicis longus muscle

■ **Figure 25-7   Anterior view of the intermediate muscles of the right forearm**

Anterior Muscles of the Forearm – Deep Layer

■ To reveal the deepest layer of muscles in the anterior region of the forearm, you may follow either of these instructions:

1. Flex the hand at the wrist and flex the forearm at the elbow; the intervening muscles will bow and therefore may be pushed away from the deeper muscles

2. Transect the flexor digitorum superficialis m. to see the deep layer of muscles in the anterior forearm

■ Be careful to leave vessels and nerves intact

■ Identify the following muscles in the deep layer of the forearm (Figure 25–8):

- **Flexor digitorum profundus m.** – deep to the flexor digitorum superficialis m.; you will not study the tendinous attachments on the fingers until the hand dissection

- **Flexor pollicis longus m.** – lateral to the flexor digitorum profundus m.; the radial a. is superficial to this muscle; avoid damaging this artery

- **Pronator quadratus m.** – deep and distal to the flexor pollicis longus m.; the muscle fibers course horizontally and attach to both the radius and ulna

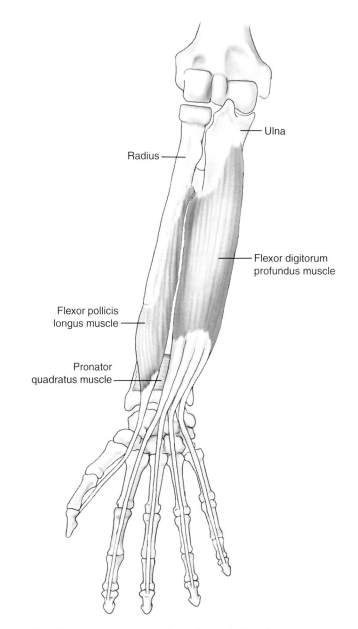

■ **Figure 25–8   Anterior view of the deep muscles of the right forearm**

Posterior Muscles of the Forearm – Superficial Layer

- Turn the forearm prone

- Identify the lateral epicondyle of the humerus; this is a common attachment for many of the posterior forearm muscles

- Identify the following forearm muscles (Figure 25–9):

  - **Brachioradialis m.** – has the most superior attachment of the extensor group of forearm muscles

  - **Extensor carpi radialis longus m.** – parallels the brachioradialis m. on the lateral aspect of the forearm

  - **Anconeus m.** – short triangular muscle; closely associated with the distal end of the triceps brachii m.

  - **Extensor carpi radialis brevis m.** – lies deep to and is somewhat shorter than the extensor carpi radialis longus m.

  - **Extensor digitorum m.** – lies medial to the extensor carpi radialis brevis m.; four tendons arise from the muscle belly to attach to digits 2–5

  - **Extensor digiti minimi m.** – medial to the extensor digitorum m. and courses to the little finger

  - **Extensor carpi ulnaris m.** – most medial of the superficial posterior muscles of the forearm

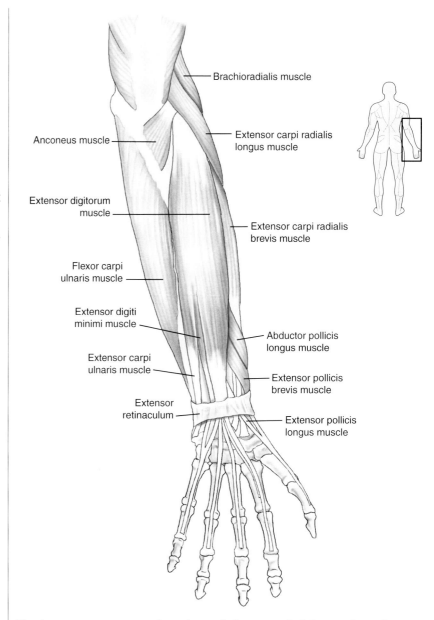

- **Figure 25–9** **Posterior view of the superficial muscles of the right forearm**

Posterior Muscles of the Forearm – Deep Layer

■ To reveal the deeper muscles of the posterior region of the forearm, you may follow either of these instructions:

1. Extend the hand at the wrist and extend the forearm at the elbow; the intervening muscles will bow and therefore may be pushed away from the deeper muscles

2. Transect the extensor digitorum m. about 6 cm proximal to the wrist and reflect the cut muscle superiorly

■ Be careful to leave vessels and nerves intact

■ Identify the following (Figure 25–10A):

● **Supinator m.** – deep to the extensor carpi radialis brevis and extensor digitorum mm.; attached to both the ulna and radius

● **Abductor pollicis longus m.** – lateral and parallel to the extensor pollicis longus m.; distal to the supinator m.

● **Extensor pollicis brevis m.** – distal to the abductor pollicis longus m.

● **Extensor pollicis longus m.** – distal to the extensor pollicis brevis m.

● **Extensor indicis m.** – inferior to the extensor pollicis longus m.

● **Anatomical snuff box** (Figure 25–10B) – bordered by the abductor pollicis longus, extensor pollicis brevis, and extensor pollicis longus tendons

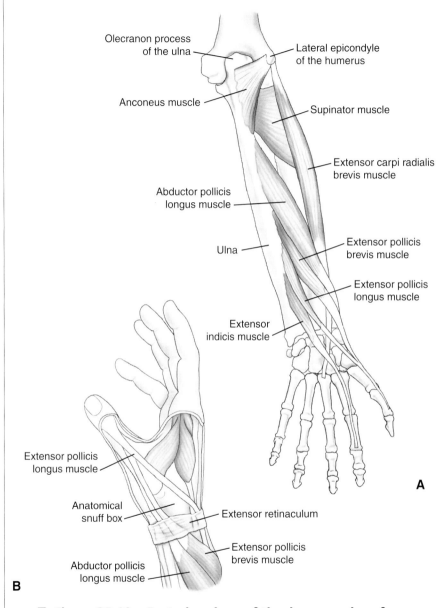

■ **Figure 25-10   Posterior views of the deep muscles of the right forearm. *A*, Posterior view. *B*, Lateral view**

Upper Limb – Nerves

Lab 26

Prior to dissection, you should familiarize yourself with the following structures:

INNERVATION

5 Roots (C5, C6, C7, C8, T1)
- Dorsal scapular n. and long thoracic n.

3 Trunks
- Superior – suprascapular n. and n. to the subclavius m.
- Middle
- Inferior

6 Divisions
- 3 anterior and 3 posterior divisions

3 Cords
- Lateral
  - Lateral pectoral n.
- Medial
  - Medial pectoral n.

- Medial cutaneous n. of the arm*
- Medial cutaneous n. of the forearm*
- Posterior
  - Upper subscapular n.
  - Thoracodorsal n.
  - Lower subscapular n.

5 Terminal Branches
- Musculocutaneous n.
  - Muscular branches
  - Lateral cutaneous n. of the forearm
- Median n.
  - Muscular branches
  - Palmar branch
  - Recurrent branch of the median n.
  - Digital branches of the median n. (common and proper palmar digital nn.)

- Ulnar n.
  - Muscular branches
  - Superficial and deep branches of the ulnar n.
  - Digital branches of the ulnar n. (common and proper palmar digital nn.)
- Radial n.
  - Muscular branches
  - Posterior cutaneous nn. of the arm and forearm
  - Inferior lateral cutaneous n. of the arm
  - Superficial and deep branches of the radial n.
- Axillary n.
  - Muscular branches
  - Superior lateral cutaneous n. of the arm

Note: Arm and brachium refer to the same region of the upper limb, as do the forearm and antebrachium. For example, the medial cutaneous n. of the forearm is the same as the medial antebrachial cutaneous n.

Proximal Innervation of the Upper Limb

- The brachial plexus is a weave of nerves, much like a "Jacob's ladder" woven of string

- The brachial plexus is somewhat symmetrical in its design (Figure 26–1):

  - five roots
  - three trunks
  - six divisions (three anterior and three posterior)
  - three cords
  - five terminal branches

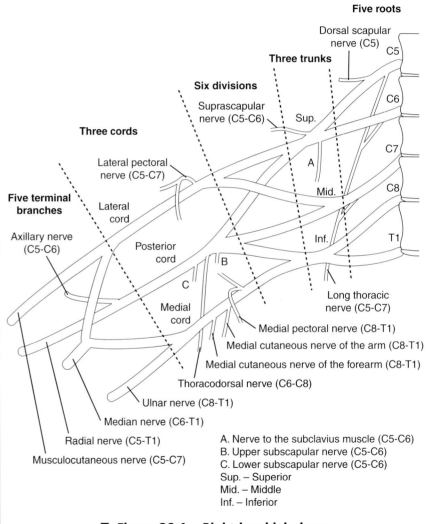

■ Figure 26–1   Right brachial plexus

Brachial Plexus – Roots

- The roots are the **ventral rami of C5–T1** spinal segments

  - Find the roots as they pass between the anterior and middle scalene mm., and eventually flank the subclavian a.

- Identify the following branches from the roots (Figure 26–2):

  - **Dorsal scapular n. (C5)** – pierces the middle scalene m., descends deep to the levator scapulae m., and enters the deep surface of the rhomboid mm.; accompanied by the dorsal scapular a. (deep branch of the transverse cervical a.)

  - **Long thoracic n. (C5–C7)** – descends posterior to C8 and T1 rami, and courses distally on the external surface of the serratus anterior m.

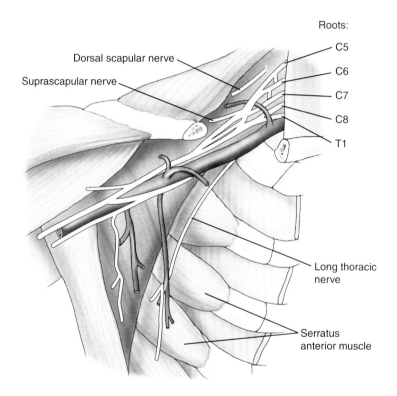

Roots:

C5

C6

C7

C8

T1

Dorsal scapular nerve

Suprascapular nerve

Long thoracic nerve

Serratus anterior muscle

■ **Figure 26–2   Right brachial plexus roots**

Brachial Plexus – Trunks

■ At the root of the neck, the roots of the brachial plexus unite to form three trunks

■ Identify the following (Figure 26–3):

- **Brachial plexus – Trunks**
  - **Superior trunk** – union of C5 and C6 roots
    - ◆ **Suprascapular n.** – passes posterolaterally across the posterior triangle of the neck, superior to the brachial plexus, to pass through the scapular notch
  - **Middle trunk** – continuation of the root of C7
  - **Inferior trunk** – union of C8 and T1 roots

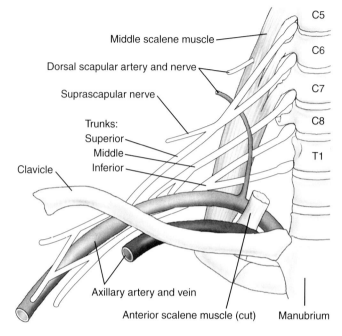

■ **Figure 26–3   Right brachial plexus trunks**

Brachial Plexus – Divisions and Cords

■ Identify the following (Figure 26–4A):

● **Brachial plexus: Divisions** – each trunk (superior, middle, and inferior) of the brachial plexus divides into an anterior and posterior division as the plexus passes posterior to the clavicle

  ● **Anterior division** – supplies the muscles of the anterior (flexor) compartment of the upper limb

  ● **Posterior division** – supplies the muscles of the posterior (extensor) compartment of the upper limb

● **Brachial plexus: Cords** (Figure 26–4B) – the three anterior divisions and three posterior divisions form three cords that are named by their anatomic position relative to the axillary a.

  ● **Lateral** – formed by the anterior division of the superior and middle trunks; positioned lateral to the axillary a.

  ● **Medial** – formed by the anterior division of the inferior trunk; positioned medial to the axillary a.

  ● **Posterior** – formed by the posterior divisions of all three trunks; positioned posterior to the axillary a.

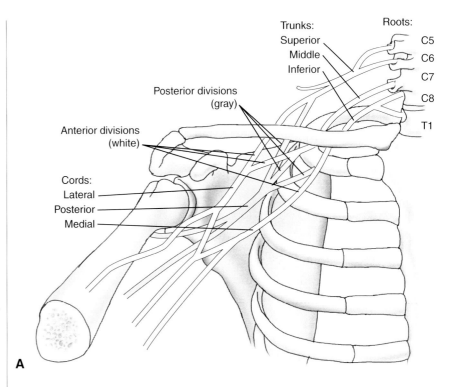

A

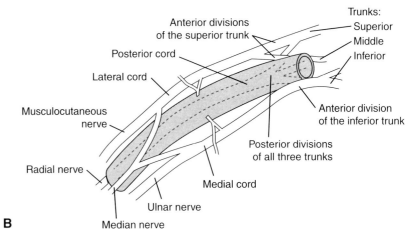

B

■ **Figure 26–4   Right brachial plexus divisions and cords.** *A,* Divisions and cords. *B,* Cords in situ

Brachial Plexus – Branches from the Lateral and Medial Cords

■ Identify the following (Figure 26–5):

- **Lateral cord** – lateral to the axillary a.

  - **Lateral pectoral n.** – innervates the pectoralis major m.; there may be a communicating branch between the lateral and medial cords

- **Medial cord** – medial to the axillary a.

  - **Medial pectoral n.** – innervates the pectoralis major and minor mm.

  - **Medial cutaneous n. of the arm** – a small branch that courses along the medial surface of the axillary v.; communicates with the intercostobrachial n.

  - **Medial cutaneous n. of the forearm** – courses between the axillary a. and v. along the side of the ulnar n.

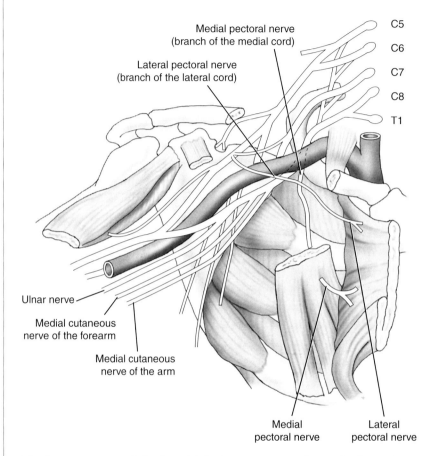

■ **Figure 26–5 Right brachial plexus nerves from the lateral and medial cords**

Brachial Plexus – Branches from the Posterior Cord

■ Bluntly remove the axillary fat to expose the following branches of the posterior cord

■ Identify the following (Figure 26–6):

- **Upper subscapular n.** – courses posteriorly to enter the subscapularis m.

- **Thoracodorsal n.** – courses along the posterior axillary wall to enter the deep surface of the latissimus dorsi m.

- **Lower subscapular n.** – passes inferolaterally, deep to the subscapular a. and v., to the subscapularis and teres major mm.

- **Radial n.** – a terminal branch of the posterior cord; enters the radial groove with the deep a. of the arm to pass between the long and medial heads of the triceps brachii m.

- **Axillary n.** – the other terminal branch of the posterior cord; courses through the quadrangular space to innervate the deltoid and teres minor mm.

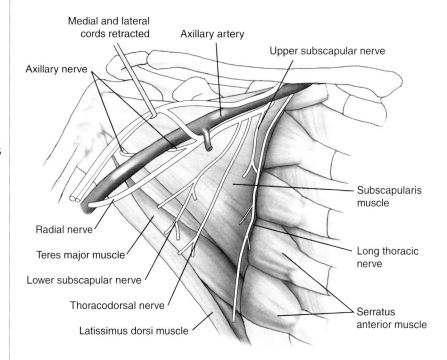

■ **Figure 26–6   Right brachial plexus nerves from the posterior cord**

Brachial Plexus – Five Terminal Branches in the Arm

■ Observe that the medial and lateral cords bifurcate and converge in a form that resembles the letter M; identify the five terminal branches (Figure 26–7):

● **Musculocutaneous n.** – terminal branch of the lateral cord; pierces the coracobrachialis m. and descends between the biceps and brachialis mm.

    ● Continues distally as the **lateral cutaneous n. of the forearm**

● **Median n.** – terminal branch formed by the convergence of the medial branch of the lateral cord and the lateral branch of the medial cord

    ● In the proximal part of the arm, the median n. courses lateral to the brachial a.

    ● At the insertion of the coracobrachialis m., the median n. crosses anterior to the brachial a. and courses distally on the medial side of the brachial a., deep to the bicipital aponeurosis

● **Ulnar n.** – terminal branch of the medial cord

    ● Courses medial to the brachial a. to the middle of the arm, where the ulnar n. pierces the medial intermuscular septum to enter the posterior compartment

    ● Courses anterior to the medial head of the triceps brachii m.

● **Radial n.** – one of two terminal branches of the posterior cord

    ● Courses posterior to the axillary and brachial aa. and enters the radial groove with the deep a. of the arm

    ● Descends between the heads of the triceps m. and pierces the lateral intermuscular septum

● **Axillary n.** – the second terminal branch of the posterior cord; passes posteriorly through the quadrangular space, with the posterior circumflex humeral a., en route to the deltoid and teres minor mm.

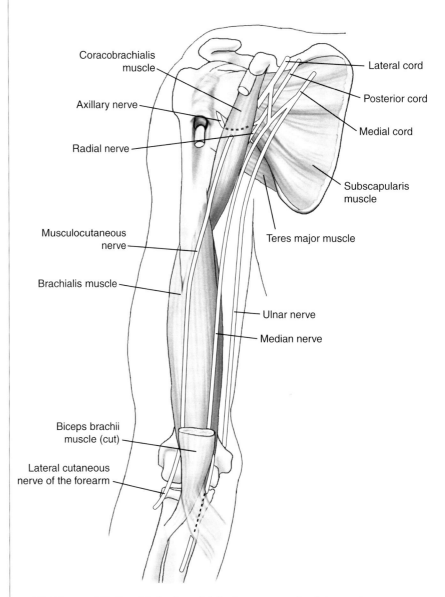

■ **Figure 26–7   Right brachial plexus terminal branches – arm**

Brachial Plexus – Branches in the Forearm

■ Identify the following terminal nerves of the brachial plexus in the forearm; follow them distally to the wrist (Figure 26–8):

- **Radial n.** – after piercing the lateral intermuscular septum, the radial n. enters the anterior compartment and descends lateral to the brachialis m., anterior to the lateral epicondyle; the radial n. branches into the following:

  - **Superficial branch** – passes superficial to the supinator m. and descends deep to the brachioradialis m.

  - **Deep branch** – wraps posteriorly around the radius and pierces the supinator m. (instructions to follow this nerve are on the next page)

- **Median n.** – enters the forearm by coursing between the humeral and ulnar heads of the pronator teres m.; descends between the flexor digitorum superficialis and profundus mm. in the forearm, entering the carpal tunnel of the wrist

  - The median n. divides into the **anterior interosseous n.** and **palmar cutaneous branches**

- **Ulnar n.** – courses posterior to the medial epicondyle of the humerus ("funny bone") to enter the forearm

  - Enters the forearm between the two heads of the flexor carpi ulnaris m.

  - Observe a **palmar branch** that courses superficially over the flexor retinaculum

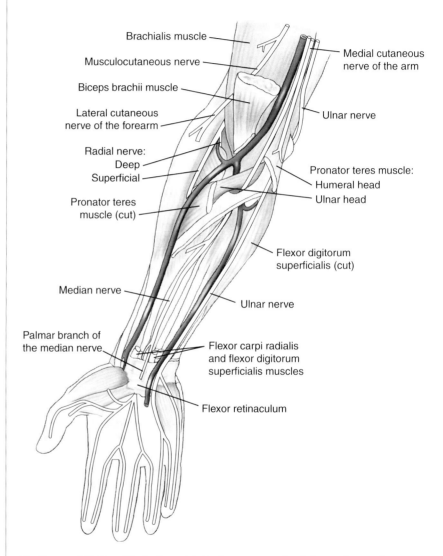

■ **Figure 26–8   Right brachial plexus terminal branches – forearm**

Brachial Plexus – Posterior Branches of the Shoulder

■ Turn the cadaver prone to identify the following (Figure 26–9):

- **Dorsal scapular n.** – after piercing the middle scalene m., the dorsal scapular n. descends with the deep branch of the transverse cervical a., deep to the levator scapulae and rhomboid mm., along the medial (vertebral) border of the scapula

- **Suprascapular n.** – courses inferior to the suprascapular ligament to enter the supraspinous fossa accompanied by the suprascapular a., which courses superior to the suprascapular ligament; the nerve and artery course around the spinoglenoid notch to enter the infraspinous fossa; make a vertical incision through the supraspinatus and infraspinatus mm. to verify their course

- **Axillary n.** – courses posteriorly, accompanied by the posterior circumflex humeral a., to enter the posterior region of the arm through the following anatomic region:

  - **Quadrangular space** – a four-sided space on the posterior aspect of the shoulder bordered by the teres minor m., teres major m., long head of the triceps brachii m., and the surgical neck of the humerus

- **Radial n.** – enters the posterior region of the arm, accompanied by deep a. of the arm, between the long and medial heads of the triceps m., inferior to the teres major m., to follow the radial groove

  - Descends between the long and medial heads of the triceps brachii m. and pierces the lateral intermuscular septum

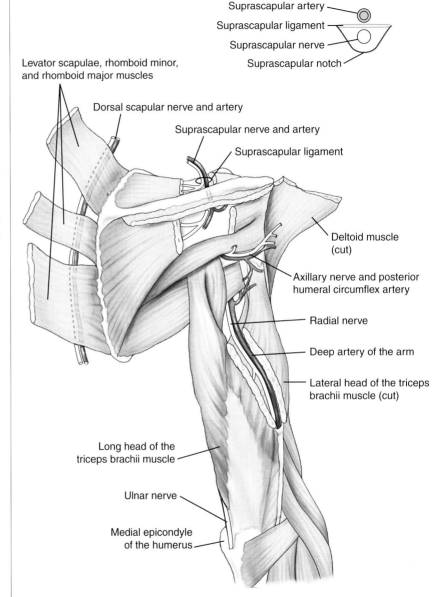

■ **Figure 26–9   Posterior view of the nerves of the right shoulder**

Brachial Plexus – Posterior Branches of the Forearm

■ Identify the following nerves of the brachial plexus in the forearm; follow them distally to the wrist (Figure 26–10):

● **Radial n.** – after piercing the lateral intermuscular septum, the radial n. descends lateral to the brachialis m.; anterior to the lateral epicondyle, the radial n. divides into the following:

 ● **Superficial radial n.** – passes superficial to the supinator m. and descends deep to the brachioradialis m.; the superficial branch enters the forearm in the same region as the abductor pollicis longus m.

 ● **Deep branch** – wraps posteriorly around the radius, piercing the supinator m.; the deep branch of the radial n. descends deep among the posterior muscles of the forearm

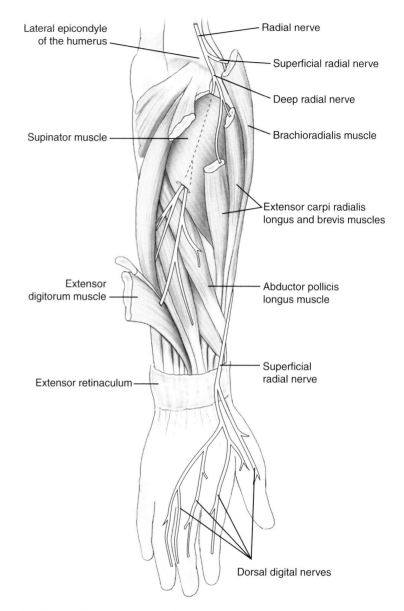

■ **Figure 26–10   Posterior view of the nerves of the right forearm**

Upper Limb – Vasculature

Prior to dissection, you should familiarize yourself with the following structures:

VASCULATURE

- Subclavian a.
  - Thyrocervical trunk
    - Transverse cervical a.
    - Superficial branch
    - Deep branch (dorsal scapular a.)
    - Suprascapular a.
- Axillary a.
  - Superior thoracic a.
  - Thoracoacromial a.
    - Acromial branch
    - Deltoid branch
  - Pectoral branch
  - Clavicular branch
  - Lateral thoracic a.
  - Subscapular a.
    - Circumflex scapular a.
    - Thoracodorsal a.
  - Anterior humeral circumflex a.
  - Posterior humeral circumflex a.
- Brachial a.
  - Deep artery of the arm
    - Radial collateral a.
    - Middle collateral a.
  - Superior ulnar collateral a.
  - Inferior ulnar collateral a.
- Radial a.
  - Radial recurrent a.
- Ulnar a.
  - Anterior ulnar recurrent a.
  - Posterior ulnar recurrent a.
  - Common interosseous a.
    - Anterior interosseous a.
    - Posterior interosseous a.
      - Interosseous recurrent a.

Notes: Deep veins are named the same as the arteries they accompany. Vasculature of the hand will be dissected in another laboratory session.

Axillary Artery

- Turn the upper limb supine

- **Veins** – the deep veins are named the same as the arteries they accompany; ask your instructor if you may remove the veins to improve visual clarity

- Identify the following (Figure 27–1):

  - **Subclavian a.** – courses from the aortic arch (left side) or the brachiocephalic a. (right side) to the external border of rib 1, where the subclavian a. continues as the axillary a.

  - **Axillary a.** – courses from the external border of rib 1 to the inferior border of the teres major m., where the axillary a. continues as the brachial a.

  - **Axillary sheath** – located about 2 cm below the coracoid process; the sheath contains the axillary a., axillary v., and the cords of the brachial plexus

  - Open the axillary sheath to follow the course of the axillary a.

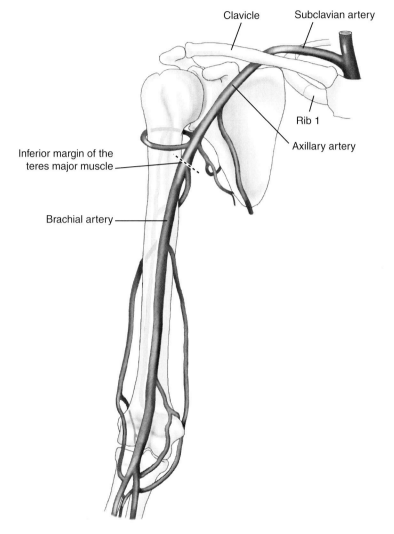

Clavicle   Subclavian artery

Rib 1

Axillary artery

Inferior margin of the teres major muscle

Brachial artery

■ **Figure 27–1   Right axillary artery**

Axillary Artery – Branches

■ Identify the following (Figure 27–2):

- **First part** – medial to the pectoralis minor m.; between rib 1 and the medial border of the pectoralis minor m.; one branch

  - ● **Superior (supreme) thoracic a.**

- **Second part** – deep to the pectoralis minor m.; two branches

  - ● **Thoracoacromial a.** – follow its four branches (each named for the region supplied):

    - ◆ **Acromial, deltoid, pectoral,** and **clavicular aa.**

  - ● **Lateral thoracic a.** – supplies the serratus anterior m.; in the female, it provides **mammary branches**

- **Third part** – lateral to the pectoralis minor m.; between the pectoralis minor and teres major mm.; three branches

  - ● **Subscapular a.** – largest branch of the axillary a.

    - ◆ After the subscapular a. gives rise to the **circumflex scapular a.** (seen on the posterior side of the scapula), the artery continues as the **thoracodorsal a.,** which supplies the latissimus dorsi m.

  - ● **Anterior circumflex humeral a.** – courses anterior to the surgical neck of the humerus

  - ● **Posterior circumflex humeral a.** – usually larger than the anterior circumflex humeral a.; courses posteriorly through the quadrangular space with the axillary n.

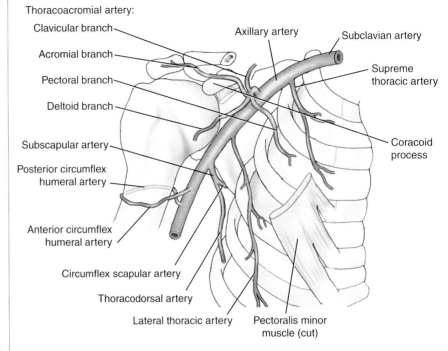

■ **Figure 27–2   Branches of the right axillary artery**

Arteries of the Posterior Region of the Shoulder

■ Turn the upper limb prone

■ Identify the following (Figure 27–3):

- **Deep branch of the transverse cervical a.** – also known as the dorsal scapular a.; descends with the dorsal scapular n., deep to the levator scapulae and rhomboid mm., along the medial (vertebral) border of the scapula

- **Suprascapular a.** – courses superior to the supra-scapular ligament; enters the supraspinous fossa accompanied by the suprascapular n., which courses inferior to the suprascapular ligament; the nerve and artery course around the spinoglenoid notch to enter the infraspinous fossa

- **Circumflex scapular a.** – a branch of the subscapular a.; courses through a triangular space bordered by the teres minor, teres major, and long head of the triceps brachii mm.

- **Posterior circumflex humeral a.** – courses through the quadrangular space (teres minor m., teres major m., long head of triceps brachii m., and the surgical neck of humerus)

  - The axillary n. also is located in this region

■ *Note:* The dorsal scapular a. anastomoses with the supra-scapular and circumflex scapular aa.

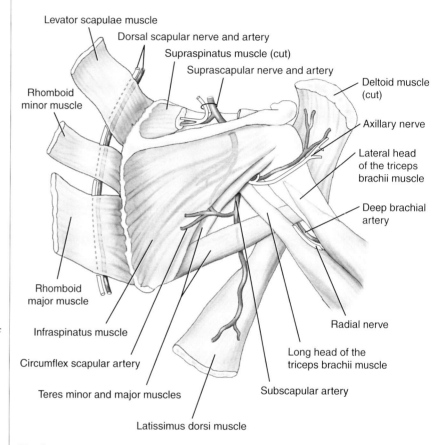

Levator scapulae muscle
Dorsal scapular nerve and artery
Supraspinatus muscle (cut)
Suprascapular nerve and artery
Deltoid muscle (cut)
Rhomboid minor muscle
Axillary nerve
Lateral head of the triceps brachii muscle
Deep brachial artery
Rhomboid major muscle
Infraspinatus muscle
Radial nerve
Circumflex scapular artery
Long head of the triceps brachii muscle
Teres minor and major muscles
Subscapular artery
Latissimus dorsi muscle

■ **Figure 27–3   Arteries of the posterior region of the right shoulder**

Brachial Artery

■ Identify the following (Figure 27–4):

- **Venae comitantes** – observe that the brachial a. is accompanied by smaller paired veins called venae comitantes; ask your instructor if you may remove the venae comitantes

- **Brachial a.** – begins at the inferior border of the teres major m. as the continuation of the axillary a.; the brachial a. courses on the medial aspect of the biceps brachii m. to the cubital fossa under the protection of the bicipital aponeurosis; the artery has the following branches:

  - **Deep a. of the arm** – courses in the radial groove (accompanied by the radial n.) on the posterior aspect of the humerus (described in more detail on the next page)

  - **Superior ulnar collateral a.** – arises midway along the brachial a.; accompanies the ulnar n.; courses posterior to the medial epicondyle of the humerus; anastomoses with the **posterior ulnar recurrent a.**

  - **Inferior ulnar collateral a.** – arises from the brachial artery about 5 cm superior to the elbow; courses anterior to the medial epicondyle of the humerus, between the median n. and the brachialis m.; anastomoses with the **anterior ulnar recurrent a.**

■ *Note:* The anastomotic connections around the elbow are often difficult to demonstrate; limit the time spent in pursuit of these anastomoses

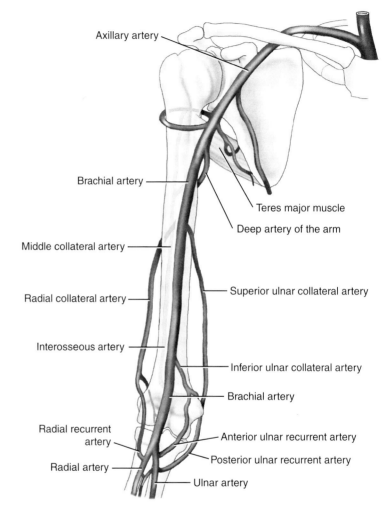

■ **Figure 27–4 Right brachial artery**

Arteries of the Posterior Region of the Arm

- The brachial a. courses within the anterior compartment of the arm; however, a few of the branches of the brachial a. course in the posterior compartment of the arm

- Identify the following (Figure 27–5):

  - **Superior ulnar collateral a.** – courses posterior to the medial epicondyle of the humerus; anastomoses with the **posterior ulnar recurrent a.**

  - **Deep a. of the arm** – courses in the radial groove along the posterior aspect of the humerus (accompanied by the radial n.), between the long head of triceps brachii m. and the medial head of triceps brachii m.; the deep a. of the arm has the following branches:

    - **Middle collateral a.** – descends in the medial head of the triceps brachii m., posterior to the lateral epicondyle of the humerus; anastomoses with the **interosseous recurrent a.** (shown on the next page)

    - **Radial collateral a.** – accompanies the radial n. between the brachioradialis and brachialis mm.; anterior to the lateral epicondyle of the humerus, the radial collateral a. anastomoses with the **radial recurrent a.**

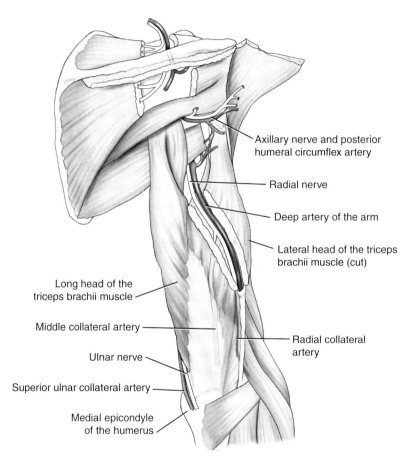

Axillary nerve and posterior humeral circumflex artery

Radial nerve

Deep artery of the arm

Lateral head of the triceps brachii muscle (cut)

Long head of the triceps brachii muscle

Middle collateral artery

Ulnar nerve

Superior ulnar collateral artery

Medial epicondyle of the humerus

Radial collateral artery

■ **Figure 27–5   Arteries of the posterior region of the right arm**

Proximal Portions of the Radial and Ulnar Arteries

■ Identify the following (Figure 27–6):

- **Radial a.** – the lateral terminal branch of the brachial a.

  - **Radial recurrent a.** – arises in the cubital fossa; crosses anterior to the biceps tendon, between the superficial and deep branches of the radial n.; anastomoses with the radial collateral branch of the deep a. of the arm

- **Ulnar a.** – the medial terminal branch of the brachial a.

  - **Anterior** and **posterior ulnar recurrent aa.** – located anterior and posterior, respectively, to the medial epicondyle of the humerus

  - **Common interosseous a.** – about 1 cm long; arises from the ulnar a., medial to the radial tuberosity; bifurcates into the anterior and posterior interosseous aa.

    ◆ **Anterior interosseous a.** – courses along the anterior surface of the **interosseous membrane**

    ◆ **Posterior interosseous a.** – courses along the posterior surface of the interosseous membrane

      ◆ **Interosseous recurrent a.** – courses in a superior direction, posterior to the lateral epicondyle of the humerus, to anastomose with the middle collateral branch of the deep a. of the arm

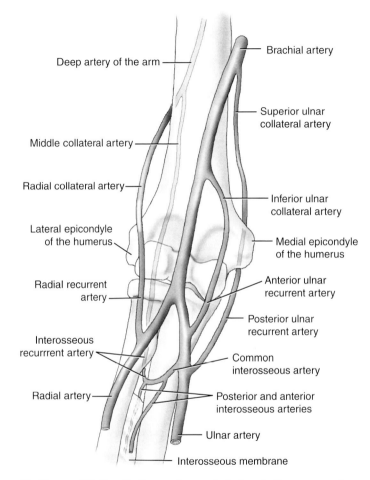

■ **Figure 27–6   Collateral arterial circulation around the right elbow**

Distal Portion of the Radial and Ulnar Arteries

■ Trace the following arteries distally through the forearm (Figure 27–7):

- **Radial a.** – smaller than the ulnar a.

  - Passes along the radial side of the forearm (deep to brachioradialis m.) to the wrist

  - In the wrist, the radial a. is lateral to the flexor carpi radialis tendon and anterior to the styloid process of the radius

  - Winds laterally around the wrist, deep to the abductor pollicis longus m. and extensor pollicis brevis and longus mm.; principal contributor to the deep palmar arterial arch of the hand

  - Observe the radial a. within the anatomical snuff box

- **Ulnar a.** – larger than the radial a.; courses medial to the biceps brachii tendon

  - Accompanied by the median n., passes between the ulnar and radial heads of the flexor digitorum superficialis m.

  - Midway down the forearm, crosses posterior to the median n. to reach the superficial side of the flexor digitorum profundus m.

  - Along the distal region of the forearm, lies lateral to the ulnar n.; exits the forearm lateral to the pisiform bone and superficial to the flexor retinaculum

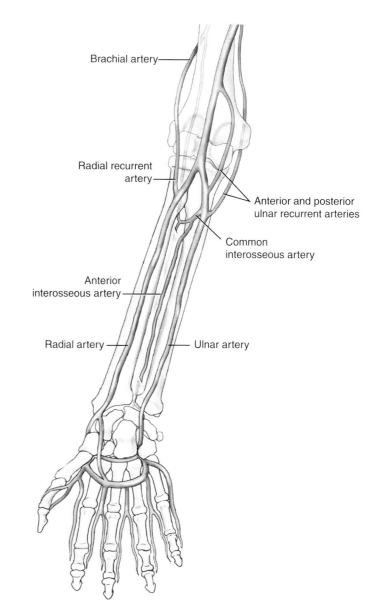

■ **Figure 27–7   Distal portion of the right radial and ulnar arteries**

Upper Limb – Hand

Prior to dissection, you should familiarize yourself with the following structures:

OSTEOLOGY

- Hand
  - Carpal bones (8)
    - Proximal row
      - Pisiform
      - Triquetrum
      - Lunate
      - Scaphoid
    - Distal row
      - Trapezium
      - Trapezoid
      - Capitate
      - Hamate
        - Hook of the hamate
  - Metacarpals (5)
  - Phalanges (14)
    - Proximal, middle, and distal phalanges

VASCULATURE

- Radial and ulnar aa.
  - Superficial palmar arch
    - Common palmar digital aa.
    - Proper palmar digital aa.
  - Deep palmar arch
    - Palmar metacarpal aa.
    - Princeps pollicis a.

HAND

- Tendons and muscles
  - Flexor digitorum superficialis tendons
  - Flexor digitorum profundus tendons
  - Palmaris longus m.
  - Palmaris brevis m.
  - Thenar mm.
    - Abductor pollicis brevis m.
    - Opponens pollicis m.
    - Flexor pollicis brevis m.
  - Adductor pollicis m.
  - Hypothenar mm.
    - Abductor digiti minimi m.
    - Opponens digiti minimi m.
    - Flexor digiti minimi m.
  - Lumbrical mm. (4)
  - Dorsal interossei mm. (4)
  - Palmar interossei mm. (3)

NERVES

- Median n.
  - Superficial palmar branch of the median n.
  - Recurrent branch of the median n.
  - Common palmar digital nn.
    - Proper palmar digital nn.
- Ulnar n.
  - Superficial branch of the ulnar n.
  - Common palmar digital nn.
    - Proper palmar digital nn.
  - Deep branch of the ulnar n.
- Radial n.

MISCELLANEOUS

- Flexor retinaculum
- Anatomical snuff box
- Palmar aponeurosis
- Carpal tunnel
- Fibrous digital sheaths
- Extensor expansion hood

TABLE 28-1 Muscles of the Hand

Muscle	Proximal Attachment	Distal Attachment	Action	Innervation
Palmaris brevis	Palmar aponeurosis and flexor retinaculum	The dermis on the ulnar side of the hand	Tenses the skin over the hypothenar mm.	Ulnar n., superficial branch (C8–T1)
Thenar muscles of the hand				
• Abductor pollicis brevis	Flexor retinaculum and tubercles of the scaphoid and trapezium bones	Base of the proximal phalanx of digit 1	Abducts and opposes the thumb	Median n., recurrent branch (C8–T1)
• Flexor pollicis brevis			Flexes the thumb	
• Opponens pollicis		Base of the first metacarpal bone	Opposes the thumb to the other digits	
Adductor pollicis	Oblique head: base of second and third metacarpal bones, capitate bone	Base of the proximal phalanx of digit 1	Adducts the thumb	Ulnar n., deep branch (C8–T1)
	Transverse head: palmar surface of metacarpal 3			
Hypothenar muscles of the hand				
• Abductor digit minimi	Pisiform bone	Proximal phalanx of digit 5	Abducts digit 5	Ulnar n., deep branch (C8–T1)
• Flexor digiti minimi brevis	Hook of the hamate bone and flexor retinaculum		Flexes the proximal phalanx of digit 5	
• Opponens digiti minimi		Fifth metacarpal	Opposes digit 5 to the thumb	
Lumbricals 1 and 2	Lateral two tendons of the flexor digitorum profundus m.	Lateral sides of the extensor expansion for digits 2–5	Flexes the metacarpophalangeal joints and extends the interphalangeal joints	Median n. (C8–T1)
Lumbricals 3 and 4	Medial two tendons of the flexor digitorum profundus m. in each space			Ulnar n. (C8–T1)
Dorsal interossei 1–4	Adjacent sides of two metacarpals	Extensor expansion and base of the proximal phalanges of digits 2–4	Abducts the digits	Ulnar n., deep branch (C8–T1)
Palmar interossei 1–3	Anterior surface of metacarpals 2, 4, and 5	Extensor expansion of the digits and proximal phalanges 2, 4, and 5	Adducts the digits	

Hand – Skin Removal

- Make the following incisions to remove the skin from the hand; leave the palmar aponeurosis intact (Figure 28–1):

  - If the hand is clenched, have a partner hold it open while you make a longitudinal incision along the middle of the palm (A) and transversely across the palm (B)

  - Remove the skin from the hand; note that the skin is bound to the palmar aponeurosis (a tough fascia), which may require time and patience to separate

  - Make a longitudinal incision along the length of each finger and the thumb (C)

  - Remove the skin, being careful to not damage the underlying nerves, vessels, and fibrous sheaths

  - Ask your instructor about the number of fingers to dissect

- Remove the skin on the dorsal surface of the hand, leaving the superficial fascia intact

- The palmar structures are customarily grouped into the following compartments:

  - **Thenar** – lateral compartment

  - **Hypothenar** – medial compartment

  - **Intermediate** – central compartment

  - **Adductor** – deep compartment

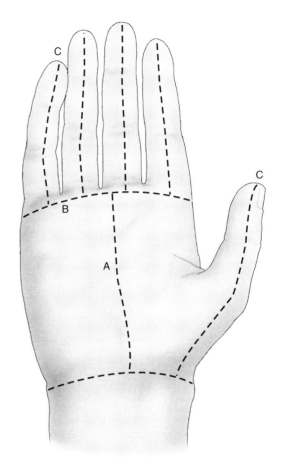

■ **Figure 28–1   Right hand skin incisions**

Palmar Aponeurosis

■ Identify the following structures (Figure 28–2):

• **Palmar aponeurosis** – a thick, well-defined part of the deep fascia of the palm

  • The proximal end is continuous with the flexor retinaculum and the palmaris longus tendon

  • **Fibrous digital sheaths of the hand** – ligamentous tubes that enclose the synovial sheaths of tendons en route to the digits

• **Thenar eminence** – the lateral eminence on the palmar surface; the palmar aponeurosis is thin over the thenar eminence

• **Hypothenar eminence** – the medial eminence on the palmar surface; the palmar aponeurosis is also thin over the hypothenar eminence

• **Palmaris brevis m.** – muscle fibers course transversely over the hypothenar eminence

■ Detach the palmaris brevis m. from the palmar aponeurosis and then carefully remove the palmar aponeurosis; do not damage the **common** and **proper digital nn., aa.,** and **vv.** deep to the palmar aponeurosis; removal requires lifting up the palmar aponeurosis and cutting, with scalpel or scissors, connections to the underlying tendons and bones

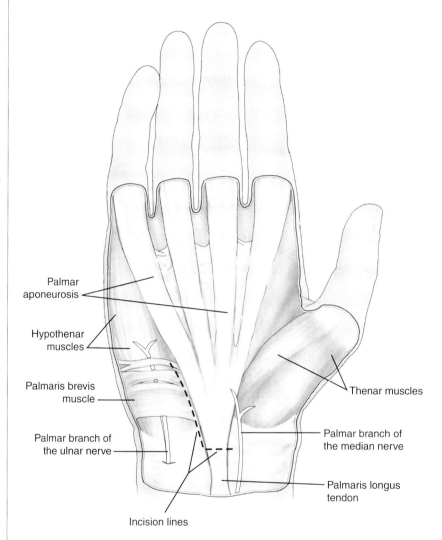

Palmar aponeurosis

Hypothenar muscles

Palmaris brevis muscle

Palmar branch of the ulnar nerve

Thenar muscles

Palmar branch of the median nerve

Palmaris longus tendon

Incision lines

■ **Figure 28–2   Palmar aponeurosis of the right hand**

Carpal Tunnel

■ Identify the following (Figure 28–3):

● **Flexor retinaculum** (transverse carpal ligament) – attached medially to the pisiform bone and hook of the hamate bone; attached laterally to the trapezium and scaphoid bones (Figure 28–3*A*)

● **Carpal tunnel** – the carpal bones form an anterior concavity, which is converted into a tunnel by the overlying flexor retinaculum (Figure 28–3*B*)

  ○ Contains the flexor pollicis longus tendon, all of the flexor digitorum tendons, and the median n.

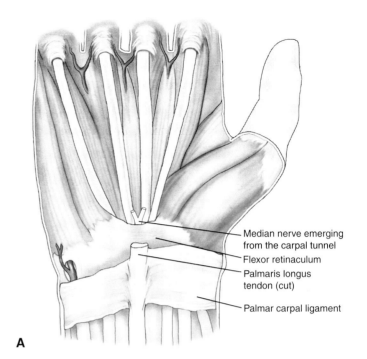

Median nerve emerging from the carpal tunnel
Flexor retinaculum
Palmaris longus tendon (cut)
Palmar carpal ligament

A

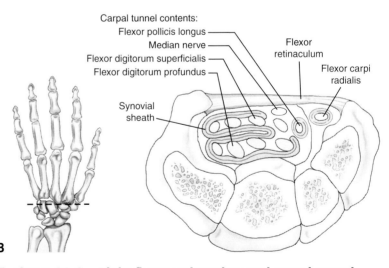

Carpal tunnel contents:
Flexor pollicis longus
Median nerve
Flexor digitorum superficialis
Flexor digitorum profundus
Flexor retinaculum
Flexor carpi radialis
Synovial sheath

B

■ **Figure 28–3   Right flexor retinaculum and carpal tunnel. *A*, Anterior view. *B*, Cross-section through the distal row of carpal bones**

Vessels of the Palm of the Hand

■ Identify the following arteries; the veins have the same names and may be removed (Figure 28–4):

● **Ulnar a.** – lateral to the pisiform bone and superficial to the flexor retinaculum; major contributor to the superficial palmar arch

  ○ **Deep branch of the ulnar a.** – courses deep to the hypothenar mm.; do not follow the deep branch at this time

● **Superficial palmar branch of the radial a.** – courses deep to the thenar mm.

  ○ **Superficial palmar arch** – formed by the superficial anastomosis between the radial and ulnar aa.

    ◆ **Common palmar digital aa.**

      ◆ **Proper palmar digital aa.** – bifurcation of the common palmar digital aa.; course along the medial and lateral surfaces of each finger

      ◆ **Princeps pollicis a.** – arterial supply to the thumb

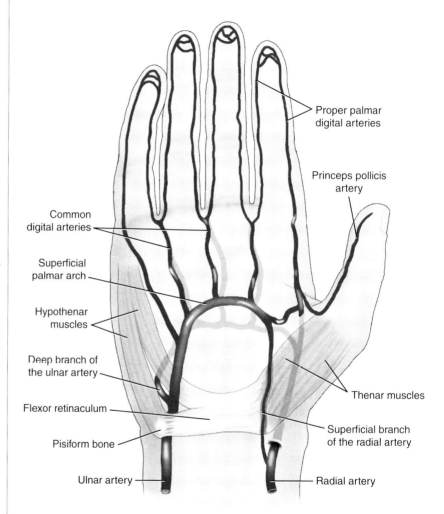

Labels on figure:
Proper palmar digital arteries
Princeps pollicis artery
Common digital arteries
Superficial palmar arch
Hypothenar muscles
Deep branch of the ulnar artery
Flexor retinaculum
Pisiform bone
Ulnar artery
Thenar muscles
Superficial branch of the radial artery
Radial artery

■ **Figure 28-4   Vessels of the palm of the right hand**

Nerves of the Palm of the Hand

■ Identify the following nerves:

- **Median n.** – enters the palmar surface of the hand, deep to the flexor retinaculum, and divides into a number of branches (Figure 28–5*A*)

  - **Recurrent median n.** – supplies the thenar mm.

  - **Common palmar digital nn.** – course distally through the hand and terminate by dividing into proper palmar digital nn.

    ◆ **Proper palmar digital nn.** – supply digits 1, 2, 3, and the lateral (radial) side of digit 4 (Figure 28–5*B*); accompany the proper palmar digital aa. and vv.

- **Ulnar n.** – enters the palmar surface of the hand with the ulnar a., superficial to the flexor retinaculum, and divides into superficial and deep branches (Figure 28–5*A*)

  - **Superficial branch of the ulnar n.**

    ◆ **Common palmar digital nn.** – course distally through the hand and terminates by dividing into proper palmar digital nn.

      ◆ **Proper palmar digital nn.** – supply digit 5 and the medial (ulnar) side of digit 4 (Figure 28–5*B*); accompany the proper palmar digital aa. and vv.

  - **Deep branch of the ulnar n.** – disappears through the hypothenar mm., in company with the deep ulnar a.; do not follow the deep branch at this time

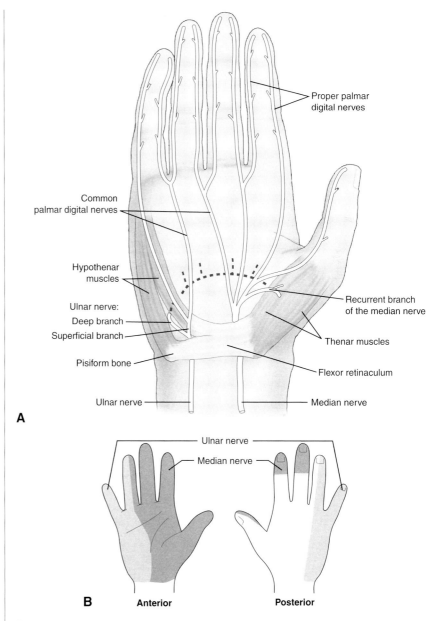

A

B   Anterior   Posterior

■ **Figure 28–5   Nerves of the palm of the right hand. *A*, Nerves. *B*, Cutaneous fields**

Thenar Compartment of the Hand

- **Thenar compartment** – lateral compartment of the palmar surface of the hand; identify the three muscles of the thumb that form the thenar eminence (Figure 28–6):

  - **Abductor pollicis brevis m.** – forms the anterolateral portion of the thenar eminence

  - **Flexor pollicis brevis m.** – medial to the abductor pollicis brevis m.; the recurrent branch of the median n. crosses superficial to the flexor pollicis brevis m.

  - **Opponens pollicis m.** – quadrangular-shaped muscle deep to the abductor pollicis m. and lateral to the flexor pollicis brevis m.

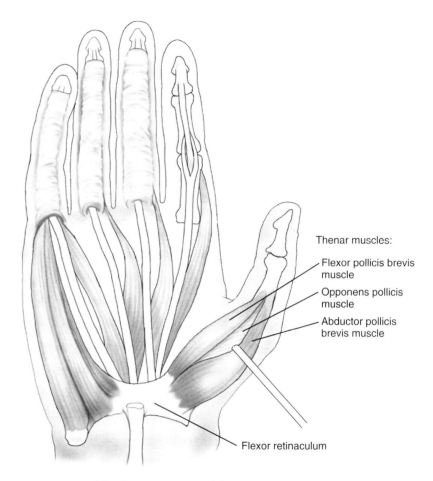

Thenar muscles:

Flexor pollicis brevis muscle

Opponens pollicis muscle

Abductor pollicis brevis muscle

Flexor retinaculum

■ **Figure 28–6   Right thenar muscles**

Hypothenar Compartment of the Hand

- **Hypothenar compartment** – medial compartment of the palmar surface of the hand; identify the three muscles of digit 5 that form the hypothenar eminence (Figure 28–7):

  - **Abductor digiti minimi m.** – most superficial and medial of the three muscles

  - **Flexor digiti minimi m.** – lateral to the abductor digiti minimi m.

  - **Opponens digiti minimi m.** – quadrangular-shaped muscle deep to the abductor and flexor digiti minimi mm.

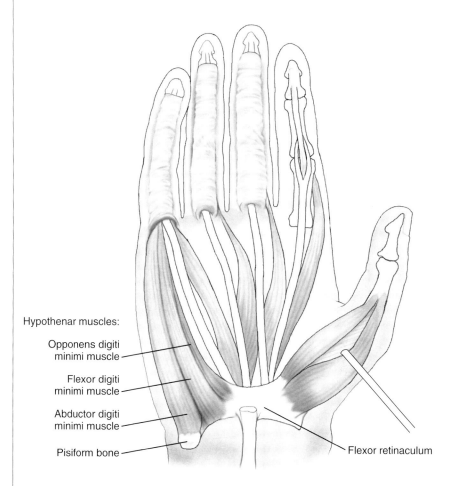

Hypothenar muscles:

Opponens digiti minimi muscle

Flexor digiti minimi muscle

Abductor digiti minimi muscle

Pisiform bone

Flexor retinaculum

■ **Figure 28–7   Right hypothenar muscles**

Central Compartment of the Palm of the Hand

■ Make a longitudinal incision through the flexor retinaculum to identify the following (Figure 28–8):

• **Flexor digitorum tendons** – tendons of both the flexor digitorum superficialis and profundus mm. course through the central compartment

• **Lumbrical mm.** (4) – arise from the flexor digitorum profundus tendons and insert on the radial side of the **extensor expansion hood** of the corresponding digit

• **Fibrous digital sheaths** – cover each of the fingers

  ● Make an incision along the length of one or two fibrous digital sheaths to expose their contents

  ● Observe that each tendon of the **flexor digitorum superficialis m.** bifurcates to attach to the sides of the middle phalanx of digits 2–5

  ● Observe that each **flexor digitorum profundus tendon** passes deep to the bifurcation of the flexor digitorum superficialis tendon to attach on the distal phalanx of digits 2–5

■ The following instructions will enable you to see the deep structures of the palm:

• Either flex the hand at the wrist to relieve tension from the tendons, which can be pushed aside . . .

• Or, cut horizontally through the bellies of both the flexor digitorum superficialis and profundus mm. at the wrist of one hand only (ask instructor)

  ● Reflect the flexor digitorum superficialis and profundus tendons from the wrist toward the fingertips

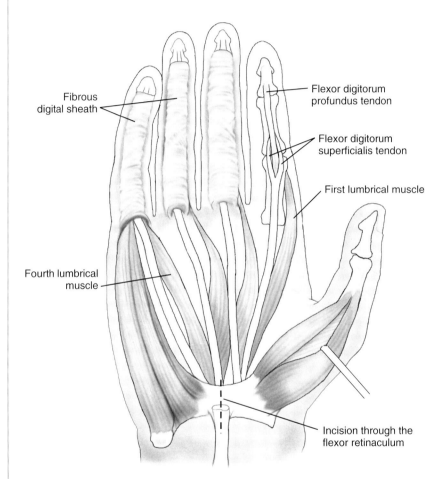

Fibrous digital sheath

Flexor digitorum profundus tendon

Flexor digitorum superficialis tendon

First lumbrical muscle

Fourth lumbrical muscle

Incision through the flexor retinaculum

■ **Figure 28–8   Central compartment of the right palm**

Adductor Compartment

■ Identify the following (Figure 28–9):

- **Adductor pollicis m.** – a fan-shaped muscle; has two heads (oblique and transverse) that are separated by the radial a. as the artery enters the palm to form the deep palmar arterial arch

  - Follow the muscle's tendon to the base of the proximal phalanx of the thumb

- **Deep palmar arch** – direct continuation of the radial a.; continues with the deep branch of the ulnar a.; deep to the long flexor tendons and in contact with the bases of the metacarpal bones

  - **Palmar metacarpal aa.** – anastomose with the common palmar digital aa. from the superficial palmar arch

- **Princeps pollicis a.** – courses along the medial side of the thumb

- **Deep ulnar n.** – courses deep to the abductor digiti minimi and flexor digiti minimi mm.; perforates the opponens digiti minimi m.

- **Palmar interossei mm.** (3) – three muscles located between the metacarpal bones on the palmar surface of the hand (adduct digits 2, 4, and 5)

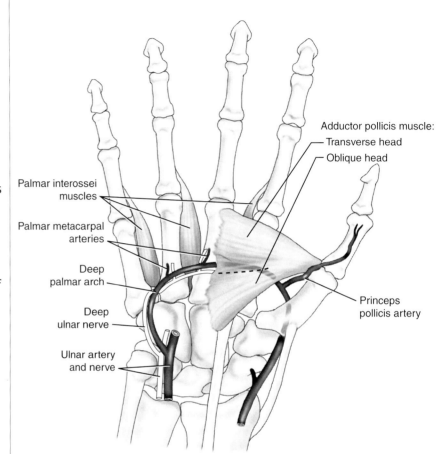

Adductor pollicis muscle:
— Transverse head
— Oblique head

Palmar interossei muscles

Palmar metacarpal arteries

Deep palmar arch

Deep ulnar nerve

Ulnar artery and nerve

Princeps pollicis artery

■ **Figure 28–9   Deeper structures of the central compartment of the right palm**

Dorsal Surface of the Hand

■ Turn the hand prone and identify the following:

- **Dorsal interossei mm.** (4) – the four muscles that occupy the intervals between the dorsal metacarpal bones (abduct digits 2, 3, and 4) (Figure 28–10*A*)

  - **First dorsal interosseus m.** – the most obvious; the radial a. separates its two proximal attachments between metacarpals 1 and 2

- **Anatomical snuff box** – region bounded by the tendons of the abductor pollicis longus, extensor pollicis brevis, and extensor pollicis longus mm.

  - **Radial a.** – courses through the floor of the anatomical snuff box to enter the palmar surface of the hand; courses between the two heads of the adductor pollicis m.

- **Superficial radial n.** – usually exits the forearm en route to the dorsum of the hand; courses between the extensor carpi radialis brevis and abductor pollicis longus mm. (Figure 28–10*B*)

  - Provides cutaneous innervation to much of the dorsum of the hand, thumb, and proximal regions of digits 2, 3, and the radial half of 4

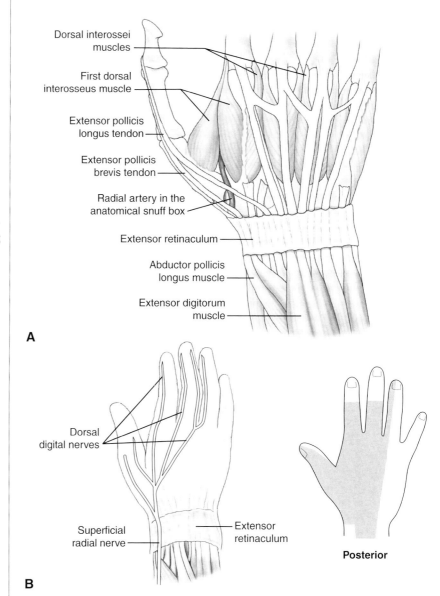

■ **Figure 28-10   Dorsal surface of the right hand. *A*, Structures. *B*, Superficial radial nerve**

Dorsal Hood and Extensor Expansion of the Digits

■ Identify the following (Figure 28–11):

- **Dorsal hood** – associated with the metacarpophalangeal joint; stabilizes the lumbrical and interossei tendons and provides leverage to the extensor digitorum tendons

- **Extensor expansion** – the four tendons from the extensor digitorum m. spread out over the metacarpophalangeal joint to form the extensor expansion

  - Triangular shaped, with the base facing proximally and the apex facing distally

  - The extensor expansions of digit 2 (index finger) and digit 5 are joined by the tendons of the extensor indicis and extensor digiti minimi mm., respectively

  - The lumbrical and interosseous mm. attach to the proximal sides of the extensor expansion

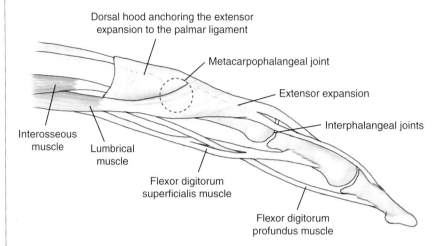

■ **Figure 28–11   Lateral view of digit three**

Upper Limb – Joints

Prior to dissection, you should familiarize yourself with the following structures:

SHOULDER

- Sternoclavicular joint
  - Articular disc
- Costoclavicular ligament
- Acromioclavicular joint
- Coracoclavicular ligaments
  - Conoid ligament
  - Trapezoid ligament
- Glenohumeral joint capsule
  - Posterior region of the joint capsule
  - Superior glenohumeral ligament
  - Middle glenohumeral ligament
  - Inferior glenohumeral ligament

- Subscapular bursa
- Coracohumeral ligament
- Transverse humeral ligament
- Glenoid labrum
- Articular cartilage
- Glenoid cavity
- Tendon of the long head of the biceps brachii m.

ELBOW

- Joint capsule
- Radial collateral ligament of the elbow

- Ulnar collateral ligament of the elbow
- Annular ligament

WRIST

- Radial collateral ligament of the wrist
- Ulnar collateral ligament of the wrist
- Intercarpal ligaments

HAND AND DIGITS

- Metacarpophalangeal joints
- Interphalangeal joints
- Medial collateral digital ligaments
- Lateral collateral digital ligaments

Upper Limb – Joint Dissection Overview

■ Dissection will be done in several steps:

- First, muscles that surround the joints will be removed from the joint to reveal the joint capsule

- Second, the joint capsule and tendons that traverse the capsule will be cut to reveal the articulation

Sternoclavicular Joint

■ Articulation of the clavicle and the manubrium of the sternum

■ Identify the following (Figure 29–1):

- **Sternoclavicular joint** – clean and open the joint capsule

  - **Articular disc** – located between the manubrium, the clavicle, and rib 1

- **Costoclavicular ligament** – attached to rib 1 and the inferior border of the clavicle

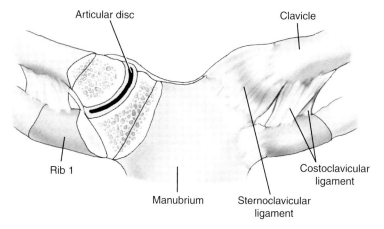

Articular disc

Clavicle

Rib 1

Manubrium

Sternoclavicular ligament

Costoclavicular ligament

■ **Figure 29–1** **Sternoclavicular joint**

Shoulder Joint

■ Identify the following (Figure 29–2):

- **Coracoclavicular ligament** – consists of two ligaments:

  - **Conoid ligament** – courses between the conoid tubercle and superior aspect of the coracoid process

  - **Trapezoid ligament** – attached to the superior aspect of the coracoid process and the trapezoid line on the inferior aspect of the clavicle

- **Acromioclavicular joint** – open with a scalpel

  - **Articular disc** – difficult to see; the disc separates the joint surfaces

- **Coracoacromial ligament** – ligamentous attachment between the coracoid process and the anterior tip of the acromion

- **Coracohumeral ligament** – a strong, ligamentous band that attaches the coracoid process to the anterior region of the greater tubercle of the humerus

- **Transverse humeral ligament** – covers the intertubercular groove (between the greater and lesser tubercles); crosses over the tendon of the long head of the biceps brachii m.

- **Subscapular bursa** – deep to the subscapularis m.; attempt to identify the communicating openings between the subscapular bursa and the synovial joint capsule

- **Rotator cuff** – observe the tendons of the teres minor, infraspinatus, supraspinatus, and subscapularis mm. where they insert into the glenohumeral joint capsule

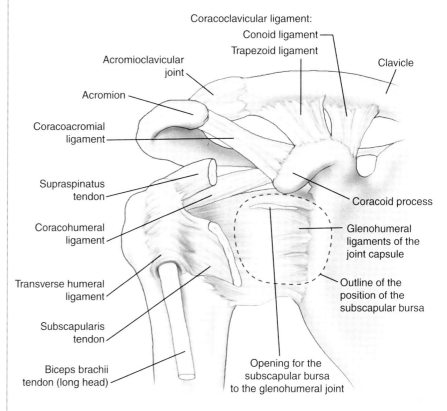

■ **Figure 29–2 Right shoulder joint**

Glenohumeral Joint Capsule

- The glenohumeral joint capsule is attached from the periphery of the glenoid fossa of the scapula to the anatomic neck of the humerus

- To identify the ligaments of the glenohumeral joint capsule, cut the tendon of the supraspinatus m. from its attachment and cut the subscapularis tendon from the lesser tubercle attachment; reflect both muscles

  - **Glenohumeral ligaments** – intrinsic thickenings in the joint capsule anteriorly; there are **superior, middle, and inferior glenohumeral ligaments** (Figure 29–3A) and one posterior glenohumeral ligament

- To better study the structures composing the glenohumeral joint, make a vertical incision through the joint capsule anteriorly (see Figure 29–3A) and identify the following (Figure 29–3B):

  - **Glenoid labrum** – fibrocartilaginous ring that adds depth to the shoulder joint

  - **Articular cartilage** – hyaline cartilage covering the articular surfaces

  - **Glenoid cavity** – the shallow socket of the glenoid fossa into which rests the head of the humerus

  - **Long head of the biceps brachii m.** – trace the tendon of this muscle to its attachment on the supraglenoid tubercle of the scapula; also trace the tendon of the long head of the triceps brachii m. to the infraglenoid tubercle of the scapula

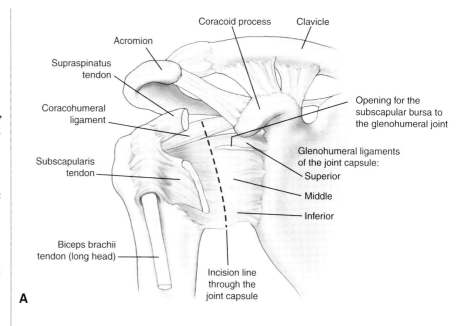

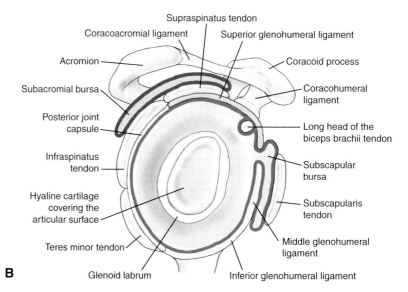

- **Figure 29–3   Right shoulder joint. A, Anterior view of the joint capsule. B, Lateral view of the opened joint capsule**

Elbow Joint

- Palpate the olecranon process of the ulna, and the medial and lateral epicondyles of the humerus

- Remove the muscles, arteries, veins, and nerves from the anterior, medial, and lateral surfaces of one elbow joint

- Clean the underlying capsule of the elbow joint

- Identify the following ligaments (Figure 29–4):

  - **Ulnar collateral ligament:**
    - Anterior portion – between the medial epicondyle of the humerus and coronoid process of the ulna
    - Posterior portion – between the medial epicondyle of the humerus and olecranon process of the ulna

  - **Radial collateral ligament** – between the lateral epicondyle of the humerus and neck of the radius

  - **Annular ligament** – courses around the head of the radius; attaches to the anterior and posterior margins of the radial notch of the ulna; this ligament and its associated capsule form the proximal radioulnar joint

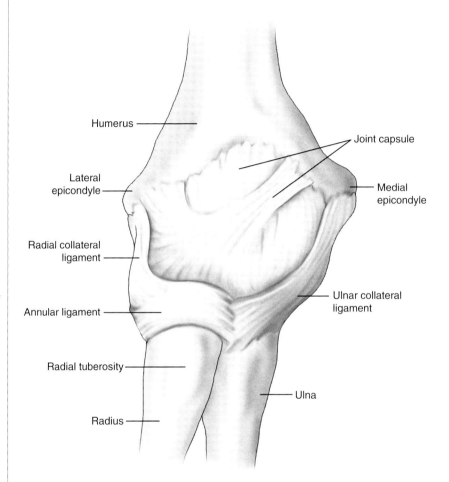

■ **Figure 29–4   Right elbow joint**

Wrist Joint

- If you desire to dissect the wrist joint, you will need to remove all muscles, arteries, veins, and nerves from the anterior region of one wrist and hand

- Using a skeleton, identify the following bones (Figure 29–5A):

  - Proximal row
    - **Pisiform (1)**
    - **Triquetrum (2)**
    - **Lunate (3)**
    - **Scaphoid (4)**
  - Distal row
    - **Trapezium (5)**
    - **Trapezoid (6)**
    - **Capitate (7)**
    - **Hamate (8)**
    - **Metacarpals (9)**

- Identify the following (Figure 29–5B):

  - **Radiocarpal joint** – the distal end of the radius has two joint surfaces to articulate with only the scaphoid and lunate carpal bones

  - **Radial collateral ligament** – attached to the styloid process of the radius and the scaphoid

  - **Ulnar collateral ligament** – attached to the styloid process of the ulna and the triquetrum

  - **Palmar radiocarpal ligaments** – attached to the radius and the two rows of carpal bones

  - **Palmar ulnocarpal ligament** – attached to the ulna and the two rows of carpal bones

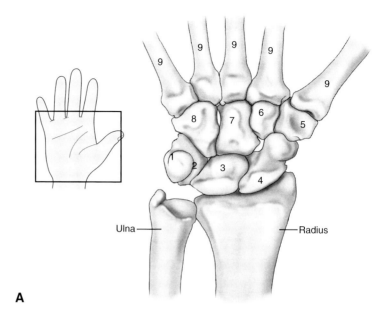

A

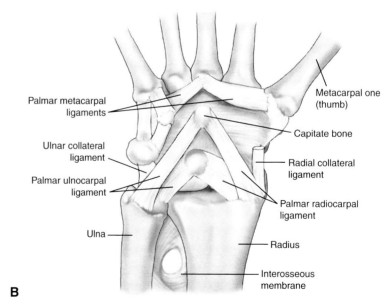

B

■ **Figure 29–5   Carpal bone joints of the right hand. *A*, Osteology. *B*, Ligaments**

Joints of the Hand and Digits

■ Remove muscles and tendons to observe the joint capsules of the thumb and one digit

■ Identify the following (Figure 29–6):

- **Carpometacarpal ligaments** – attached to the distal row of carpal bones and the base of the metacarpal bones

- **Saddle joint** – articulation of the trapezium and base of the first metacarpal bones (thumb only)

- **Hinge joints** – the interphalangeal joints of the thumb and digits 2–5

- **Joint capsule** – each **metacarpophalangeal** (MP) and **interphalangeal joint** (IP) is surrounded by its own joint capsule

- **Medial** and **lateral collateral digital ligaments** – medial and lateral thickenings of the joint capsule

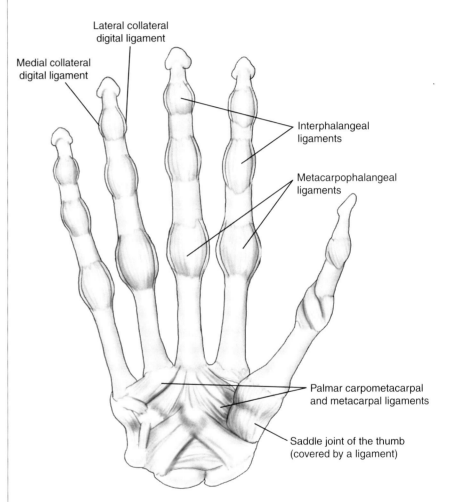

■ Figure 29–6   Joints of the right hand

Unit 4 Upper Limb Overview

At the end of Unit 4, you should be able to identify the following structures on cadavers, skeletons, and/or radiographs:

Osteology

- Scapula
  - Subscapular fossa
  - Spine
  - Supraspinous fossa
  - Infraspinous fossa
  - Acromion
  - Medial and lateral borders
  - Superior angle
    - Suprascapular notch
  - Inferior angle
  - Glenoid cavity
  - Supraglenoid and infraglenoid tubercles
  - Coracoid process
- Clavicle
  - Acromial and sternal ends
  - Conoid tubercle
- Humerus
  - Head
  - Anatomic and surgical necks
  - Greater and lesser tubercles
  - Intertubercular (bicipital) groove
  - Radial groove
  - Deltoid tuberosity
  - Lateral and medial supracondylar ridges
  - Lateral and medial epicondyles
  - Trochlea
  - Capitulum

- Coronoid fossa
- Olecranon fossa
- Radius
  - Head
  - Neck
  - Radial tuberosity
  - Interosseous border
  - Radial styloid process
  - Ulnar notch
- Ulna
  - Olecranon
  - Coronoid process
    - Radial notch
  - Trochlear notch
  - Interosseous border
  - Supinator crest
  - Ulnar head
    - Ulnar styloid process
- Hand
  - Carpal bones (8)
    - Scaphoid
    - Lunate
    - Triquetrum
    - Pisiform
    - Hamate
      - Hook of the hamate
    - Capitate
    - Trapezoid

  - Trapezium
  - Metacarpal bones (5)
  - Phalanges (14) Proximal, middle, and distal phalanges

Muscles

- Shoulder
  - Deltoid m.
  - Trapezius m.
  - Levator scapulae m.
  - Serratus anterior m.
  - Pectoralis minor m.
  - Rhomboid major m.
  - Rhomboid minor m.
  - Subclavius m.
  - Rotator cuff muscles
    - Supraspinatus m.
    - Infraspinatus m.
    - Teres minor m.
    - Subscapularis m.
  - Teres major m.
  - Pectoralis major m.
  - Latissimus dorsi m.
- Arm
  - Coracobrachialis m.
  - Biceps brachii m.
    - Long head
    - Short head

- Brachialis m.
- Triceps brachii m.
  - Long head
  - Lateral head
  - Medial head
- Anterior compartment of the forearm
  - Pronator teres m.
  - Flexor carpi radialis m.
  - Flexor carpi ulnaris m.
  - Palmaris longus m.
  - Flexor digitorum superficialis m.
  - Flexor digitorum profundus m.
  - Flexor pollicis longus m.
  - Pronator quadratus m.
- Posterior compartment of the forearm
  - Brachioradialis m.
  - Extensor carpi radialis longus m.
  - Extensor carpi radialis brevis m.
  - Extensor digitorum m.
  - Extensor digiti minimi m.
  - Extensor carpi ulnaris m.
  - Extensor indicis m.
  - Extensor pollicis longus and brevis mm.
  - Abductor pollicis longus m.
  - Anconeus m.
  - Supinator m.
- Hand
  - Palmaris brevis m.
  - Thenar mm.
    - Abductor pollicis brevis, opponens pollicis, and flexor pollicis brevis mm.
  - Adductor pollicis m.
  - Hypothenar mm.
    - Abductor digiti minimi, opponens digiti minimi, and flexor digiti minimi mm.
  - Lumbrical mm.
  - Dorsal interossei mm.
  - Palmar interossei mm.

Arteries
- Aorta
  - Subclavian a.
    - Thyrocervical trunk
      - Transverse cervical a.
        - Superficial branch
        - Deep branch (dorsal scapular a.)
      - Suprascapular a.
  - Axillary a.
    - Superior thoracic a.
    - Thoracoacromial a.
      - Acromial, deltoid, pectoral, and clavicular branches
    - Lateral thoracic a.
    - Subscapular a.
      - Circumflex scapular a.
      - Thoracodorsal a.
    - Anterior humeral circumflex a.
    - Posterior humeral circumflex
  - Brachial a.
    - Deep a. of the arm
      - Radial collateral a.
      - Middle collateral a.
    - Superior ulnar collateral a.
    - Inferior ulnar collateral a.
  - Radial a.
    - Radial recurrent a.
    - Palmar carpal branch
    - Superficial palmar arch
    - Dorsal metacarpal aa.
    - Dorsal digital aa.
    - Princeps pollicis a.
    - Deep palmar arch
      - Palmar metacarpal aa.
- Ulnar a.
  - Anterior ulnar recurrent a.
  - Posterior ulnar recurrent a.
  - Common interosseous a.
    - Anterior interosseous a.
    - Posterior interosseous a.
      - Interosseous recurrent a.
  - Palmar carpal branch
  - Deep palmar branch
  - Superficial palmar arch
  - Common palmar digital aa.
    - Proper palmar digital aa.

Veins
- Axillary v.
  - Subscapular v.
  - Circumflex scapular v.
    - Thoracodorsal v.
    - Posterior circumflex humeral v.
    - Anterior circumflex humeral v.
  - Lateral thoracic v.
- Superficial veins of the upper limb
  - Cephalic v.
    - Thoracoacromial v.
  - Basilic v.
  - Median cubital v.
  - Dorsal venous network of the hand
- Deep veins of the upper limb
  - Brachial v.
  - Ulnar v.
  - Radial v.

Lymphatics
- Axillary lymph nodes

Nerves
- *Five Roots (C5, C6, C7, C8, T1)*
  - Dorsal scapular n. and long thoracic n.
- *Three Trunks*
  - Superior – suprascapular n. and n. to the subclavius m.
  - Middle
  - Inferior
- *Six Divisions*
  - Three anterior and three posterior

Three Cords
- Lateral
  - Lateral pectoral n.
- Medial
  - Medial pectoral n.
  - Medial cutaneous n. of the arm
  - Medial cutaneous n. of the forearm
- Posterior
  - Upper subscapular n.
  - Thoracodorsal n.
  - Lower subscapular n.

Five Terminal Branches
- Musculocutaneous n.
  - Muscular branches
  - Lateral cutaneous n. of the forearm
- Median n.
  - Muscular branches
  - Palmar branch
  - Recurrent branch of the median n.
  - Digital branches of the median n. (common and proper palmar digital nn.)
- Ulnar n.
  - Muscular branches
  - Superficial branch of the ulnar n.
  - Deep branch of the ulnar n.
  - Digital branches of the ulnar n. (common and proper palmar digital nn.)
- Radial n.
  - Muscular branches
  - Posterior cutaneous nn. of the arm and forearm

- Inferior lateral cutaneous n. of the arm
- Deep branch (posterior interosseous n.)
- Superficial branch
- Axillary n.
  - Muscular branches
  - Superior lateral cutaneous n. of the arm

Joints
- Shoulder
  - Sternoclavicular joint
    - Articular disc
  - Costoclavicular joint
  - Acromioclavicular joint
  - Coracoclavicular ligaments
    - Conoid ligament
    - Trapezoid ligament
  - Glenohumeral capsule
    - Posterior region of the joint capsule
    - Superior glenohumeral ligament
    - Middle glenohumeral ligament
    - Inferior glenohumeral ligament
  - Subscapular bursa
  - Coracohumeral ligament
  - Transverse humeral ligament
  - Glenoid labrum
  - Articular cartilage
  - Tendon of the long head of the biceps brachii m.
- Elbow
  - Joint capsule
  - Radial collateral ligament of the elbow

- Ulnar collateral ligament of the elbow
- Annular ligament
- Wrist
  - Radial collateral ligament of the wrist
  - Ulnar collateral ligament of the wrist
  - Intercarpal ligaments
- Digits
  - Metacarpophalangeal joints
  - Interphalangeal joints
  - Lateral collateral digital ligaments
  - Medial collateral digital ligaments

Fascia
- Supraspinous fascia
- Infraspinous fascia
- Axillary fascia
- Suspensory ligament of the axilla
- Brachial fascia
- Medial intermuscular septum of the arm
- Lateral intermuscular septum of the arm
- Antebrachial fascia
- Flexor and extensor retinacula
- Carpal tunnel
- Fibrous digital sheaths
- Extensor expansion

Miscellaneous
- Deltopectoral triangle
- Anatomical snuff box

Lower Limb – Osteology and Superficial Structures

Prior to dissection, you should familiarize yourself with the following structures:

OSTEOLOGY

- Features of the os coxa
  - Acetabulum
  - Obturator foramen
  - Ischiopubic ramus
- Bones of the os coxae
  - Ilium
    - Ala (wing) of the ilium
    - Iliac crest
      - Anterior superior iliac spine
      - Anterior inferior iliac spine
      - Posterior superior iliac spine
      - Posterior inferior iliac spine
      - Iliac fossa
    - Anterior, posterior, and inferior gluteal lines
    - Auricular surface
  - Ischium
    - Ischial tuberosity, ramus, and spine
    - Greater sciatic notch
    - Lesser sciatic notch
  - Pubis
    - Pubic tubercle and crest
    - Superior pubic ramus
      - Pecten pubis (pectineal line)

- Inferior pubic ramus
- Obturator groove and crest
- Pubic arch

- Pelvis
  - Pelvic cavity
  - Symphysis pubis
- Femur
  - Head
    - Fovea for the ligament of the head
  - Neck
    - Greater trochanter
      - Trochanteric fossa
    - Lesser trochanter
    - Intertrochanteric line and crest
  - Linea aspera
  - Pectineal line
  - Gluteal tuberosity
  - Medial condyle and epicondyle
    - Adductor tubercle
  - Lateral condyle and epicondyle
  - Popliteal and patellar surfaces
- Patella
  - Base and apex
  - Articular and anterior surfaces

- Tibia
  - Medial and lateral condyles
  - Intercondylar eminence
  - Tibial tuberosity
  - Soleal line
  - Interosseous border
  - Medial malleolus
  - Fibular notch
  - Inferior articular surface
- Fibula
  - Head and neck
  - Interosseous border
  - Lateral malleolus
- Foot
  - Tarsal bones (7)
    - Talus
    - Calcaneus – sustentaculum tali
    - Navicular
    - Medial cuneiform
    - Intermediate cuneiform
    - Lateral cuneiform
    - Cuboid
  - Metatarsal bones (5)
  - Phalanges (14)
    - Proximal, middle, and distal

SUPERFICIAL STRUCTURES

- Superficial veins
  - Great saphenous v.
  - Small saphenous v.
  - Dorsal venous arch
- Superficial nerves
  - Lateral cutaneous n. of the thigh
  - Medial and intermediate cutaneous nn. of the thigh
  - Posterior cutaneous n. of the thigh
  - Cutaneous branch of the obturator n.
  - Medial and lateral sural cutaneous nn.
  - Saphenous n.
  - Superficial fibular n.
  - Dorsal digital nn.
- Lymphatic system
  - Femoral lymph nodes
  - Inguinal lymph nodes

TABLE 30-1 Osteology of the Lower Limb

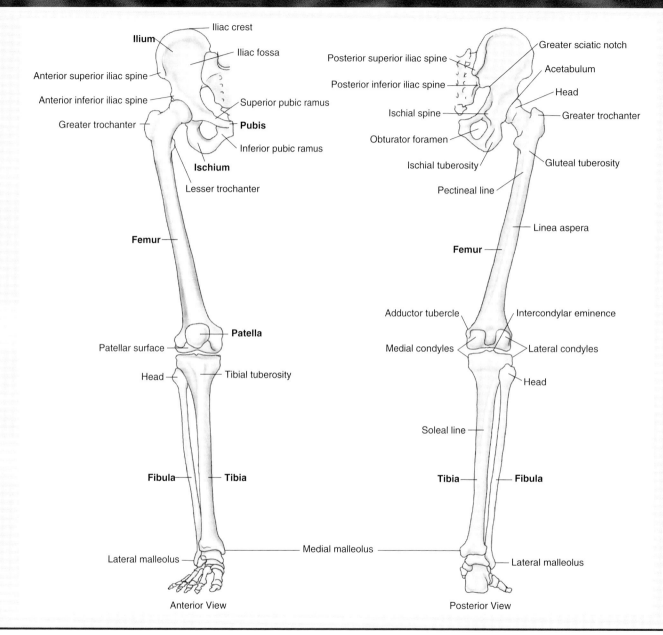

Anterior View

Posterior View

Lower Limb – Skin Removal

- Make the following circular incisions on the lower limb (Figure 30–1):

  - Proximal on the thigh (*A*)

  - Inferior to the patella (*B*)

  - Ankle (*C*)

  - About 10 cm distal to the ankle on the dorsal surface of the foot along the metatarsophalangeal joints (*D*)

- Join the three circular and one semilunar incisions with a longitudinal incision on the anterior aspect of the lower limb (*A–D*)

- Remove the skin of the lower limb, leaving the superficial fascia intact

  - Do *not* cut the superficial nerves and veins in the superficial fascia

  - Leave the flaps of skin between *C* and *D* attached at this time, but reflect the flaps from the dorsum of the foot

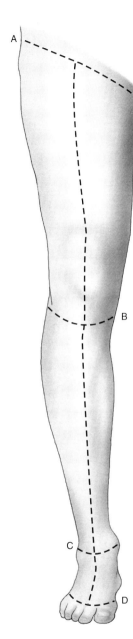

■ **Figure 30–1   Right lower limb skin incisions**

Lower Limb – Superficial Veins

■ Identify the following veins in the superficial fascia (Figure 30–2):

- **Dorsal venous arch** – identify the veins on the dorsal (superior) aspect of the foot; main tributary to the great saphenous v. (see Figure 30–2A)

- **Great saphenous v**. – the longest vein in the body; begins at the medial side of the dorsal venous arch of the foot and ascends anterior to the medial malleolus, then ascends along the medial aspect of the leg; at the knee

  - Ascends posterior to the medial condyle of the femur, then courses along the anteromedial aspect of the thigh

  - Terminates in the thigh by passing through the deep fascia (saphenous opening), inferior to the inguinal ligament, as a tributary of the femoral v.

- **Small saphenous v**. – begins on the lateral surface of the foot (see Figure 30–2B); ascends posterior to the lateral malleolus and penetrates the deep fascia in the popliteal fossa to terminate in the popliteal v., between the two heads of the gastrocnemius m.

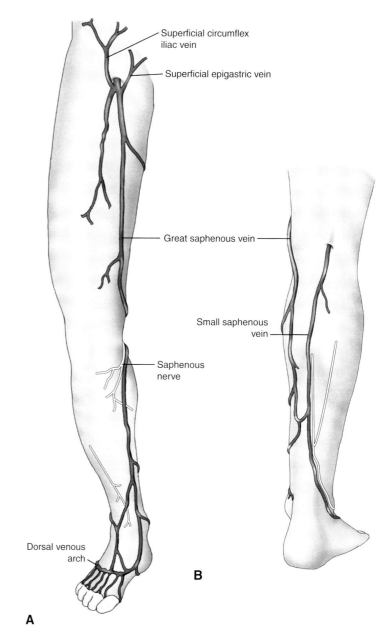

■ **Figure 30–2   Superficial veins of the right lower limb. A, Anterior view. B, Posterior view**

Lower Limb – Cutaneous Nerves

■ Identify the following superficial cutaneous nerves (Figure 30–3):

- **Lateral cutaneous n. of the thigh** – passes deep to the inguinal ligament, inferior to the anterior superior iliac spine, before piercing the deep fascia to supply the skin on the lateral aspect of the thigh

- **Cutaneous branch of the obturator n.** – located on the medial surface of the thigh; inferior to the saphenous ring and medial to the great saphenous v.

- **Medial and intermediate cutaneous nn. of the thigh** – branches emerge along the medial border of the sartorius m. and course along the anteromedial surface of the thigh

- **Saphenous n.** – courses along the deep surface of the sartorius m. and pierces the fascia distal and medial to the knee; follows the great saphenous v. in the leg

- **Posterior cutaneous n. of the thigh** – emerges from under the inferior border of the gluteus maximus m.; courses along the posterior surface of the thigh

- **Sural n.** – emerges from the deep fascia near the middle of the posterior aspect of the leg; follow this n. inferiorly as it courses together with the small saphenous v., posterior to the lateral malleolus

- **Superficial fibular n.** – pierces the deep fascia in the lateral distal third of the leg; follow it distally to the dorsum of the foot

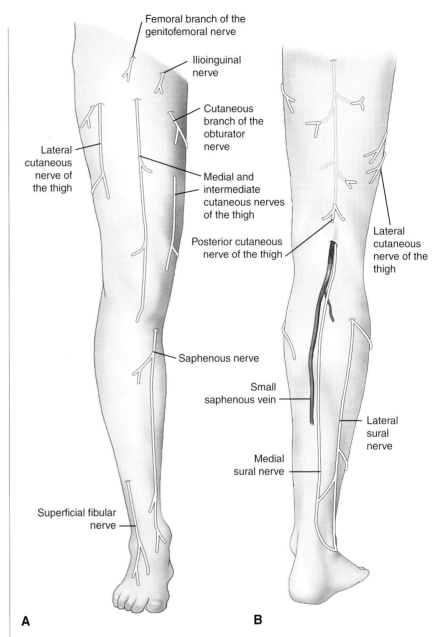

A

B

■ **Figure 30–3   Cutaneous nerves of the right lower limb. A, Anterior view. B, Posterior view**

Lower Limb – Superficial Lymph Nodes

■ Identify the following lymph nodes of the thigh (Figure 30–4):

● **Superficial inguinal lymph nodes**

  ● Horizontal group – inferior to the inguinal ligament

  ● Vertical group – bilaterally flank the great saphenous v. where the vein penetrates the deep fascia

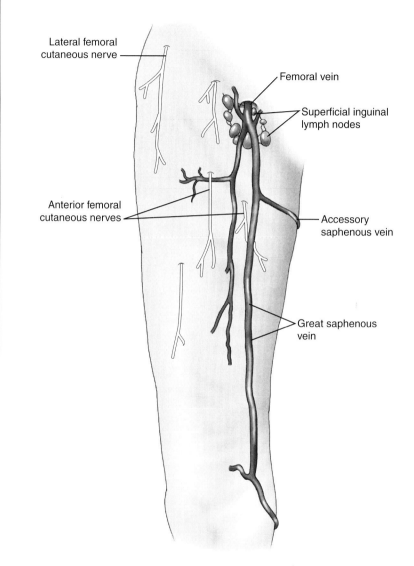

Lateral femoral cutaneous nerve

Femoral vein

Superficial inguinal lymph nodes

Anterior femoral cutaneous nerves

Accessory saphenous vein

Great saphenous vein

■ **Figure 30–4   Superficial lymph nodes of the right thigh**

Lower Limb – Musculature

Lab 31

Prior to dissection, you should familiarize yourself with the following structures:

MUSCLES

- Gluteal region
  - Tensor fascia lata m.
  - Gluteus maximus m.
  - Gluteus medius m.
  - Gluteus minimus m.
  - Piriformis m.
  - Superior gemellus m.
  - Obturator internus m.
  - Inferior gemellus m.
  - Quadratus femoris m.
- Iliopsoas m.
  - Iliacus m.
  - Psoas major m.
- Psoas minor m.
- Anterior muscles of the thigh
  - Quadriceps femoris m.
    - Rectus femoris m.
      - Vastus medialis m.
      - Vastus intermedius m.
      - Vastus lateralis m.
  - Sartorius m.
- Medial muscles of the thigh
  - Pectineus m.
  - Adductor brevis m.
  - Adductor longus m.
  - Adductor magnus m.
  - Gracilis m.
  - Obturator externus m.
- Posterior muscles of the thigh
  - Biceps femoris m.
  - Semitendinosus m.
  - Semimembranosus m.
- Lateral muscles of the leg
  - Fibularis (peroneus) longus m.
  - Fibularis (peroneus) brevis m.
- Anterior muscles of the leg
  - Tibialis anterior m.
  - Extensor digitorum longus m.
  - Extensor digitorum brevis m.
  - Extensor hallucis longus m.
  - Extensor hallucis brevis m.
  - Fibularis (peroneus) tertius m.
- Posterior muscles of the leg
  - Superficial layer
    - Gastrocnemius m.
    - Soleus m.
  - Deep layer
    - Tibialis posterior m.
    - Flexor hallucis longus m.
    - Flexor digitorum longus m.
    - Plantaris m.
    - Popliteus m.

TABLE 31-1 Muscles of the Lower Limb

Muscle	Proximal Attachment	Distal Attachment	Action	Innervation
Iliopsoas				
• Psoas major	T12–L5 vertebrae	Lesser trochanter of the femur	Flexes the thigh and vertebral column	Lumbar nn. (L1–L4)
• Iliacus	Iliac fossa and crest	Lesser trochanter of the femur	Flexes the thigh	Femoral n. (L2–L3)
• Psoas minor	T12–L1 vertebrae	Pectineal line of the femur	Weak flexor of the pelvis	Lumbar n. (L1)
Muscles of the anterior region of the thigh				
• Tensor fascia lata	Anterior superior iliac spine and ventral iliac crest	Lateral condyle of the tibia via the iliotibial tract	Abducts, medially rotates, and flexes the thigh	Superior gluteal n. (L4–L5)
• Sartorius	Anterior superior iliac spine	Proximal and medial surfaces of the tibia	Flexes, abducts, and laterally rotates the thigh; flexes the leg	Femoral n. (L2–L3)
Quadriceps femoris • Rectus femoris	Anterior inferior iliac spine	All four muscles of the quadriceps femoris form a common tendon that envelopes the patella and inserts on the tibial tuberosity	Flexes the thigh; extends the leg	Femoral n. (L2–L4)
• Vastus medialis	Intertrochanteric line and the linea aspera		Extend the leg	
• Vastus lateralis	Linea aspera			
• Vastus intermedius	Anterior and lateral surfaces of the femur			
Muscles of the medial region of the thigh				
• Pectineus	Pectineal line of the superior pubic ramus	Pectineal line of the femur	Adducts and flexes the thigh	Femoral n. (L2–L3); may receive a branch from the obturator n.
• Adductor longus	Pubis	Linea aspera (medial third)	Adducts the thigh	Obturator n. (L2–L4)
• Adductor brevis	Inferior ramus of the pubis	Linea aspera (superior part)		
• Adductor magnus	Ramus of the ischium and ischial tuberosity	Linea aspera and adductor tubercle of the femur		Obturator n. (L2–L3); tibial division of the sciatic n. (L2–L4)

TABLE 31-1   Muscles of the Lower Limb—cont'd

Muscle	Proximal Attachment	Distal Attachment	Action	Innervation
• Gracilis	Inferior ramus and body of the pubis	Proximal, medial surface of the tibia	Adducts and flexes the thigh	Obturator n. (L2–L3)
• Obturator externus	Margins of the obturator foramen and obturator membrane	Trochanteric fossa of the femur	Laterally rotates the thigh	Obturator n. (L3–L4)
Muscles of the posterior region of the thigh				
• Gluteus maximus	Ilium and sacrum	Iliotibial tract and gluteal tuberosity	Extends and laterally rotates the thigh; extends the trunk when the distal attachment is fixed	Inferior gluteal n. (L5–S2)
• Gluteus medius	External surface of the ilium	Greater trochanter of the femur	Abducts and medially rotates the thigh	Superior gluteal n. (L5–S1)
• Gluteus minimus				
• Piriformis	Anterior surface of the sacrum			Ventral rami S1–S2
• Obturator internus	Pelvic surface of the obturator membrane	Greater trochanter of the femur	Rotate the thigh laterally	Nerve to the obturator internus (L5–S1)
• Superior gemellus	Ischial spine			
• Inferior gemellus				Nerve to the quadratus femoris (L5–S1)
• Quadratus femoris				
• Semitendinosus	Ischial tuberosity	Medial surface of the proximal end of the tibia	Extends the thigh; flexes the leg and medially rotates the leg	Tibial n. (L5–S2)
• Semimembranosus		Posterior medial surface of the condyle of the tibia	Extends the thigh; flexes the leg and medially rotates the leg	
• Biceps femoris	Long head: Ischial tuberosity Short head: Lateral lip of the linea aspera of the femur	Head of the fibula	Long head: Extends the thigh Long and short heads: Flex the leg and laterally rotate the leg	Long head: Tibial n. (S1–S3) Short head: Fibular n. (L5–S2)
Muscles of the anterior region of the leg				
• Tibialis anterior	Lateral proximal surface of the tibia	Base of the first metatarsal bone	Dorsiflexes and inverts the foot	Deep fibular n. (L4–L5)

Continued

TABLE 31-1 Muscles of the Lower Limb—cont'd

Muscle	Proximal Attachment	Distal Attachment	Action	Innervation
• Extensor digitorum longus	Lateral condyle of the tibia	Middle and distal phalanx of digits 2–5	Extends digits 2–5 and dorsiflexes the foot	
• Extensor hallucis longus	Fibula	Distal phalanx of the great toe	Extends the great toe and dorsiflexes the foot	Deep fibular n. (L5–S1)
• Fibularis tertius		Base of the fifth metatarsal bone	Dorsiflexes and everts the foot	
Muscles of the lateral region of the leg				
• Fibularis longus	Head and proximal surface of the fibula	Base of the first metatarsal and medial cuneiform bones	Evert and plantar flex the foot	Superficial fibular n. (L4–S1)
• Fibularis brevis	Distal two-thirds of the lateral surface of the fibula	Dorsal base of the fifth metatarsal bone		
Muscles of the posterior region of the leg				
• Gastrocnemius	Medial and lateral condyles of the femur		Flexes the leg and plantar flexes the foot	
• Soleus	Posterior aspect of the fibula and tibia	Posterior surface of the calcaneus via the calcaneal tendon	Plantar flexes the foot	Tibial n. (S1–S2)
• Plantaris	Lateral supracondylar line of the femur		Assists the gastrocnemius; flexes the leg and plantar flexes the foot	
• Popliteus	Lateral condyle of the femur	Posterior surface of the proximal end of the tibia	Weakly flexes the leg; rotates the tibia on the femur and unlocks the leg	Tibial n. (L4–S1)
• Flexor hallucis longus	Distal two-thirds of the posterior surface of the fibula and the interosseous membrane	Base of the distal phalanx of the great toe	Flexes the great toe and plantar flexes the foot	Tibial n. (S2–S3)
• Flexor digitorum longus	Medial posterior surface of the tibia	Base of the distal phalanx of digits 2–5	Flexes digits 2–5 and plantar flexes the foot	
• Tibialis posterior	Posterior surface of the fibula, interosseus membrane, and tibia	Navicular, cuneiform, and cuboid tarsal bones and metatarsal bones 2–4	Plantar flexes and inverts the foot	Tibial n. (L4–L5)

Anterior Muscles of the Thigh

■ Identify the following (Figure 31–1):

- **Femoral triangle**

  - Borders – sartorius m., adductor longus m., and inguinal ligament

- **Sartorius m.** – longest muscle in the body; crosses two joints (hip and knee)

- **Quadriceps femoris mm.** (4 heads); inferolateral to the sartorius m.

  - **Rectus femoris m**. – the most superficial and medial muscle of the quadriceps femoris group

  - **Vastus lateralis m.** – lateral to the rectus femoris m.

  - **Vastus medialis m.** – medial to the rectus femoris m.

  - **Vastus intermedius m.** – deep to the rectus femoris m.

- **Tensor fascia lata m.** – most lateral of the thigh muscles; its tendon becomes the iliotibial tract

- **Iliopsoas m.** – union of the iliacus and psoas major mm.; posterolateral to the **femoral sheath** and **femoral n.**

- **Pectineus m.** – posteromedial to the femoral sheath and femoral n.

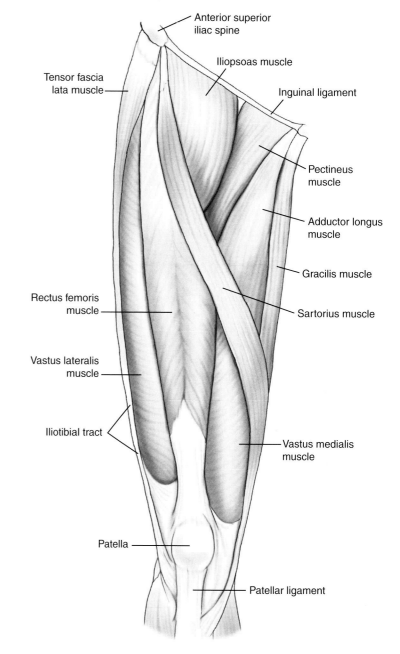

■ **Figure 31-1   Muscles of the anterior region of the right thigh**

Medial Muscles of the Thigh

■ Reflect the sartorius and rectus femoris mm. to identify the following (Figure 31–2A):

- **Gracilis m.** – the most medial muscle in the medial region of the thigh; courses vertically between the pubis and tibia

- **Pectineus m.** – posteromedial to the femoral sheath

- **Adductor longus m.** – inferior to the pectineus m.; reflect this muscle from the femoral attachment to see the deeper muscles

- **Adductor magnus m.** (Figure 31–2B) – inferior to the adductor brevis m. and deep to the adductor longus m.

  - **Adductor canal** – muscular channel that transmits the femoral a. and v. from the femoral sheath to the adductor hiatus

    ◆ Boundaries – sartorius m. (roof), vastus medialis m. (lateral wall and floor), and adductor magnus m. (medial wall and floor)

  - **Adductor hiatus** – an opening in the adductor magnus m. between the adductor tubercle on the medial condyle of the femur and the linea aspera; provides a passageway for the femoral a. and v. to reach the popliteal fossa

- **Adductor brevis m.** – deep to the pectineus and adductor longus mm.

- **Obturator externus m**. – deep to the iliopsoas and pectineus mm.; you may cut the iliopsoas and pectineus mm. on one side to better identify the obturator externus m.

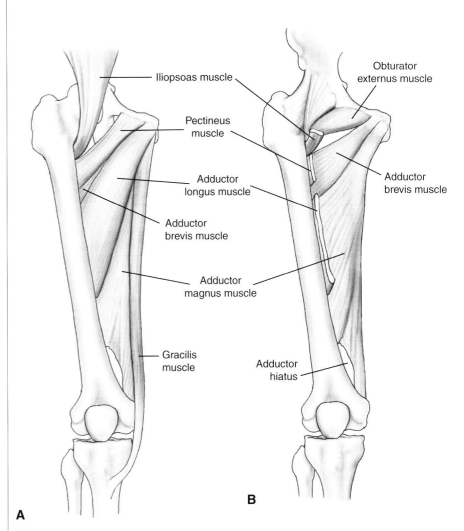

■ **Figure 31-2 Muscles of the medial region of the right thigh.** *A*, **Intermediate layer.** *B*, **Deep layer**

Muscles of the Gluteal Region

■ Turn the limb prone; identify the following muscles:

■ As you reveal the following muscles, preserve the nerve and blood supplies, which are identified in subsequent laboratory dissections

- **Gluteus maximus m.** – cut its attachment along the ilium, sacrum, and coccyx to reflect the muscle belly laterally (Figure 31–3A)

- **Gluteus medius m.** – deep to the gluteus maximus m.; cut the gluteus medius m. along the ilium to reflect the muscle belly laterally (Figure 31–3B)

- **Gluteus minimus m.** – deep to the gluteus medius m.

- **Piriformis m.** – observe that the **sciatic n.** exits the pelvis inferior to the piriformis m.; the sciatic n. and piriformis m. pass through the greater sciatic foramen

- **Superior gemellus m.** – inferior to the piriformis m.

- **Obturator internus m.** – inferior to the superior gemellus m.

- **Inferior gemellus m.** – inferior to the obturator internus m.

- **Quadratus femoris m**. – inferior to the inferior gemellus m.

- **Obturator externus m.** – search, with a probe, near the greater trochanter of the femur in the interval between the quadratus femoris and inferior gemellus mm. to find the tendon of the obturator externus m. (not shown)

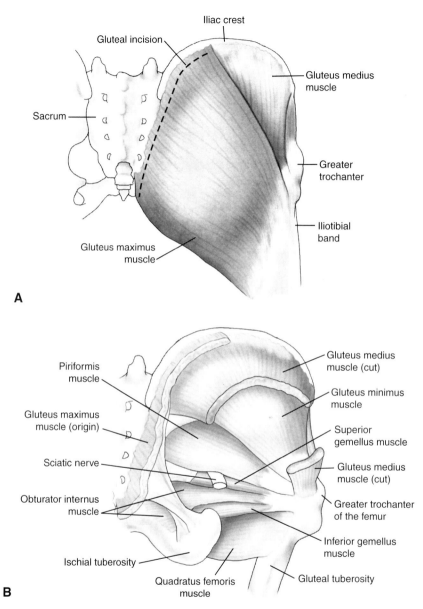

■ **Figure 31–3   Muscles of the gluteal region. A, Posterior view. B, Posterior view (gluteus maximus m. is removed and the gluteus medius m. is cut to show the gluteus minimus m.)**

Posterior Muscles of the Thigh

■ Identify the following (Figure 31–4):

● **Hamstrings** – a group of muscles that attach proximally to the ischial tuberosity; span both the hip and knee joints, and are innervated by the tibial division of the sciatic n.

 ● **Biceps femoris m. (long head only)** – the most lateral of the three hamstring muscles; attached distally to the fibular head

 ● **Semitendinosus m.** – middle of the three hamstring muscles; attached distally to the same region as the sartorius and gracilis mm., medial to the tibial tuberosity

 ● **Semimembranosus m.** – most medial of the three hamstring muscles; attached distally to the medial condyle of the tibia

● **Biceps femoris m. (short head)** – on the lateral side of the thigh (linea aspera) and deep to the long head of the biceps femoris m.

 ● Because the short head of the biceps femoris m. does not originate on the ischial tuberosity, span two joints, or have the same innervation as the hamstring muscles, it is not considered a hamstring muscle

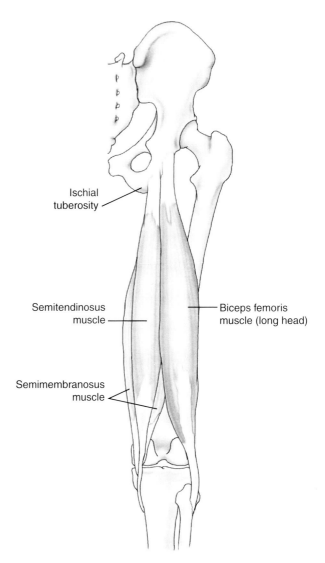

Ischial tuberosity

Semitendinosus muscle

Semimembranosus muscle

Biceps femoris muscle (long head)

■ **Figure 31–4   Hamstring muscles of the posterior region of the right thigh**

Muscles of the Popliteal Fossa

■ Greater access to the popliteal fossa may be gained by flexing the knee to ease the tension of the deep fascia and tendons; keep vessels and nerves intact

■ Identify the following (Figure 31–5):

● **Popliteal fossa** – a diamond-shaped region on the posterior aspect of the knee

  ◦ Boundaries – superior angle:

  ◆ Lateral border – biceps femoris tendon

  ◆ Medial border – semimembranosus and semitendinosus tendons

  ◦ Boundaries – inferior angle:

  ◆ Both heads of the gastrocnemius m.

● **Tibial n.** – pull the two heads of the gastrocnemius m. apart at the popliteal fossa to observe the tibial n.

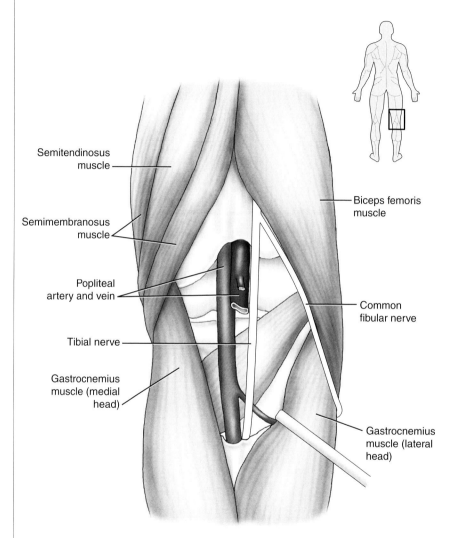

Semitendinosus muscle

Semimembranosus muscle

Popliteal artery and vein

Tibial nerve

Gastrocnemius muscle (medial head)

Biceps femoris muscle

Common fibular nerve

Gastrocnemius muscle (lateral head)

■ **Figure 31–5   Posterior view of the muscles of the right popliteal fossa**

Posterior Muscles of the Leg – Superficial Layer

- Turn the leg prone

- Identify the following muscles that converge on the calcaneal (Achilles) tendon (Figure 31–6):

  - **Gastrocnemius m.** – the most superficial muscle in the posterior region of the leg; attached to the condyles of the femur and the calcaneus; thus, this muscle spans two joints

  - **Plantaris m.** – the smallest of the three muscles in the superficial layer; the muscle and its tendon span two joints

- Cut both heads of the gastrocnemius m., where the two heads unite into one belly; keep vessels and nerves intact

- As you reflect the gastrocnemius m. inferiorly, identify the following:

  - **Soleus m.** – deep to the gastrocnemius m.; atttached to the posterior aspect of the tibia and fibula; inserted with the gastrocnemius m. on the calcaneus

  - **Plantaris tendon** – sometimes referred to as the "freshman's nerve" because it looks like a nerve

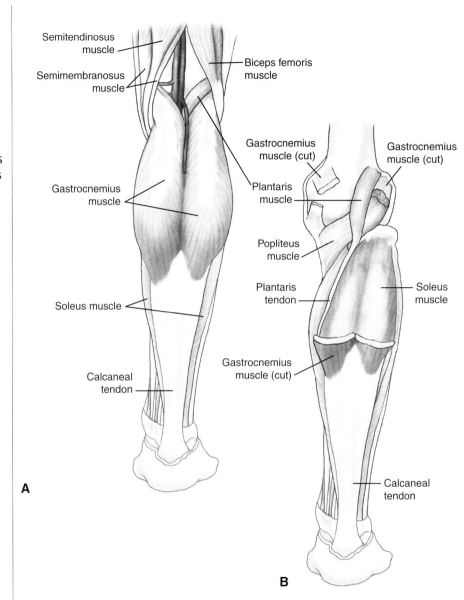

- **Figure 31–6   Posterior muscles of the right leg. *A*, Superficial muscles. *B*, Deep muscles**

Posterior Muscles of the Leg – Deep Layer

■ To better identify the deep muscles in the posterior aspect of the leg, transect the soleus m. (in a similar fashion to the gastrocnemius m.), then reflect the cut muscles superiorly and inferiorly

  ● Keep vessels and nerves intact during the reflection

■ Identify the following (Figure 31–7):

  ● **Popliteus m.** – deepest muscle in the popliteal fossa; passes obliquely downward and medially from the femur to the tibia

  ● **Tibialis posterior m.** – thick, flat muscle between the flexor hallucis longus and flexor digitorum longus mm.

  ● **Flexor digitorum longus m.** – medial to and partially overlies the tibialis posterior m.

  ● **Flexor hallucis longus m.** – its muscle belly is lateral to the flexor digitorum longus m.; its tendon crosses the medial side of the ankle joint

  ● **Flexor retinaculum** – fibrous band that tethers tendons that cross the ankle joint on its medial side; spans the interval between the medial malleolus to the medial surface of the calcaneus; secures the tendons of the tibialis posterior, flexor digitorum longus, and flexor hallucis longus mm.

    ● The plantar attachments will be studied in a future laboratory session

  ● A helpful way to remember the three deep muscles is Tom, Dick and Harry (from anterior to posterior at the medial malleolus):

    ● Tibialis posterior, flexor Digitorum longus, flexor Hallucis longus mm.

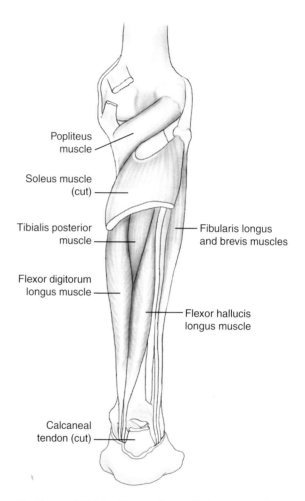

■ **Figure 31-7   Deep view of the posterior muscles of the right leg**

Popliteus muscle

Soleus muscle (cut)

Tibialis posterior muscle

Flexor digitorum longus muscle

Calcaneal tendon (cut)

Fibularis longus and brevis muscles

Flexor hallucis longus muscle

Anterior and Lateral Muscles of the Leg

- Identify the following structures at the ankle (Figure 31–8):

  - **Superior extensor retinaculum** – courses between the anterior borders of the tibia and fibula

  - **Inferior extensor retinaculum** – Y-shaped structure; attached to the calcaneus, medial malleolus, and plantar fascia

- Cut both of the retinacula vertically with scissors to free the underlying tendons; dorsiflex the foot to relieve some tension from the tendons; keep vessels and nerves intact

  - **Tibialis anterior m.** – located along the lateral border of the anterior surface of the tibia

  - **Extensor digitorum longus m.** – visible inferior and lateral to the tibialis anterior m.; follow the tendon to digits 2–5

  - **Fibularis tertius m.** – attached to the fibula; inserts on the base of the fifth metatarsal bone; your cadaver may not have this muscle

  - **Extensor hallucis longus m.** – deep between the tibialis anterior and extensor digitorum longus mm.; follow the tendon to digit 1

  - **Fibularis longus m.** – the most superficial muscle on the lateral aspect of the fibula

  - **Fibularis brevis m.** – deep to the fibularis longus m.

  - **Extensor digitorum brevis m.** – located on the dorsal surface of the foot, deep to the extensor digitorum longus m.

  - **Extensor hallucis brevis m.** – the tendon is located on the dorsum of the foot; located deep to the extensor digitorum longus m.

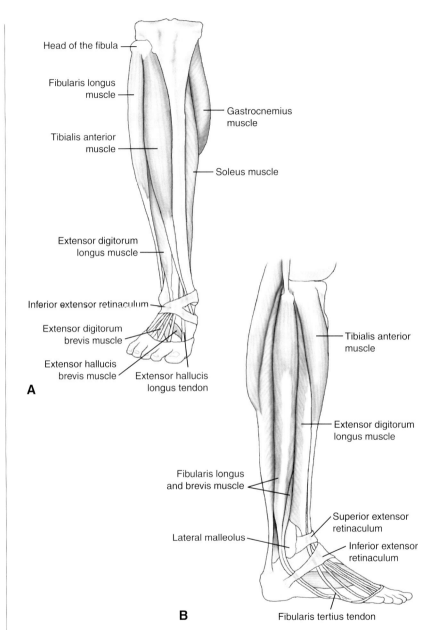

■ **Figure 31-8   Superficial views of the anterior and lateral muscles of the right leg. *A*, Anterior view. *B*, Lateral view**

Lower Limb – Nerves

Lab 32

Prior to dissection, you should familiarize yourself with the following structures:

LUMBAR PLEXUS

- Subcostal n. (T12)
- Iliohypogastric n. (L1)
- Ilioinguinal n. (L1)
- Genitofemoral n. (L1–L2)
  - Genital branch
  - Femoral branch
- Lateral cutaneous n. of the thigh (L2–L3)
- Obturator n. (L2–L4)
  - Anterior branch
    - Cutaneous branch
  - Posterior branch
- Accessory obturator n.
- Femoral n. (L2–L4)
  - Muscular branches

- Medial cutaneous n. of the thigh (L2–L4)
- Intermediate cutaneous n. of the thigh (L2–L4)
- Saphenous n. (L2–L4)
- Lumbosacral trunk (L4–L5)

SACRAL PLEXUS

- Lumbosacral trunk (L4–L5)
- Nerve to the obturator internus m. (L5–S2)
- Nerve to the piriformis m. (S1–S2)
- Nerve to the quadratus femoris m. (L4–S1)
- Superior gluteal n. (L4–S1)

- Inferior gluteal n. (L5–S2)
- Posterior cutaneous n. of the thigh (S1–S3)
- Sciatic n. (L4–S3)
  - Common fibular (peroneal) n. (L4–S3)
    - Lateral sural cutaneous n.
    - Superficial fibular (peroneal) n.
    - Deep fibular (peroneal) n.
  - Tibial n. (L4–S3)
    - Interosseous n.
    - Medial sural cutaneous n.
    - Medial plantar n.
      - Common plantar digital nn.
        - Proper plantar digital nn.
    - Lateral plantar n.
      - Common plantar digital nn.
        - Proper plantar digital nn.

Proximal Innervation of the Lower Limb

- During this laboratory period, you will dissect and study the innervation of the lower limb (Figure 32–1)

- The nerves to the lower limb arise from the lumbar and sacral regions of the spinal cord

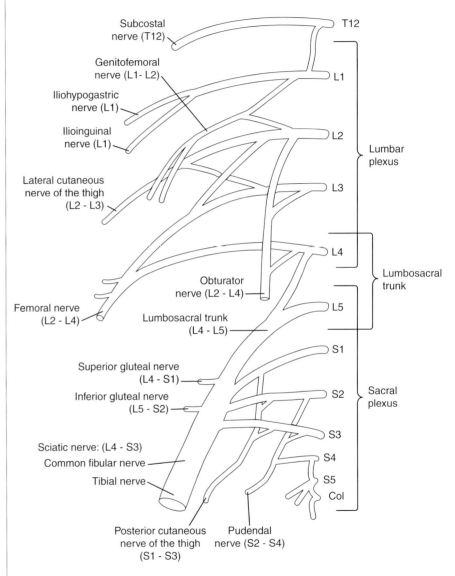

■ **Figure 32–1 Diagram of the right lumbar and sacral plexuses**

Dissection of the Lumbar Plexus

■ Verify with your instructor that you may cut and reflect the psoas major m.

■ To study the ventral rami that contribute the nerves of the lumbar plexus, remove the psoas major m. from its attachments on one side of your cadaver in the following three steps (Figure 32–2):

- Isolate the obturator, genitofemoral, and femoral nn.; bisect the distal end of the belly of the psoas major m., without cutting the isolated nerves (*A*)

- Grasp the proximal end of the bisected psoas major m. firmly with your hand (*B*)

- Pull anteriorly and superiorly, tearing the proximal attachment of the psoas major m. from the ilium and lumbar vertebrae to reveal the lumbar plexus (*C*)

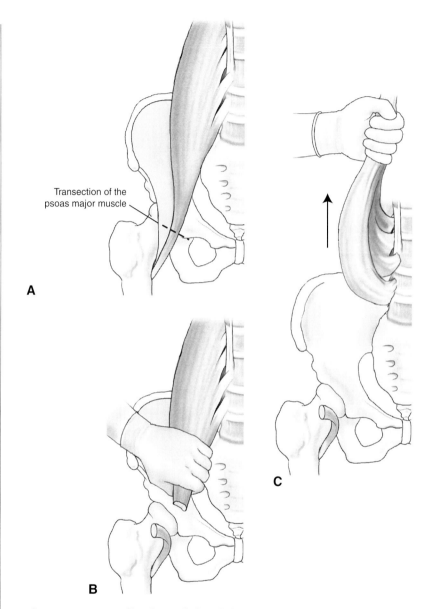

Transection of the psoas major muscle

A

B

C

■ **Figure 32–2   Reflection of the right psoas major muscle. *A*, Transect the distal end of the psoas major muscle. *B*, Grasp the proximal end of the cut psoas major muscle. *C*, Pull superiorly on the psoas major muscle to peel it from its attachments**

Lumbar Plexus – Posterior Abdominal Wall

■ Identify the following (Figure 32–3):

- **Subcostal n.** – lateral to the psoas major m.; courses along the inferior border of rib 12 in the same fashion as the intercostal nn.

- **Iliohypogastric n.** – lateral to the psoas major m.; parallels the iliac crest to reach the inguinal and pubic regions

- **Ilioinguinal n.** – lateral to the psoas major m.; courses through the inguinal canal and terminates as the anterior scrotal/labial branches

- **Genitofemoral n.** – descends on the anterior surface of the psoas major m.; divides into genital and femoral branches

- **Lateral cutaneous n. of the thigh** – lateral to the psoas major m.; courses obliquely toward the anterior superior iliac spine and then passes inferior to the inguinal ligament to supply skin on the lateral aspect of the thigh

- **Femoral n.** – lateral to the psoas major m.; arises from the ventral rami of L2, L3, and L4 lumbar nn.; exits the pelvis lateral to the femoral a. and v., inferior to the inguinal ligament

- **Obturator n.** – medial to the psoas major m.; also originates from the ventral rami of L2, L3, and L4; exits the pelvis through the obturator foramen and pierces the obturator internus and externus mm.

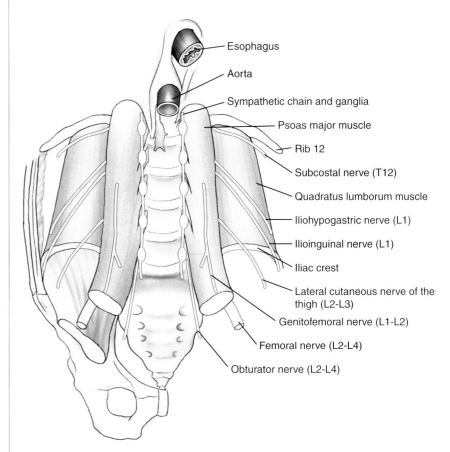

Esophagus
Aorta
Sympathetic chain and ganglia
Psoas major muscle
Rib 12
Subcostal nerve (T12)
Quadratus lumborum muscle
Iliohypogastric nerve (L1)
Ilioinguinal nerve (L1)
Iliac crest
Lateral cutaneous nerve of the thigh (L2-L3)
Genitofemoral nerve (L1-L2)
Femoral nerve (L2-L4)
Obturator nerve (L2-L4)

■ **Figure 32-3 Posterior abdominal wall, demonstrating the lumbar plexus of nerves (the psoas major muscle is shown for orientation)**

Lumbar and Sacral Plexuses – Pelvic Cavity

■ Identify the following nerves along the lateral wall of the pelvis (Figure 32–4):

- **Obturator n.** (L2–L4) – courses along the lateral wall of the pelvis en route to the obturator foramen

- **Lumbosacral trunk** (L4–L5) – branches from ventral primary rami at the level of L4 and L5; join to become the lumbosacral trunk; the trunk descends into the pelvic cavity to contribute to the sacral plexus

- **Superior gluteal n.** (L4–L5) – exits the pelvis through the greater sciatic foramen, superior to the piriformis m.

- **Inferior gluteal n.** (L5–S1) – exits the pelvis through the greater sciatic foramen, inferior to the piriformis m.

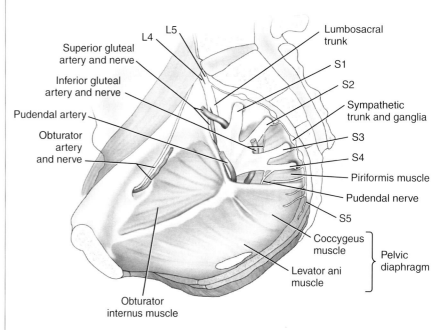

■ **Figure 32–4   Medial view of a sagittal section through the right pelvis**

Sacral Plexus – Nerves of the Gluteal Region

■ Turn the lower limb prone and identify the following (Figure 32–5):

- **Superior gluteal n.** – exits the pelvis (accompanied by the superior gluteal a. and v.) through the greater sciatic foramen, superior to the piriformis m.

- **Inferior gluteal n.** – exits the pelvis (accompanied by the inferior gluteal a. and v.) through the greater sciatic foramen, inferior to the piriformis m.

- **Posterior cutaneous n. of the thigh** – exits the pelvis with the sciatic and inferior gluteal nn.

- **Sciatic n.** – exits the pelvis with the inferior gluteal n., inferior to the piriformis m.; descends between the greater trochanter of the femur and ischial tuberosity, between the hamstring muscles; the largest nerve in the body

  - Identify tibial branches of the sciatic n. to the hamstring musculature and hamstring division of the adductor magnus m.

■ Attempt to identify the nerves that supply the quadratus femoris, inferior gemellus, obturator internus, superior gemellus, and piriformis mm.; these nerves are small

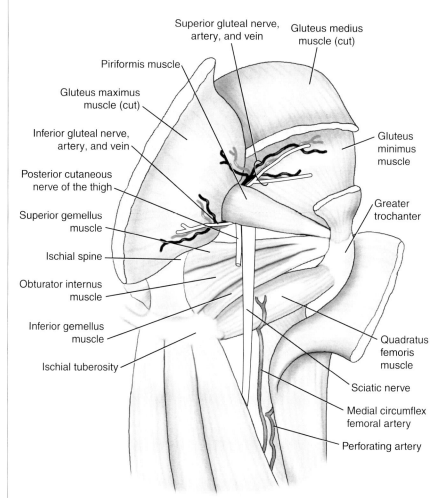

■ **Figure 32–5   Posterior view of the nerves of the right gluteal region**

Femoral Triangle

■ Turn the lower limb supine and identify the following:

● **Femoral triangle**

  ● **Borders** – sartorius m. (lateral), adductor longus m. (medial), and inguinal ligament (superior); iliopsoas and pectineus mm. (floor) (Figure 32–6*A*)

  ● **Contents** (from lateral to medial) (Figure 32–6*B*):

    ◆ **Lateral cutaneous n. of the thigh** – enters the thigh inferior to the inguinal ligament, inferomedial to the anterior superior iliac spine

    ◆ **Femoral n.** – located superficially in the furrow between the iliopsoas and pectineus mm.; not housed in the femoral sheath

      ◆ Follow branches of the femoral n. to the pectineus, sartorius, and quadriceps mm.

      ◆ Observe the intermediate and medial cutaneous nn. of the thigh branching from the femoral n.

      ◆ Observe the saphenous n. originating from the femoral n.

  ● **Femoral sheath** – a funnel-shaped fascial tube; extension of the transversalis fascia and iliopsoas fascia; the sheath ends by becoming continuous with the adventitia of the following structures:

    ◆ **Femoral a.** – continuation of the external iliac a.

    ◆ **Femoral v.** – becomes the external iliac v.

    ◆ **Femoral canal** – a closed pouch medial to the femoral v.; lymph nodes and lymphatic vessels occupy the femoral canal (potential space for hernias) to reach the external iliac lymph nodes

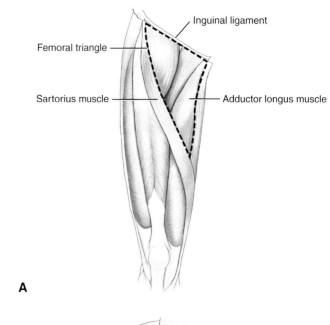

A

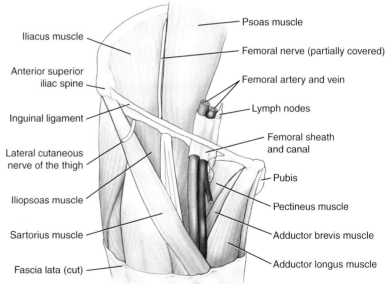

B

■ **Figure 32–6   Right femoral triangle. *A*, Borders. *B*, Contents**

Lumbar Plexus – Anterior and Medial Regions of the Thigh

■ Identify the following:

- **Saphenous n.** (Figure 32–7A) – a branch from the femoral n.

  - Accompanies the femoral vessels distally through the **adductor canal**

  - Observe that this nerve does not traverse the **adductor hiatus;** instead, it curves anterior to the adductor magnus tendon toward the medial aspect of the knee

  - Find the saphenous n. where it emerges from the deep fascia inferior to the knee, proximal to the attachment of the gracilis and sartorius mm.; trace the nerve proximally to the adductor canal

- **Obturator n.** (Figure 32–7B) – exits the pelvis and courses through a foramen at the top of the obturator membrane and pierces the obturator internus and externus mm.

  - **Anterior division of the obturator n.** – superficial to the adductor brevis

  - **Posterior division of the obturator n.** – deep to the adductor brevis m.

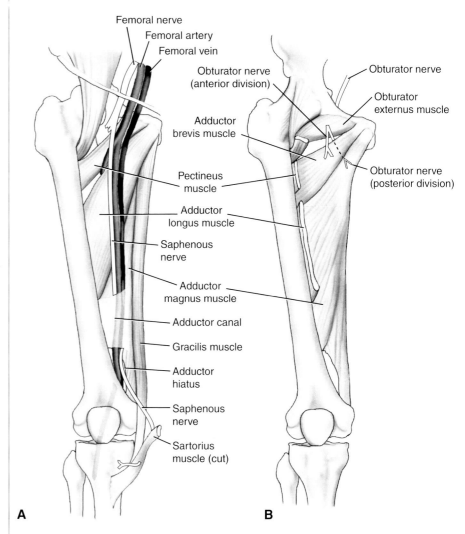

■ Figure 32–7   Nerves of the anterior and medial regions of the right thigh. *A*, Femoral and saphenous nn. *B*, Obturator n.

Nerves of the Popliteal Fossa and Leg

■ Identify the following (Figure 32–8):

● **Tibial n.** – courses through the middle of the popliteal fossa, next to the popliteal a. and v.

  ● In the leg, the tibial n. descends in company with the posterior tibial vessels between the superficial and deep group of muscles to the region between the heel and medial malleolus, where it branches into the **medial** and **lateral plantar nn.,** both of which course to the sole of the foot

● **Common fibular n.** – smaller than the tibial n.; descends to the head of the fibula, between the tendon of the biceps femoris m. and lateral head of the gastrocnemius m.; winds around the lateral surface of the neck of the fibula, deep to the fibularis longus m., where it bifurcates into the following:

  ● **Superficial fibular n.** – courses anteriorly around the neck of the fibula, between the fibularis mm. and extensor digitorum longus m.

  ● **Deep fibular n.** – courses anterior to the interosseous membrane, deep to the extensor digitorum longus m., accompanied by the anterior tibial a. in the proximal part of the leg

■ *Note:* The plantar nerves of the foot will be dissected in another laboratory session

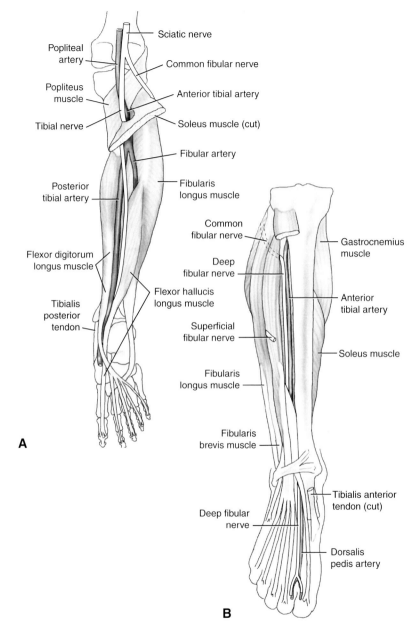

■ **Figure 32-8 The right popliteal fossa and leg. *A*, Posterior view. *B*, Anterior view**

Lower Limb – Vasculature

Lab 33

Prior to dissection, you should familiarize yourself with the following structures:

VASCULATURE

- ■ Superior and inferior gluteal aa.
- ■ Obturator a.
- ■ Femoral a.
  - Deep a. of the thigh
  - Medial circumflex femoral a.

- Lateral circumflex femoral a.
  - Ascending, transverse, and descending branches
  - Perforating branches
- ■ Popliteal a.
  - Superior medial genicular a.
  - Superior lateral genicular a.

- Inferior medial genicular a.
- Inferior lateral genicular a.
- ■ Anterior tibial a.
  - Dorsalis pedis a.
- ■ Posterior tibial a.
  - Fibular a.

Notes: Deep veins are named the same as the arteries they accompany. Vasculature of the foot will be dissected in another laboratory session.

Arteries of the Gluteal Region

■ Veins are not shown but accompany each artery; ask your instructor if you may remove the veins

■ Place the thigh prone

■ Identify the following (Figure 33–1):

● **Superior gluteal a.** – exits the pelvis superior to the piriformis m.

● **Inferior gluteal a.** – exits the pelvis inferior to the piriformis m.

● **Medial circumflex femoral a.** – courses between the proximal parts of the adductor magnus and quadratus femoris mm.

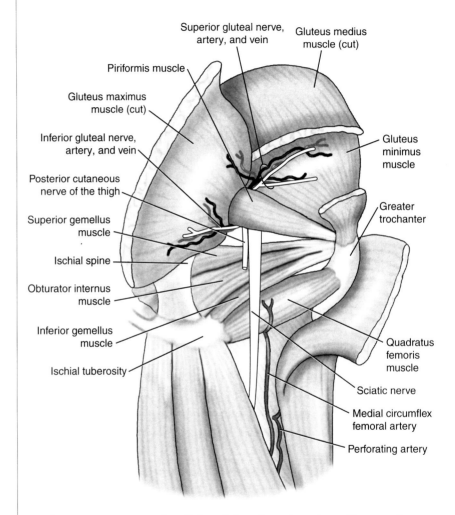

Superior gluteal nerve, artery, and vein
Gluteus medius muscle (cut)
Piriformis muscle
Gluteus maximus muscle (cut)
Inferior gluteal nerve, artery, and vein
Gluteus minimus muscle
Posterior cutaneous nerve of the thigh
Greater trochanter
Superior gemellus muscle
Ischial spine
Obturator internus muscle
Inferior gemellus muscle
Quadratus femoris muscle
Ischial tuberosity
Sciatic nerve
Medial circumflex femoral artery
Perforating artery

■ **Figure 33–1** **Vessels of the right gluteal region. (Some veins are not shown.)**

Vessels of the Thigh

■ Turn the lower limb supine; identify the **femoral triangle** bordered by the sartorius m., adductor longus m., and inguinal ligament

■ Identify the following structures (Figure 33–2):

- **Obturator a.** – emerges through the obturator canal (accompanied by the obturator n.) and divides into anterior and posterior branches

- **Femoral sheath** – surrounds the femoral a., v., and lymph nodes; note that the femoral n. is outside the femoral sheath

- **Femoral a. and v.** – open the anterior wall of the femoral sheath by making a vertical incision to expose the femoral a. and v.

  - **Deep a. of the thigh** – arises from the lateral surface of the femoral a.; gives rise to two proximal circumflex aa.

    - ◆ **Medial circumflex femoral a.** – passes between the iliopsoas and pectineus mm.; courses around the posterior aspect of the femoral neck

    - ◆ **Lateral circumflex femoral a.** – passes superficial to the iliopsoas tendon; courses around the anterior aspect of the femoral neck

      - ◆ Attempt to identify its three branches: **ascending, transverse,** and **descending branches**

  - ◆ **Venae comitantes** – observe that the femoral a. is accompanied by smaller paired veins called venae comitantes; ask your instructor if you may remove the venae comitantes

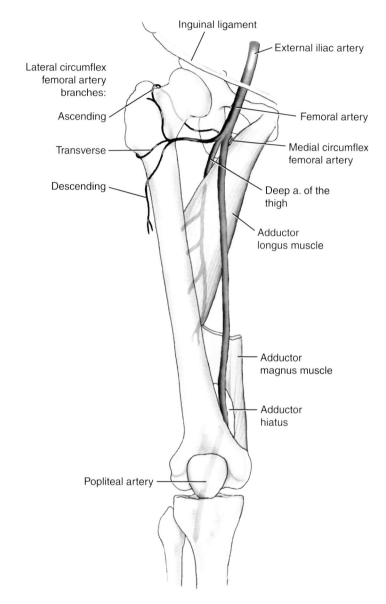

■ **Figure 33-2   Right femoral artery and its branches**

Vessels of the Thigh—cont'd

■ Identify the following structures (Figure 33–3*A*, *B*):

- **Femoral a.**

  - **Deep a. of the thigh**

    - ◆ **Perforating branches** – trace the deep a. of the thigh distally, as it passes posterior to the adductor longus m.; identify the perforating branches of the deep a. of the thigh that pierce the adductor mm.; the perforating branches terminate in the posterior compartment of the thigh

- **Adductor canal** – a narrow fascial tunnel that courses from the apex of the femoral triangle to the adductor hiatus; the canal lies deep to the sartorius m. (see Figure 33–3*A*)

  - Trace the femoral a. and v. and the saphenous n. as they descend through the adductor canal

- **Adductor hiatus** – an opening through the distal end of the adductor magnus m., through which pass the femoral a. and v. to enter the popliteal fossa

  - The saphenous n. does not course through the adductor hiatus (see Figure 33–3*A*)

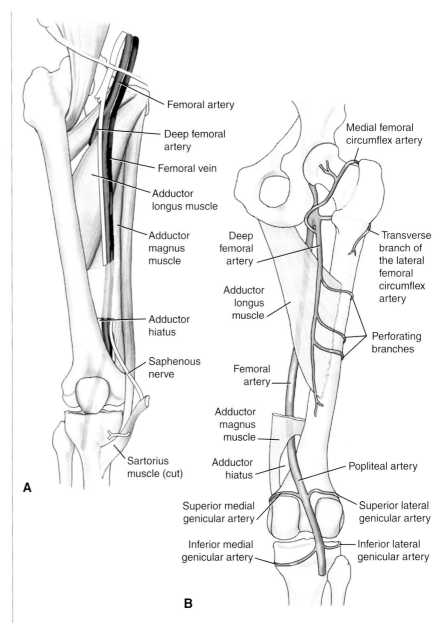

A

Femoral artery

Deep femoral artery

Femoral vein

Adductor longus muscle

Adductor magnus muscle

Adductor hiatus

Saphenous nerve

Sartorius muscle (cut)

Deep femoral artery

Adductor longus muscle

Femoral artery

Adductor magnus muscle

Adductor hiatus

Superior medial genicular artery

Inferior medial genicular artery

Medial femoral circumflex artery

Transverse branch of the lateral femoral circumflex artery

Perforating branches

Popliteal artery

Superior lateral genicular artery

Inferior lateral genicular artery

B

■ **Figure 33–3   Vasculature of the right thigh. *A*, Anterior view. *B*, Posterior view. (Some veins are not shown.)**

Vessels of the Popliteal Fossa

■ Turn the lower limb prone

■ Identify the following (Figure 33–4A, B):

- **Popliteal a.** – trace the femoral a. and v. distally through the adductor hiatus, at which location their names change to the popliteal a. and v.

  - Ask your instructor if you may remove the popliteal v. and its tributaries, which accompany the following arteries

- Identify the following branches of the popliteal a.:

  - **Superior medial genicular a.** – passes superior to the medial condyle of the femur and anterior to the semitendinosus and semimembranosus mm.

  - **Superior lateral genicular a.** – passes superior to the lateral condyle of the femur and anterior to the biceps femoris tendon

  - **Lateral and medial inferior genicular aa.** – deep to the two cut heads of the gastrocnemius m.

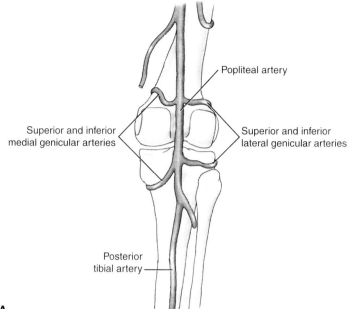

A

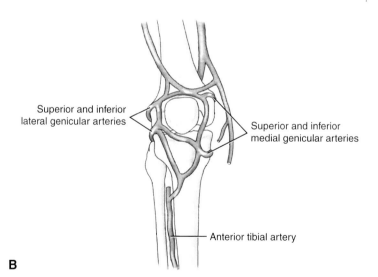

B

■ **Figure 33–4   Genicular arteries of the right knee.**
A, **Posterior view.** *B,* **Anterior view**

Arteries of the Leg

■ Identify the following structures (Figure 33–5):

- **Posterior tibial a.** – trace the popliteal a. distally to where it crosses the soleus m.; at this location the name of the artery changes to the posterior tibial a.

- **Anterior tibial a.** – pierces the belly of the tibialis posterior m. and the interosseus membrane to enter the anterior compartment of the leg

- **Fibular a.** – branch of the posterior tibial a. that courses to the lateral compartment of the leg

- **Dorsal pedis a**. – follow the anterior tibial a. distally to the foot where the artery's name changes to the dorsalis pedis a. on the dorsal surface of the foot; located between the tendons of the extensor hallucis longus and extensor digitorum longus mm.; this artery is used for assessing peripheral arterial pulse

- Attempt to locate the **arcuate a., dorsal metatarsal aa.,** and **dorsal digital aa.**

■ Plantar vessels of the foot will be dissected in another laboratory session

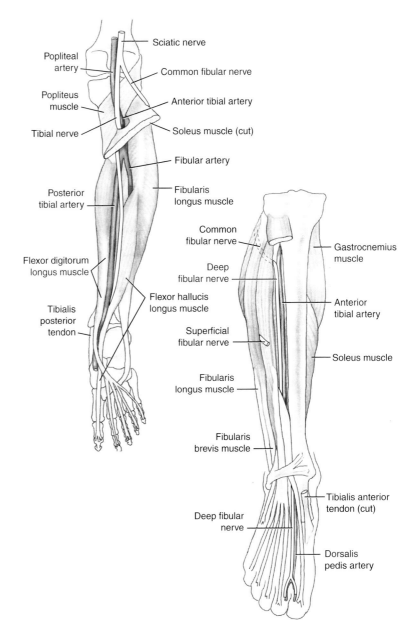

■ **Figure 33–5   Vessels of the right leg. *A*, Posterior view. *B*, Anterior view**

Lower Limb – Foot

Lab 34

Prior to dissection, you should familiarize yourself with the following structures:

OSTEOLOGY

- Tarsal bones (7)
  - Calcaneus
    - Sustentaculum tali
  - Cuboid
  - Navicular
  - Cuneiform bones (lateral, intermediate, medial)
- Metatarsal bones (5)
- Phalanges (14)

LAYERS OF THE PLANTAR SURFACE OF THE FOOT

- Plantar aponeurosis

- First layer
  - Abductor hallucis m.
  - Abductor digiti minimi m.
  - Flexor digitorum brevis m.
  - Medial and lateral plantar nn.
  - Medial and lateral plantar a.
- Second layer
  - Lateral plantar n.
  - Quadratus plantae m.
  - Flexor digitorum longus tendon
  - Lumbrical mm. (4)
  - Flexor hallucis longus tendon
- Third layer
  - Adductor hallucis m. (oblique and transverse heads)

- Flexor hallucis brevis m.
- Common plantar digital nn.
- Proper plantar digital nn.
- Flexor digiti minimi brevis m.
- Deep plantar arch
- Plantar metatarsal aa.
- Common plantar digital aa.
- Fourth layer
  - Dorsal interosseus mm. (4)
  - Plantar interosseus mm. (3)
  - Tibialis posterior tendon
  - Peroneus longus tendon
  - Medial and lateral plantar aa.

TABLE 34-1   Muscles of the Foot

Muscle	Proximal Attachment	Distal Attachment	Action	Innervation
Plantar compartment of the foot (Layer 1)				
Abductor hallucis		Proximal phalanx of the great toe (digit 1)	Abducts and flexes the great toe	Medial plantar n. (S2–S3)
Flexor digitorum brevis	Calcaneus bone	Lateral surfaces of the middle phalanx of digits 2–5	Flexes digits 2–5	
Abductor digiti minimi		Lateral side of the base of the proximal phalanx for digit 5	Abducts and flexes digit 5	Lateral plantar n. (S2–S3)
Plantar compartment of the foot (Layer 2)				
Quadratus plantae	Calcaneus bone	Tendon of the flexor digitorum longus	Flexes digits 2–5	Lateral plantar n. (S2–S3)
Lumbricals	Tendons of the flexor digitorum longus	Expansion over digits 2–5	Flex the proximal phalanges and extend the middle and distal phalanges of digits 2–5	Medial one: medial plantar n. (S2–S3) Lateral three: lateral plantar n. (S2–S3)
Plantar compartment of the foot (Layer 3)				
Flexor hallucis brevis	Cuboid and lateral cuneiform bones	Base of the proximal phalanx for the great toe	Flexes the great toe	Medial plantar n. (S2–S3)
Adductor hallucis	Oblique head: base of the metatarsals 2–4 Transverse head: metatarsophalangeal joints	Proximal phalanx of the great toe	Adducts the great toe	Deep branch of the lateral plantar n. (S2–S3)
Flexor digiti minimi brevis	Base of the fifth metatarsal bone	Base of the proximal phalanx of digit 5	Flexes the proximal phalanx of digit 5	Superficial branch of the lateral plantar n. (S2–S3)
Plantar compartment of the foot (Layer 4)				
Plantar interossei (3 muscles)	Base and medial side of metatarsals 3–5	Medial side of the base of the proximal phalanx of digits 3–5	Adduct digits 2–4 and flex the metatarsophalangeal joints	Lateral plantar n. (S2–S3)
Dorsal interossei (4 muscles)	Adjacent sides of metatarsals 1–5	First: medial side of the proximal phalanx of digit 2 Second to fourth: lateral side of digits 2–4	Abduct digits 2–4 and flex the metatarsophalangeal joints	

Skin Removal – Foot

■ Make the following skin incisions (Figure 34–1):

- Remove the flaps between *A* and *B* from the ankle and dorsal and plantar surfaces of the foot

- Make longitudinal incisions (*C*) along the dorsal and plantar surfaces of the great toe and along the top of the other digits connected to the incision at *B*

- Remove the remaining skin from the toes

  - Ask your instructor about the number of toes to dissect

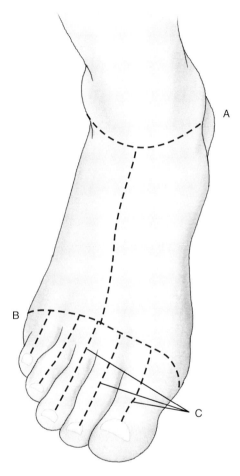

■ **Figure 34–1   Skin incisions of the right foot**

Dissection of the Plantar Surface of the Foot

- Plantar aponeurosis (Figure 34–2)

  - Use a scalpel handle (with the blade off) to scrape the remaining superficial fascia from the **plantar aponeurosis**

  - Identify digital bands of the plantar aponeurosis that extend to each toe; look for **proper digital n., a., and v.** lateral and medial to each band

  - Cut the digital bands of the aponeurosis from the plantar aponeurosis proximal to the base of the toes; do not damage the neurovascular bundles

  - Make a longitudinal incision through the plantar aponeurosis, from the base of the toes to the **calcaneus**

- Plantar structures are customarily grouped into four layers:

  - First layer
  - Second layer
  - Third layer
  - Fourth layer

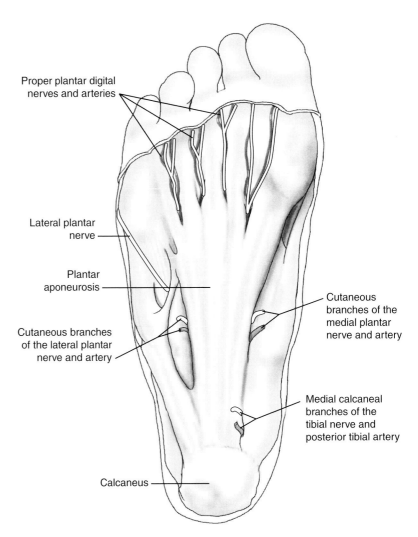

Proper plantar digital nerves and arteries

Lateral plantar nerve

Plantar aponeurosis

Cutaneous branches of the lateral plantar nerve and artery

Cutaneous branches of the medial plantar nerve and artery

Medial calcaneal branches of the tibial nerve and posterior tibial artery

Calcaneus

■ **Figure 34-2   Plantar surface of the right foot. (Veins are not shown.)**

Plantar Surface of the Foot – First Layer

■ Identify the following (Figure 34–3):

- **_Abductor hallucis m._** – located on the medial surface of the calcaneus; trace the muscle's tendon to its attachment on the base of the great toe's proximal phalanx

  - You may want to either reflect or transect the abductor hallucis m. at its calcaneal attachment to see the course of the medial plantar n., a., and v.

- **_Abductor digiti minimi m._** – located on the lateral side of the calcaneus; trace its tendon to the attachment on the proximal phalanx of digit 5

- **Flexor digitorum brevis m.** – reflect the flaps of the plantar aponeurosis medially and laterally to expose the underlying flexor digitorum brevis m.; you may want to cut the flexor digitorum brevis m. from its attachment to the calcaneus and reflect it toward the toes

- **Medial** and **lateral plantar nn., aa., and vv.** – enter the sole of the foot from the deep surface of the abductor hallucis m.

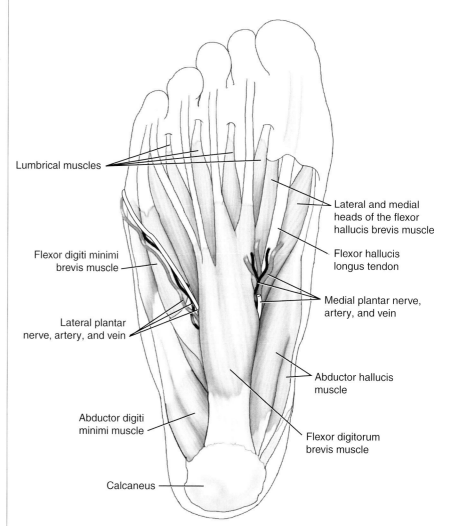

■ **Figure 34–3   First layer of the plantar surface of the right foot**

Plantar Surface of the Foot – Second Layer

■ Identify the following (Figure 34–4):

- **Quadratus plantae m.** – arises by two heads from the calcaneus and inserts into the tendon of the flexor digitorum longus m.

- **Flexor digitorum longus tendons** – trace the tendons to digits 2–5; note that these tendons pass deep to the tendons of the flexor digitorum brevis m.

- **Lumbrical mm.** – the four lumbrical mm. arise from the tendons of the flexor digitorum longus m.

- **Flexor hallucis longus tendon** – trace the tendon as it crosses deep to the tendon of the flexor digitorum longus m. to attach to digit 1

- **Tibial n.** and **posterior tibial a. and v.** – before entering the plantar surface of the foot, the nerve and vessels branch into the following:

    - **Medial plantar n., a., and v.** – course between the abductor hallucis and flexor digitorum brevis mm.

    - **Lateral plantar n., a., and v.** – course laterally across the sole of the foot between the quadratus plantae m. and the reflected flexor digitorum brevis m.

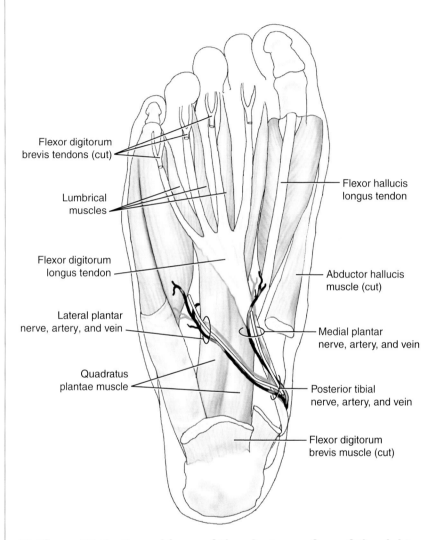

Figure 34–4   Second layer of the plantar surface of the right foot

Labels on figure:
- Flexor digitorum brevis tendons (cut)
- Lumbrical muscles
- Flexor digitorum longus tendon
- Lateral plantar nerve, artery, and vein
- Quadratus plantae muscle
- Flexor hallucis longus tendon
- Abductor hallucis muscle (cut)
- Medial plantar nerve, artery, and vein
- Posterior tibial nerve, artery, and vein
- Flexor digitorum brevis muscle (cut)

Plantar Surface of the Foot – Third Layer

- To see the structures in the third layer, you may want to make the following transections:
  - The quadratus plantae m. from the calcaneus; reflect the muscle toward the toes
  - The tendon of the flexor digitorum longus m., inferior to the tendon of the flexor hallucis longus m.; reflect the cut muscle toward the toes
- Identify the following (Figure 34–5):
  - **Flexor hallucis brevis m.** – two heads, one on each side of the flexor hallucis longus tendon; note that sesamoid bones may be within the tendon
    - Detach the lateral head of the flexor hallucis brevis m. from its origin on the bases of metatarsals 2–4 and reflect the muscle medially
  - ***Add*uctor hallucis m.** lateral to the flexor hallucis brevis m.; has oblique and transverse heads
    - Reflect the oblique head of the adductor hallucis m. toward the great toe
  - **Flexor digiti minimi m.** – medial to the abductor digiti minimi m.
  - **Deep plantar arterial arch** – at the interface between the third and fourth layers; observe that the medial and lateral plantar aa. anastomose to form the deep plantar arterial arch, deep to the flexor hallucis brevis and oblique head of the adductor hallucis mm.
    - **Plantar metatarsal aa.** – the deep plantar arterial arch gives rise to four plantar metatarsal aa. located in the region of the metatarsal bones
      - ◆ **Plantar digital aa. proper** — supply the digits
    - **Deep plantar a.** – joins the deep plantar arterial arch with the dorsalis pedis a. by piercing interosseous space 1
  - **Lateral and medial plantar nn.** – branches of the posterior tibial n.
    - **Common plantar digital nn.**
    - **Proper plantar digital nn.**

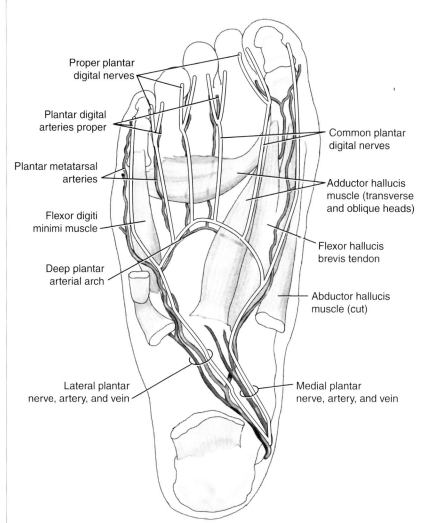

Labels:
Proper plantar digital nerves
Plantar digital arteries proper
Plantar metatarsal arteries
Flexor digiti minimi muscle
Deep plantar arterial arch
Lateral plantar nerve, artery, and vein
Common plantar digital nerves
Adductor hallucis muscle (transverse and oblique heads)
Flexor hallucis brevis tendon
Abductor hallucis muscle (cut)
Medial plantar nerve, artery, and vein

■ **Figure 34–5 Third layer of the plantar surface of the right foot**

Plantar Surface of the Foot – Fourth Layer

- Identify the following (Figure 34–6):

  - **Plantar interosseous mm.** (3) – located on the medial side of the metatarsals for digits 3–5; *add*uct digits 3–5

  - **Dorsal interosseous mm.** (4) – located on the lateral side of the metatarsals for digits 3 and 4 and both sides of the metatarsals for digit 2; *abd*uct digits 2–4

  - **Tibialis posterior tendon** – trace the tendon of the tibialis posterior m. to its multiple insertions on the tarsal bones

  - **Tibialis anterior tendon** – note its attachment on metatarsal 1

  - **Fibularis longus tendon** – follow the tendon of the fibularis longus m. across the deep plantar surface of the foot to the base of metatarsal 1 and the medial cuneiform bone

  - **Long plantar ligament** – courses from the plantar surface of the calcaneus to the cuboid and base of the metatarsals; a tunnel is formed deep to the ligament for the tendon of the fibularis longus m.; helps maintain the arches of the foot

  - **Plantar calcaneonavicular ("spring") ligament** – courses from the sustentaculum tali to the posterior and inferior surface of the navicular bone; helps maintain the longitudinal arch of the foot

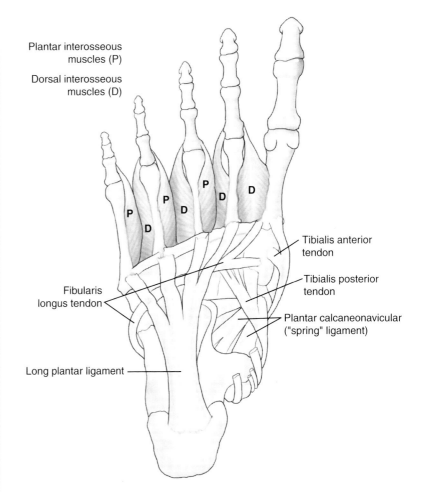

Plantar interosseous muscles (P)

Dorsal interosseous muscles (D)

Tibialis anterior tendon

Tibialis posterior tendon

Fibularis longus tendon

Plantar calcaneonavicular ("spring" ligament)

Long plantar ligament

■ **Figure 34–6** **Fourth layer of the plantar surface of the right foot**

Lower Limb – Joints

Prior to dissection, you should familiarize yourself with the following structures:

HIP
- Iliofemoral ligament
- Pubofemoral ligament
- Ischiofemoral ligament
- Round ligament of the femur
- Acetabular ligament
- Articular cartilage
- Transverse acetabular ligament

KNEE
- Fibular (lateral) collateral ligament

- Tibial (medial) collateral ligament
- Medial meniscus
- Lateral meniscus
- Anterior cruciate ligament
- Posterior cruciate ligament

ANKLE
- Medial deltoid ligament
- Anterior talofibular ligament
- Posterior talofibular ligament
- Calcaneofibular ligament

- Plantar calcaneonavicular ("spring") ligament
- Long plantar ligament

FOOT AND DIGITS
- Plantar metatarsal ligaments
- Plantar ligaments
- Metatarsophalangeal joints
- Interphalangeal joints
- Joint capsule
- Medial and lateral collateral digital ligaments

Lower Limb – Joint Dissection

- Dissection will be done in several steps:

  - First, muscles that surround the joints will be removed from the joint to reveal the joint capsule

  - Second, the joint capsule and tendons that traverse the capsule will be cut to reveal the articulation

Hip Joint

- The joint capsule attaches distally from the acetabular rim to the neck of the femur at the intertrochanteric line and root of the greater trochanter

- Thick parts of the joint capsule form the ligaments of the hip joint, which pass in a spiral fashion from the os coxae to the femur

- To identify the three capsular ligaments, cut all muscles that cross the hip joint; leave the joint capsule intact:

  - **Iliofemoral ligament** – attached to the anterior inferior iliac spine and the acetabular rim and courses inferiorly to the intertrochanteric line of the femur; reinforces the anterior aspect of the hip joint capsule (Figure 35–1 *A* and *B*)

  - **Pubofemoral ligament** – attached to the obturator crest of the pubis and courses laterally and inferiorly to the femur; reinforces the inferior and anterior aspects of the hip joint capsule (see Figure 35–1 *A*)

  - **Ischiofemoral ligament** – attached to the ischial part of the acetabular rim; spirals superior and lateral to the neck of the femur and base of the greater trochanter; reinforces the posterior aspect of the hip joint capsule (see Figure 35–1 *B*)

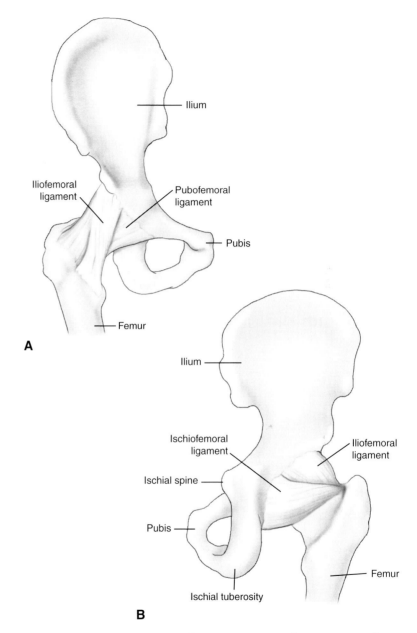

A

B

- **Figure 35–1   Right hip joint. *A*, Anterior view. *B*, Posterior view**

Hip Joint—cont'd

■ Identify the following:

- Cut the ligaments of the capsule of the one joint around its circumference with a scalpel (Figure 35–2*A*)

- Open the hip joint to break the round ligament of the femur, either by twisting the femur or cutting the ligament with a scalpel

- Identify the following (Figure 35–2*B*, *C*):

  - **Round ligament of the femur** – attached to the margins of the acetabular notch, the transverse acetabular ligament, and the fovea in the head of the femur; provides little support to the hip joint

    ◆ **The a. to the head of the femur** – a branch of the obturator a.

  - **Acetabular labrum** – a fibrocartilaginous rim that is attached to the acetabulum and the transverse acetabular ligament; increases the depth of the hip joint

  - **Transverse acetabular ligament** – forms part of the acetabular labrum; crosses the notch, forming a foramen through which vessels and nerves enter the joint

  - **Articular cartilage** – hyaline cartilage that converges with the head of the femur and fills the acetabulum, except for the fovea for attachment of the round ligament of the femoral head

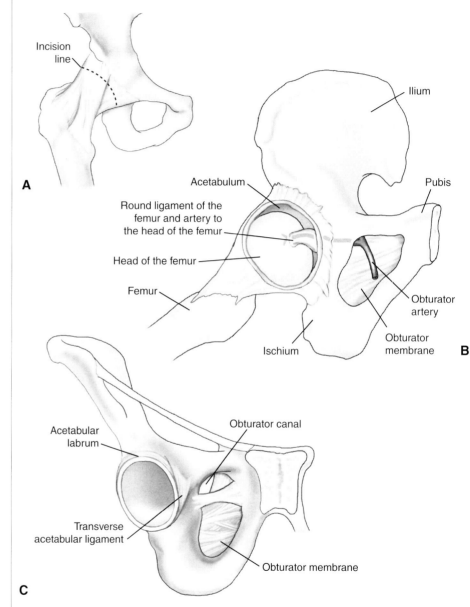

■ **Figure 35–2   Right hip joint. *A*, Capsule incision. *B*, Capsule open. *C*, Hip joint**

Knee Joint

- Remove all muscles that cross one knee joint

- Identify the following (Figure 35–3*A*, *B*):

  - **Fibular (lateral) collateral ligament** – round, cordlike ligament that extends from the lateral epicondyle of the femur to the lateral surface of the head of the fibula; the tendon of the popliteus m. passes deep to the fibular collateral ligament, separating it from the lateral meniscus; the tendon of the biceps femoris splits the ligament into two parts

  - **Tibial (medial) collateral ligament** – extends from the medial epicondyle of the femur to the medial surface of the tibia; at its midpoint, the deep fibers are firmly attached to the medial meniscus

- Use a scalpel to cut away the joint capsule but leave the collateral ligaments intact

- Identify the following:

  - **Medial meniscus** – C-shaped fibrocartilaginous wedge on the medial aspect of the tibia (Figure 35–3*C*); its anterior end is attached to the anterior cruciate ligament and posterior end to the posterior cruciate ligament

  - **Lateral meniscus** – nearly circular and smaller than the medial meniscus (see Figure 35–3*C*); the tendon of the popliteus m. separates the lateral meniscus from the fibular collateral ligament

  - **Anterior cruciate ligament** (ACL) – attached to the anterior intercondylar eminence and the posterior, medial surface of the lateral condyle of the femur

  - **Posterior cruciate ligament** (PCL) – attached to the posterior intercondylar eminence and the anterior, lateral surface of the medial condyle of the femur

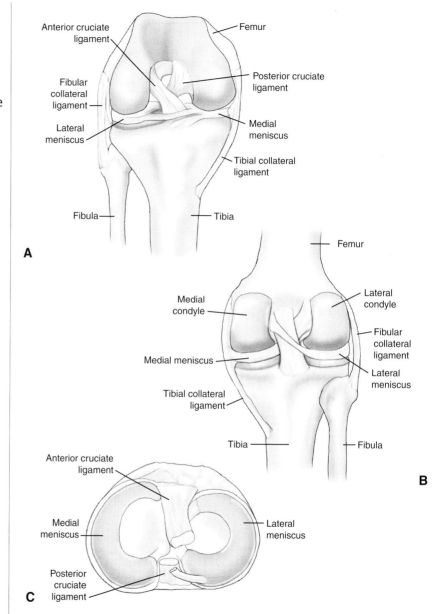

- **Figure 35–3   Right knee joint. *A*, Anterior view (flexed). *B*, Posterior view (extended). *C*, Superior view of the tibia**

Ankle and Foot Joints

■ Remove all muscles, arteries, veins, nerves, and tendons that cross the ankle joint of one lower limb to identify the following:

● **Tibiotalar joint** – allows dorsal and plantar flexion

  ● **Collateral ligaments**

    ◆ Medial side of the ankle (Figure 35–4*A*)

      ◆ **Deltoid ligament** – attached to the medial malleolus of the tibia; consists of four parts that attach to the navicular, calcaneus, and two parts of the talus

    ◆ Lateral side of the ankle (Figure 35–4*B*)

      ◆ **Posterior** and **anterior *talo*fibular ligaments** – attached between the talus and the lateral malleolus of the fibula

      ◆ **Calcaneofibular ligament** – attached at the calcaneus and the apex of the lateral malleolus of the fibula

  ● **Posterior** and **anterior *tibio*fibular ligaments** – hold the tibia and fibula together, along with the interosseous membrane

● **Talocalcaneonavicular joint** – allows inversion and eversion

  ● **Plantar calcaneonavicular ("spring") ligament** – attached from the calcaneus to the navicular bone

● **Long plantar ligament** – courses from the plantar surface of the calcaneus to the cuboid and base of the metatarsals

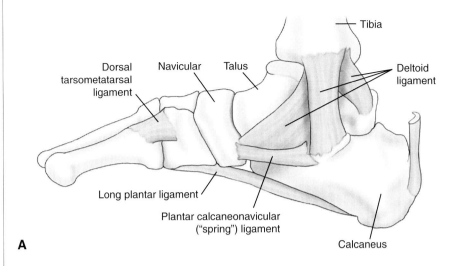

A

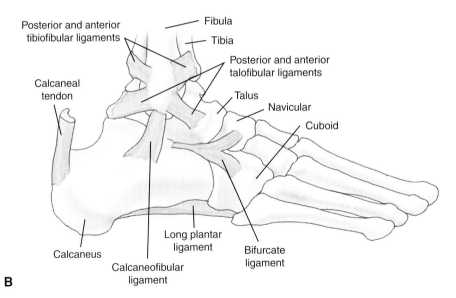

B

■ **Figure 35–4   Right ankle and foot joints. *A*, Medial view. *B*, Lateral view**

Foot and Digit Joints

■ Identify the following (Figure 35–5):

- **Plantar metatarsal ligaments** – attached between the bases of the metatarsal bones

- **Plantar ligaments** – attached to the distal end of the metatarsal bones and the proximal phalanges

- **Metatarsophalangeal joints** – articulation between the metatarsal bones and the proximal phalanges

- **Interphalangeal joints** – articulation between the phalanges

- **Joint capsule** – each metatarsophalangeal and interphalangeal joint is surrounded by its own joint capsule

  - **Medial** and **lateral collateral digital ligaments** – medial and lateral thickenings of the joint capsule

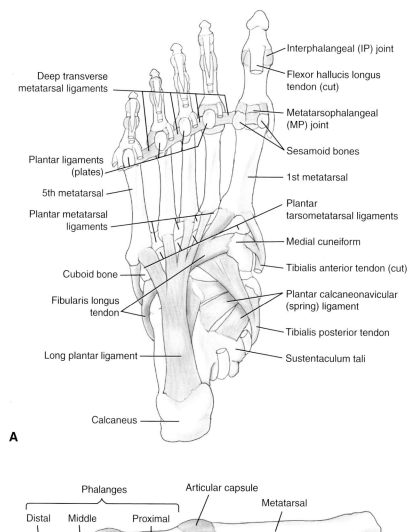

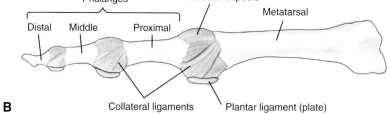

■ **Figure 35–5   Right foot and digit joints.** *A,* Plantar view. *B,* Lateral view

Arches of the Foot

- Identify the following (Figure 35–6):

  - **Arches of the foot** – the tarsal and metatarsal bones are arranged in longitudinal and transverse arches that enable the weightbearing capabilities of the foot

    - **Longitudinal arch of the foot** – composed of both the medial and lateral longitudinal arches

      - ◆ **Medial longitudinal arch** – higher than the lateral arch; composed of the calcaneus, talus, navicular, three cuneiforms, and three metatarsal bones; the talar head is the keystone of the medial longitudinal arch

      - ◆ **Lateral longitudinal arch** – flatter than the medial longitudinal arch; composed of the calcaneus, cuboid, and lateral two metatarsals

    - **Transverse arch of the foot** – courses from side to side; formed by the cuboid, cuneiforms, and bases of the metatarsals

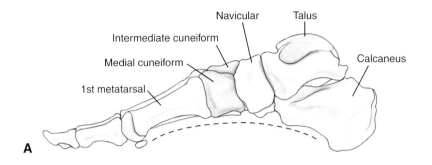

A

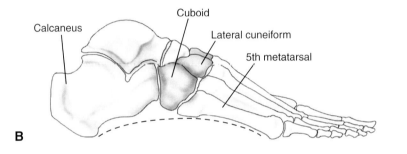

B

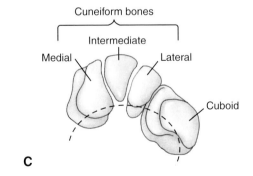

C

■ **Figure 35–6   Arches of the foot. *A*, Medial longitudinal arch. *B*, Lateral longitudinal arch. *C*, Anterior view of the transverse arch**

Unit 4 Lower Limb Overview

At the end of Unit 4, you should be able to identify the following structures on cadavers, skeletons, and/or radiographs:

Osteology

- Features of the os coxae
  - Acetabulum
  - Obturator foramen
  - Ischiopubic ramus
- Bones of the os coxae
  - Ilium
    - Ala (wing) of the ilium
    - Iliac crest
      - Anterior superior iliac spine
      - Anterior inferior iliac spine
      - Posterior superior iliac spine
      - Posterior inferior iliac spine
    - Iliac fossa
    - Anterior, posterior, and inferior gluteal lines
    - Auricular surface
  - Ischium
    - Ischial tuberosity, ramus, and spine
    - Greater sciatic notch
    - Lesser sciatic notch
  - Pubis
    - Pubic tubercle and crest
    - Superior pubic ramus
      - Pecten pubis (pectineal line)
    - Inferior pubic ramus
    - Obturator groove and crest
    - Pubic arch

- Pelvis
  - Pelvic cavity
  - Symphysis pubis
- Femur
  - Head
    - Fovea for the ligament of the head
  - Neck
    - Greater trochanter
      - Trochanteric fossa
    - Lesser trochanter
    - Intertrochanteric line and crest
  - Linea aspera
  - Pectineal line
  - Gluteal tuberosity
  - Medial condyle and epicondyle
    - Adductor tubercle
  - Lateral condyle and epicondyle
  - Popliteal and patellar surfaces
- Patella
  - Base and apex
  - Articular and anterior surfaces
- Tibia
  - Medial and lateral condyles
  - Intercondylar eminence
  - Tibial tuberosity
  - Soleal line
  - Interosseous border
  - Medial malleolus

- Fibular notch
- Inferior articular surface
- Fibula
  - Head and neck
  - Interosseous border
  - Lateral malleolus
- Foot
  - Tarsal bones (7)
    - Talus
    - Calcaneus – sustentaculum tali
    - Navicular
    - Medial cuneiform
    - Intermediate cuneiform
    - Lateral cuneiform
    - Cuboid
  - Metatarsal bones (5)
    - Phalanges (14)
      - Proximal, middle, and distal

Muscles

- Gluteal region
  - Tensor fascia lata m.
  - Gluteus maximus m.
  - Gluteus medius m.
  - Gluteus minimus m.
  - Piriformis m.
  - Superior gemellus m.
  - Obturator internus m.

- Inferior gemellus m.
- Quadratus femoris m.
- Iliopsoas m.
  - Iliacus m.
  - Psoas major m.
- Psoas minor m.
- Anterior muscles of the thigh
  - Quadratus femoris m.
    - Rectus femoris m.
    - Vastus medialis m.
    - Vastus intermedius m.
    - Vastus lateralis m.
  - Sartorius m.
- Medial muscles of the thigh
  - Pectineus m.
  - Adductor brevis m.
  - Adductor longus m.
  - Adductor magnus m. (adductor hiatus)
  - Gracilis m.
  - Obturator externus m.
- Posterior muscles of the thigh
  - Biceps femoris m.
  - Semitendinosus m.
  - Semimembranosus m.
- Anterior muscles of the leg
  - Tibialis anterior m.
  - Extensor digitorum longus m.
  - Extensor digitorum brevis m.
  - Extensor hallucis longus m.
  - Extensor hallucis brevis m.
  - Fibularis (peroneus) tertius m.
- Lateral muscles of the leg
  - Fibularis (peroneus) longus m.
  - Fibularis (peroneus) brevis m.
- Posterior muscles of the leg
  - Superficial layer
    - Gastrocnemius m.
    - Soleus m.

- Deep layer
  - Tibialis posterior m.
  - Flexor hallucis longus m.
  - Flexor digitorum longus m.
  - Plantaris m.
  - Popliteus m.
- Plantar aponeurosis
- Plantar muscles of the foot
  - First layer
    - Abductor hallucis m.
    - Flexor digitorum brevis m.
    - Abductor digiti minimi m.
  - Second layer
    - Quadratus plantae m.
    - Lumbrical mm. (4)
  - Third layer
    - Flexor hallucis brevis m.
    - Adductor hallucis m.
      - ◆ Oblique head
      - ◆ Transverse head
    - Flexor digiti minimi brevis m.
  - Fourth layer
    - Plantar interossei mm. (3)
    - Dorsal interossei mm. (4)

Nerves

Lumbar Plexus

- Subcostal n. (T12)
- Iliohypogastric n. (L1)
- Ilioinguinal n. (L1)
- Genitofemoral n. (L1–L2)
  - Genital branch
  - Femoral branch
- Lateral cutaneous n. of the thigh (L2–L3)
- Obturator n. (L2–L4)
  - Anterior branch
    - Cutaneous branch
  - Posterior branch

- Accessory obturator n.
- Femoral n. (L2–L4)
  - Muscular branches
  - Medial cutaneous n. of the thigh (L2–L4)
  - Intermediate cutaneous n. of the thigh (L2–L4)
  - Saphenous n. (L2–L4)
- Lumbosacral trunk (L4–L5)

Sacral Plexus

- Lumbosacral trunk (L4–L5)
- Nerve to the obturator internus m. (L5–S2)
- Nerve to the piriformis m. (S1–S2)
- Nerve to the quadratus femoris m. (L4–S1)
- Superior gluteal n. (L4–S1)
- Inferior gluteal n. (L5–S2)
- Posterior cutaneous n. of the thigh (S1–S3)
- Sciatic n. (L4–S3)
  - Common fibular (peroneal) n. (L4–S3)
    - Lateral sural cutaneous n.
    - Superficial fibular (peroneal) n.
    - Deep fibular (peroneal) n.
  - Tibial n. (L4–S3)
    - Interosseous n.
    - Medial sural cutaneous n.
    - Medial plantar n.
      - ◆ Common plantar digital nn.
        - ◇ Proper plantar digital nn.
    - Lateral plantar n.
      - ◆ Common plantar digital nn.
        - ◇ Proper plantar digital nn.

Arteries

- Internal iliac a.
  - Superior gluteal a.
  - Inferior gluteal a.
  - Obturator a.

- External iliac a.
  - Femoral a.
    - Deep a. of the thigh
      - Medial circumflex femoral a.
      - Lateral circumflex femoral a.
        - Ascending, transverse, and descending branches
      - Perforating branches
  - Popliteal a.
    - Superior medial genicular a.
    - Superior lateral genicular a.
    - Inferior medial genicular a.
    - Inferior lateral genicular a.
  - Anterior tibial a.
    - Dorsalis pedis a.
  - Posterior tibial a.
    - Fibular a.
    - Medial plantar a.
      - Deep branch
      - Superficial branch
    - Lateral plantar a.
      - Deep plantar arch
        - Plantar metatarsal aa.
          - Common plantar digital aa.
            - Plantar digital aa. proper

Veins

- Internal iliac v.
  - Superior gluteal v.
  - Inferior gluteal v.
  - Obturator v.
- External iliac v.
  - Great saphenous v.

- Small saphenous v.
  - Dorsal venous arch of the foot
  - Plantar venous arch
  - Dorsal metatarsal vv.
  - Plantar metatarsal vv.
  - Plantar digital vv.
- Femoral v.
  - Deep femoral v. of the thigh
    - Medial and lateral circumflex femoral vv.
    - Perforating vv.
- Popliteal v.
  - Sural v.
  - Genicular vv.
  - Anterior tibial v.
  - Posterior tibial v.
  - Fibular v.

Lymphatics

- Femoral lymph nodes
- Inguinal lymph nodes

Joints

Hip

- Iliofemoral ligament
- Pubofemoral ligament
- Ischiofemoral ligament
- Round ligament of the femur
- Acetabular ligament
- Articular cartilage
- Transverse acetabular ligament

Knee

- Fibular (lateral) collateral ligament
- Tibial (medial) collateral ligament
- Medial meniscus
- Lateral meniscus
- Anterior cruciate ligament
- Posterior cruciate ligament

Ankle

- Medial deltoid ligament
- Anterior talofibular ligament
- Posterior talofibular ligament
- Calcaneofibular ligament
- Plantar calcaneonavicular ("spring") ligament
- Long plantar ligament

Foot and digits

- Plantar metatarsal ligaments
- Plantar ligaments
- Metatarsophalangeal joints
- Interphalangeal joints
- Joint capsule
- Medial and lateral collateral ligaments

Miscellaneous

- Flexor and extensor retinacula
- Iliotibial band of fascia
- Femoral triangle, sheath, and canal
- Fibrous digital sheaths

Index

Note: Page numbers followed by t represent tabular material.

Nerve(s) *(Continued)*
in brain removal procedure, 232
of posterior cranial fossa, 246
cutaneous. *See* Cutaneous nerves.
deep temporal, 281
dorsal
in perineal space
female, 165
male, 156
scapular, 13
ethmoidal
anterior, 252
external nasal branch, 267
posterior, 252
facial, 247
branches in parotid region, 271
geniculate ganglion of, in middle ear
dissection, 323
femoral, 140, 425
in femoral triangle, 411
in lumbar plexus, 422
fibular
common, 427
deep, 427
superficial, 405, 427
frontal, 250
genitofemoral, 99, 140
in lumbar plexus, 422
in spermatic cord, 152
glossopharyngeal, 247, 286, 288
gluteal, 181
inferior, 181, 424
in lumbar plexus, 423
in sacral plexus, 423
superior, 181, 424
in lumbar plexus, 423
in sacral plexus, 423
great auricular, 198
greater occipital, in suboccipital triangle
dissection, 28, 30
greater petrosal, 244, 299
hiatus for, 241, 244
in middle ear dissection, 323
humeral, posterior circumflex, 347
hypogastric, 182
hypoglossal, 247, 303
in carotid triangle, 208
in submandibular triangle, 215
iliohypogastric, 90, 140
in lumbar plexus, 422
ilioinguinal, 90, 99, 140, 160
in lumbar plexus, 422
in brachial plexus, 364

Nerve(s) *(Continued)*
axillary, 364, 365, 367
dorsal scapular, 360, 367
lateral cutaneous, of forearm, 365
long thoracic, 360
medial cutaneous, of arm and forearm,
363
median
in arm, 365
in forearm, 366
musculocutaneous, 365
pectoral, lateral and medial, 363
radial, 364
in arm, 365
in forearm, 366
to forearm, 368
to shoulder, 367
subscapular
lower, 364
upper, 364
suprascapular, 361, 367
thoracodorsal, 364
ulnar
in arm, 365
in forearm, 366
in penis, 156
infraorbital, 300
in parotid region, 268
infratrochlear, in parotid region, 267
intercostal, 37, 42
in anterior thoracic wall removal, 43,
44
in posterior mediastinum, 74
intercostobrachial, 341
interosseous, anterior, 366
lacrimal, 250
in parotid region, 267
laryngeal
external, 210, 312
internal, 210, 211, 286, 312
left recurrent, 54, 70
recurrent, 219, 286, 312, 315
superior, 211, 315
lesser occipital, 198
lingual, 281, 303
masseteric, 279
maxillary
in infratemporal fossa, 278
schematic of, 308t
median
of palm of hand, 383
palmar cutaneous branch of, 341
mental, in parotid region, 269

Nerve(s) *(Continued)*
mylohyoid, 215, 283
nasociliary, 252
nasopalatine, 294
obturator, 140, 181
cutaneous branch in lower limb,
405
in lumbar plexus, 422, 423
in sacral plexus, 423
of anterior and medial regions of
thigh, 426
oculomotor, 242, 247
divisions, of orbit, 251
in brain removal procedure, 232
of abdominal wall, 140–141
of infratemporal fossae, 283
of leg, 427
of pelvic region, 176
of pelvic wall, 181
of popliteal fossa, 427
of superficial back, 8
of urogenital region
in female, 168
in male, 160
olfactory, 247
optic, 232, 247, 252
palatine, greater and lesser, 298, 299
phrenic, 221
plantar, medial and lateral, 427, 438, 439,
440
proper digital, 437
pudendal, 145, 146, 181
branches of
in female, 168
in male, 160
radial, superficial branch of, 342
in dorsal surface of hand, 388
rectal, inferior, 146
sacral, anterior, 181
saphenous, 405
of anterior and medial regions of
thigh, 426
sciatic
in gluteal region, 145, 424
in pelvic region, 181
scrotal, posterior and anterior, 151
spinal, 27
spinal accessory, 247
in carotid triangle, 208
in posterior triangle of neck, 198
splanchnic, 74
lumbar and sacral, 182
pelvic, 182

Nerve(s) *(Continued)*
subcostal, 140
in lumbar plexus, 422
suboccipital, 31
supraclavicular, 196
supraorbital, 250
in parotid region, 267
supratrochlear, 250
in parotid region, 267
sural, 405
sympathetic. *See* Sympathetic nerves.
tibial, 427, 439
at popliteal fossa, 415
to pterygoid canal (Vidian nerve), 299
transverse cervical, 198
trochlear, 242, 247, 250
in brain removal procedure, 232
ulnar, palmar cutaneous branch of, 341, 383
vagus, 70, 247
in carotid triangle, 208, 209
larynx and, 312
left, 54
right, 54, 286
vestibulocochlear, 247, 319
Vidian, 299
zygomaticofacial, 268
zygomaticotemporal, 268
Neurovascular bundles
anterior cutaneous, 37
intercostal, 37, 74
lateral cutaneous, 37
posterior cutaneous, 8
Nipple, 35
Node(s)
atrioventricular, 63
sinoatrial, 63
Nodules, aortic valves, 67
Nose
lateral wall of, 295–297
septum of. *See* Nasal septum.
Notch
cardiac, 50
jugular, 34, 36, 195

Oblique fissure
in left lung, 50
in right lung, 51
Oblique muscles
external, 85t, 91, 94
internal, 85t, 94
of orbit
inferior, 253
superior, 250, 251, 253